AF348857

Learning Materials in Biosciences

Series Editors

Kursad Turksen, University of Ottawa
Ottawa, Canada
Pablo Carbonell, Universitat Politècnica de València
Valencia, Spain
Joan Roig, Molecular Biology Institute of Barcelona
Barcelona, Spain
Aydin Berenjian, The University of Waikato
Hamilton, New Zealand
Mia Levite, School of Pharmacy, Faculty of Medicine
The Hebrew University
Jerusalem, Israel
Anja K. Bosserhoff, Friedrich-Alexander University Erlangen-
Erlangen, Germany

Learning Materials in Biosciences textbooks compactly and concisely discuss a specific biological, biomedical, biochemical, bioengineering or cell biologic topic. The textbooks in this series are based on lectures for upper-level undergraduates, master's and graduate students, presented and written by authoritative figures in the field at leading universities around the globe.

The titles are organized to guide the reader to a deeper understanding of the concepts covered.

Each textbook provides readers with fundamental insights into the subject and prepares them to independently pursue further thinking and research on the topic. Colored figures, step-by-step protocols and take-home messages offer an accessible approach to learning and understanding.

In addition to being designed to benefit students, Learning Materials textbooks represent a valuable tool for lecturers and teachers, helping them to prepare their own respective coursework.

Ilya Digel · Aida Kistaubayeva
Irina Savitskaya · Nuraly S. Akimbekov

Introduction to Industrial Biotechnology

Second Edition

Springer

Ilya Digel
Institute for Bioengineering
Aachen University of Applied Sciences
Jülich, Nordrhein-Westfalen, Germany

Irina Savitskaya
Department of Biotechnology
Al-Farabi Kazakh National University
Almaty, Kazakhstan

Aida Kistaubayeva
Department of Biotechnology
Al-Farabi Kazakh National University
Almaty, Kazakhstan

Nuraly S. Akimbekov
Department of Biotechnology
Al-Farabi Kazakh National University
Almaty, Kazakhstan

ISSN 2509-6125   ISSN 2509-6133   (electronic)
Learning Materials in Biosciences
ISBN 978-3-032-07917-6   ISBN 978-3-032-07918-3   (eBook)
https://doi.org/10.1007/978-3-032-07918-3

Foreword

Biotechnology is where life sciences and technology converge. Today, industrial biotechnology represents a dynamic field of research and innovation, offering not only economic profitability and technological advancement but also pathways toward a more sustainable future.

This book is intended for students studying Biotechnology and aims to provide a comprehensive introduction to modern industrial biotechnology technologies, bio-based products, and the principles of a sustainable bioeconomy. It presents a structured overview of theoretical foundations and contemporary concepts related to the selection and improvement of biological agents, their cultivation, process management, and practical applications—core knowledge necessary for the development of production systems that utilize biomass and biologically active substances.

The book is organized into nine chapters, each addressing essential aspects of biotechnology based on the use of living or inactivated microbial biomass, microbial metabolites, renewable energy generation, and biotransformations using immobilized biocatalysts. Special sections are devoted to the engineering design of biotechnological processes, fermentation stages, isolation and purification techniques, and approaches to evaluating process efficiency and safety. The application of biotechnology across various sectors of the economy is also discussed, complemented by representative process flow diagrams. Each chapter includes review questions to support active learning, and a curated list of references is provided to encourage further exploration.

Although many books on biotechnology are available in print and online, the authors believe that this book, with its numerous original illustrations and a focus on industrial applications, will serve as a valuable resource for both students and educators engaged in the study and teaching of industrial biotechnology.

Prof. Dr. Ilya Digel
Institute for Bioengineering
Aachen University of Applied Sciences

Acknowledgment

The authors express their gratitude to Bhutta, Muhammad Ahmed from the FH Aachen University of Applied Sciences (Germany), master's program "Biomedical Engineering," for his valuable assistance during the textbook examination.

Competing Interests The authors have no competing interests to declare that are relevant to the content of this manuscript.

Introduction

Biotechnology is a complex concept, encompassing both science and manufacturing with specific hardware designs. It is derived from three Greek words: *"bios"* (life), *"techne"* (skill), and *"logos"* (knowledge). Biotechnology is the science of developing natural processes in artificial conditions to obtain a desired (final) product. When biological systems or agents are used as the foundational elements, the production is termed biotechnological.

A *biological agent* is a producer that synthesizes the desired product or acts as a catalyst, such as an enzyme facilitating a specific reaction. Biotechnology employs bioproducers—microorganisms, plants, higher animals, or isolated enzymes. Biocatalysts immobilized on an insoluble carrier can be used repeatedly. Modern biotechnology also utilizes innovations like artificial cell and tissue cultures. A key achievement in biotechnology is the creation of genetically engineered microorganisms, which contain recombinant DNA. These bio-factories produce substances whose structures are encoded by the introduced genes.

When a biological agent performs the entire technological process in artificial conditions, it requires optimal "comfort," which involves providing necessary nutrients and protection from external adverse effects. Therefore, the engineering and technical infrastructure plays a crucial role in supporting the bio-agent's activity, including the processes and equipment used in biotechnological production.

Industrial biotechnology is the science of essential biological processes and their practical applications to produce valuable products from organisms' vital activities, such as their biomass (the most crucial protein product), and various useful substances for national economies and medicine. This science draws from fundamental research in physiology, biochemistry, genetics, and more, turning these discoveries into practical applications.

Several industries belong to various industrial biotechnology processes, including ancient practices like baking, winemaking, brewing, vinegar production, and fermented dairy products (such as curdled milk, kefir, koumiss, yogurt, etc.). Many of these industries existed on a large scale before the concept of microorganisms as the pathogens of these processes was discovered. Unintended microorganisms, often grown accidentally in raw substrates, were initially used. However, once scientists isolated microorganisms responsible for alcoholic, lactic, and propionic acid fermentation and studied their physiological and biochemical characteristics in depth, biotechnologists took these processes under scientific control.

Industrial biotechnology provides national economies with dairy products, butter, citric acids, ethyl and butyl alcohols, acetone, and other fermentation products.

The rapid development of this field is largely due to advancements in the industrial production of penicillin. The mass production of this antibiotic and its widespread use in medicine spurred efforts to discover new antibiotics and develop more sophisticated methods for their production. Today, approximately a hundred antibiotics, used in medicine, agriculture, and the food industry, are produced via biotechnology, with a growing focus on chemically modified antibiotics.

Following the development of antibiotics, industrial biotechnology has expanded into the production of amino acids, enzymes, vitamins, hormones, and other biologically active compounds synthesized by microorganisms. These microbial products have found valuable applications in health, agriculture, and the food industry.

In the past decade, the production of microbial protein has grown significantly, with this essential product being widely used in agriculture. Microorganisms utilize non-food substances like paraffin, dairy, and meat waste to produce microbial protein. The use of single-carbon compounds, such as methane and methyl alcohol, is expected to further expand microbial protein production, as these compounds allow for the production of microbial biomass without foreign impurities, a key advantage.

The industrial use of immobilized enzymes and cells has grown in recent years. Engineering enzymology, which applies immobilized enzymes, significantly enhances biocatalyst efficiency in producing valuable substances and pure products. Immobilized cells have been tested successfully for producing organic acids, antibiotics (such as penicillin, bacitracin, and nisin), and other compounds.

Another emerging area of industrial biotechnology involves transforming natural and chemically synthesized compounds with microorganisms to obtain more valuable products, particularly in medical practice and for purging toxic compounds from industrial effluents. Examples include the biological transformation of steroids, antibiotics, and pyridine compounds.

Biotechnology has also made significant strides in traditionally non-biological industries, such as energy production (biogas) and oil and metal extraction. As microorganisms become key bio-factories for various products, the search for highly productive strains with valuable biosynthetic properties is critical. Microbiologists and geneticists work closely to address this challenge. Moreover, with the advent of genetic engineering, it is now possible to create microorganisms with specific traits tailored for industrial applications, vastly expanding the practical use of microorganisms.

To increase biosynthesis and the yield of desired products, mixed microbiological systems are often used, combining two organisms: a producer of a compound and another that stimulates its formation. Methods of mixed cultivation are employed in the industrial production of antibiotics, enzymes, vitamins, and methane, and are also commonly used in wastewater treatment. The study of complex microbial cultures for the biosynthesis of multiple valuable products is a promising area deserving further attention.

In conclusion, one of the defining characteristics of scientific and technological progress today is the integration of various research fields. This integration is driven by the need to address fundamental and national economic challenges quickly. It has led to the flourishing of industrial biotechnology, which, supported by other sciences such as chemistry, physics, mathematics, microbiology, biochemistry, molecular biology, and genetics, can solve a wide array of urgent problems. In turn, industrial biotechnology's achievements have profoundly influenced the development of these fields. This multidisciplinary approach underscores the importance of modern research methods and the collaboration of specialists from engineering, technology, and biology.

Contents

Ilya Digel, PhD, is a full professor and Executive Director of the Institute for Bioengineering at FH Aachen University of Applied Sciences, Germany. He earned his doctorate in microbiology from Al-Farabi Kazakh National University in Almaty, Kazakhstan, in 1998. Since joining FH Aachen in 2002, he has taught and conducted research in cell biology and biophysics. He teaches undergraduate and graduate courses on molecular cell biology, medical biophysics, and experimental biotechnology. In 2018, he became a full professor and head of the Institute, helping shape interdisciplinary programs and research at the intersection of biology, physics, and engineering.

His research focuses on molecular, cellular, and medical biophysics, particularly diagnostic and therapeutic technologies, including tools for cell diagnostics, biosensors, and biophotonic imaging. Beyond academia, he mentors students and junior researchers and fosters applied research with industry. He supports science-driven education and the application of bioengineering innovations.

Aida Kistaubayeva, PhD, Dr. Professor, is Head of the Biotechnology Department at Al-Farabi Kazakh National University in Almaty, Kazakhstan. She is a recognized expert in industrial and microbial biotechnology, teaches and coordinates programs in Biotechnology, Biological Engineering, and holds a double-degree in Food Biotechnology with Belgorod State University (Russia) and the University of Lille (France).

Her research focuses on microbial biotechnology, biopreparations, and innovative biomaterials, leading numerous national research projects funded by the Ministry of Science and Higher Education of Kazakhstan, and she is the principal investigator of an international project with Norway (DIKU, CPEA-LT-2017/10061) to enhance research-based education in microbial biotechnology. She represents Kazakhstan in the Organization for Women in Science for the Developing World.

She has completed international internships and training in several countries, supported by the World Bank, British Council, and the First President's Foundation. Her expertise includes food safety, ISO Codex Alimentarius standards, molecular microbiology, and academic leadership.

Irina Savitskaya, Doctor of Biological Sciences, is a Professor in the Department of Biotechnology at Al-Farabi Kazakh National University. She teaches courses such as Industrial Biotechnology, Processes and Apparatuses in Biotechnology, and Food Biotechnology for undergraduate, master's, and PhD students, and researches functional food products using polysaccharide matrices with immobilized probiotic biofilms.

She regularly participates in international scientific conferences and has authored over 100 publications, including 10 teaching manuals and two co-authored monographs—one published in Japan. Her research is protected by six Kazakhstani and four international patents. She plays a vital role in advancing biotechnology education and research in Kazakhstan, integrating academic, scientific, and methodological work.

Nuraly S. Akimbekov, PhD, is a Professor in the Department of Biotechnology at Al-Farabi Kazakh National University. He earned his bachelor's in Environmental Sciences, master's and PhD in Biotechnology from the same university, and completed postdoctoral research in Biology at Al-Farabi Kazakh National University and FH Aachen University of Applied Sciences, Germany. As a visiting professor, he taught and built research collaborations at Ural Federal University and Shaanxi University of Science and Technology. He also participated in the "500 Scientists" Program at FH Aachen.

His research focuses on geomicrobiology for green energy, environmental remediation, agriculture, micronutrients, and food systems. He has a strong academic background in Biology and Biotechnology and has authored over 10 books, including textbooks on Microbiology and Virology.

Biological Agents as Potential Resources of Biotechnological Production

1.1 Subject of Industrial Biotechnology

Biotechnology refers to the application of biological principles and processes to develop technologies that improve human life and environmental health. It harnesses cellular and biomolecular mechanisms to produce useful products and solutions. As early as 6000 years ago, humans began using microbial processes to create foods such as bread and cheese, and to preserve dairy products (Fig. 1.1).

Industrial biotechnology, also known as *white biotechnology*, applies these biological systems to industrial-scale production. It focuses on the sustainable manufacturing of valuable products from renewable resources using living cells, microorganisms, or their enzymes. It is a set of procedures that generate beneficial items or processes for the national economy and support sustainable development by utilizing living systems or their derivatives.

1.2 Characteristics of Biological Agents

Biotechnology is a key component of the modern bio-industry. Industrial applications can be broadly divided into two categories: (1) fields where biotechnological methods can replace traditional chemical technologies, and (2) fields where biotechnology plays a central, indispensable role (Fig. 1.2).

These technologies utilize the catalytic capabilities of diverse organisms and bio- systems. These include microorganisms, viruses, and plant and animal cells and tissues, as well as extracellular substances and cell components (Fig. 1.3).

Industrial biotechnology involves research, development, and production of valuable bio-based materials using biological systems. In this context, the terms product systems or

I. Digel et al., *Introduction to Industrial Biotechnology*, Learning Materials in Biosciences,
https://doi.org/10.1007/978-3-032-07918-3_1

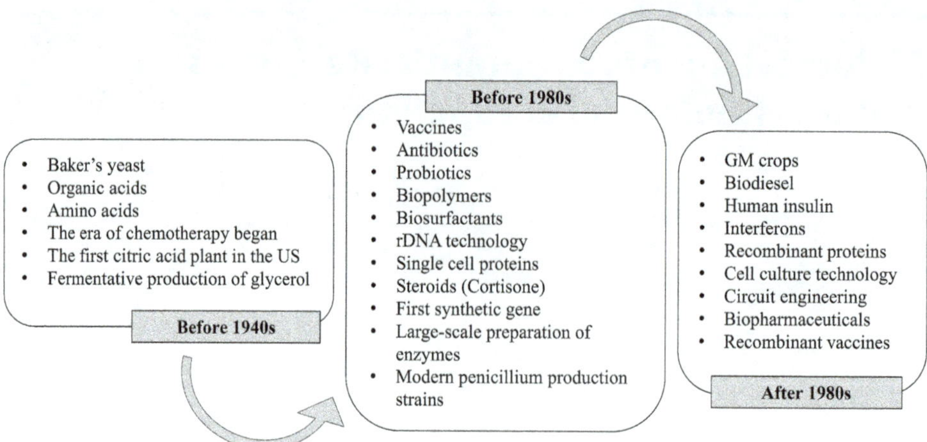

Fig. 1.1 Brief history of industrial biotechnology advancements

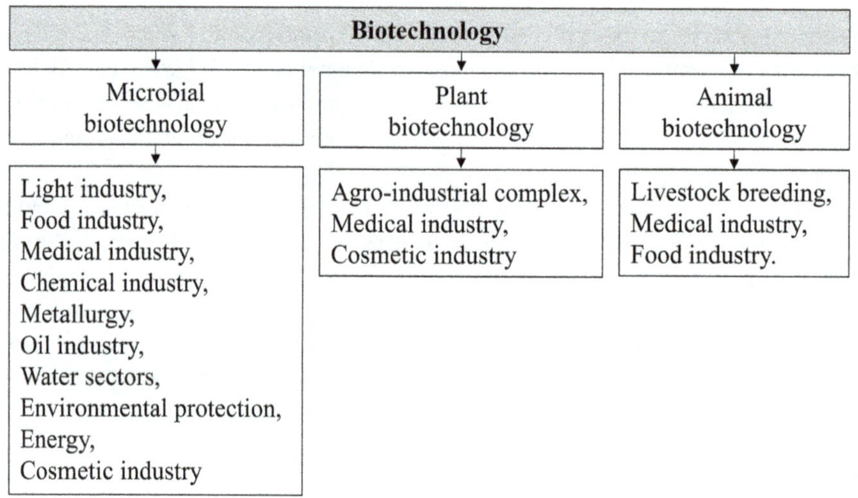

Fig. 1.2 Biotechnologies: the range of applications in various industries

biotechnological processes refer to integrated operations that combine biological agents with non-biological components. The core elements of such processes include biological agents, substrates, technical equipment, and the resulting products (Table 1.1).

1.2.1 Biological Agent

A biological agent is a central functional component in biotechnological processes and one of their most critical elements. Bio-agents can include:

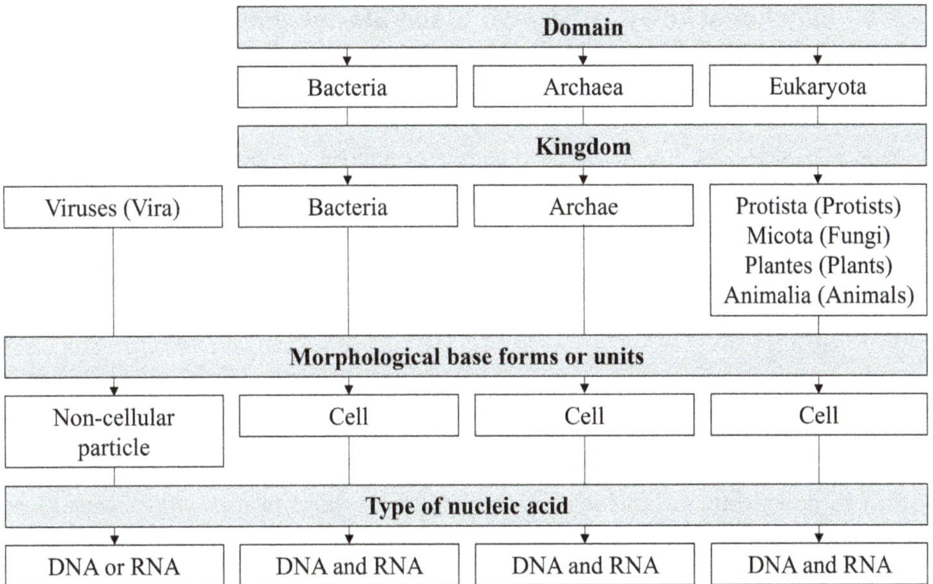

Fig. 1.3 Levels of organization in biotechnologically valuable living systems

Table 1.1 Examples of substrates, biological agents, and products generated in biotechnological processes

Substrates	Biological agents	Products
Molasses, sugar cane juice, hydrolysates of plant polymers. Sugars, alcohols, organic acids. Oil paraffin Products, precursors of biotransformation Natural gas, hydrogen. Waste from agriculture and forestry, including the processing of fruits and vegetables Household waste, wastewater. Milk serum Potatoes, grain Green biomass of plants	Microorganisms, plants, and animal cells Viruses Cell components: membranes, protoplasts, mitochondria, enzymes. Extracellular products: enzymes, coenzymes Immobilized cells of microorganisms, plants and animals, their components, and extracellular products, etc.	Biogas; clean products, medicines, diagnostics. Hormones and other biotransformation products. Organic acids; Polysaccharides. Amino acids, proteins Food products. Extracts, hydrolysates Alcohols, organic solvents. Antibiotics Enzymes, vitamins. Metals, nonmetals. Monoclonal antibodies. Biofertilizers and bioinsecticides, microbial biomass, vaccines, etc.

- Integrated holistic multicellular or unicellular organism.
- Subcellular structures, like viruses, plasmids, DNA, RNA, and proteins.
- Cell cultures of plants and animals, protoplasts, and any biological systems capable of biosynthesis and conversion.
- Multi-enzyme complexes isolated from cells.
- Individual isolated enzyme.

1.2.2   Functional Roles of Bio-agents and Bio-systems in Targeted Biosynthesis

The primary role of any bio-agent is to enable the biosynthesis of a desired product. A biological system that completes the synthesis of a target compound is referred to as a bio-producer or bio-factory. When the bio-agent is an individual enzyme catalyzing a specific reaction, it is termed a biocatalyst. Although the taxonomy of biological organisms is continuously evolving, microbial cells—whether naturally occurring or genetically engineered—remain central to industrial biotechnology due to their versatility and efficiency in production processes.

1.2.3   Properties of Microorganisms

Although prokaryotic and eukaryotic microorganisms differ in many fundamental features, they share several properties that make them especially valuable for industrial biotechnology.

- *High growth and metabolic rates:* Microorganisms have a large surface-area-to-volume ratio, allowing efficient nutrient exchange and rapid metabolism. Because bacterial cells are the smallest, they typically reproduce the fastest, followed by yeasts and fungi. For example, while a 500 kg cow produces only about 0.5 kg of protein per day, the same biomass of yeast can synthesize over 50,000 kg of protein in that time. This makes microbial biosynthesis of proteins, antibiotics, and antibodies significantly more efficient and cost-effective than chemical synthesis.
- *Metabolic flexibility:* Microorganisms can adapt to diverse environmental conditions by producing inducible (adaptive) enzymes in response to available substrates. This metabolic versatility far exceeds that of plants or animals (Table 1.2).
- *High genetic variability:* As unicellular organisms, microorganisms exhibit greater variability than multicellular organisms. Individual cells can be easily manipulated, and their short generation times and large population sizes make them ideal subjects for genetic research and biotechnology.
- *Scalability:* Microbial processes can be implemented at an industrial scale, provided the appropriate equipment, substrates, and process conditions are available.

1.2.4   Criteria for Selecting a Biotechnological Agent

When selecting a biological agent for industrial applications, the primary consideration is the efficiency of the strain. The chosen microbial cell, population, or community must retain its biochemical characteristics over prolonged fermentation periods. In addition, robust bio-producers should be resistant to genetic mutations, bacteriophages, and external

Table 1.2 Examples of microorganisms used to produce desired compounds

Biological agent	Product
Bacteria	
Acetobacter aceti, Gluconobacterium suboxidans	Vinegar
Actinoplanes missouriensis	Glucose isomerase
Azotobacter vinelandii	Alginates
Bacillus subtilis	Proteases, antibiotics
Bacillus amyloliquefaciens	α-Amylase
Bacillus licheniformis	Proteases, α-Amylase
Bacillus thuringiensis, Bacillus popilliae	Bioinsecticides
Brevibacterium flavum	Glutamate, lysine, and other amino acids
Clostridium acetobutylicum	*n*-Butanol, acetone
Corynebacterium glutamicum	Glutamate, L-lysine, inosinic acid, ribonucleotides
Escherichia coli (recombinant strain*)*	Insulin, growth hormone, interferon
Lactobacillus acidophilus, Lactococcus lactis	Yoghurt and sour-milk products Lactic acid
Leuconostoc mesenteroides	Dextran
Propionibacterium freudenreichii	Vitamin B
Propionibacterium shermanii, Streptococcus thermophilus	Yogurt, Swiss cheese
Xanthomonas campestris	Polysaccharides, Xanthan
Zymomonas mobilis	Ethanol
Yeasts	
Candida utilis	Microbial protein
Kluyveromyces fragilis	Lactase
Phaffia rhodozyma	Astaxanthin
Saccharomyces cerevisiae	Ethanol, baker's yeast, wine, ale, sake
Saccharomycopsis lipolytica	Lipase
Fungi	
Aspergillus oryzae	Amilase, Sake
Blakeslea trispora	β-carotene
Chehalosporium acremonium	Cephalosporins
Endothia parasitica	Rennet extract
Penicillium chrysogenum	Penicillins
Penicillium roqueforti	Roquefort cheese
Rhizopus nigricans	Transformation of steroids

microbial contamination. Ideal agents are safe, produce minimal waste, deliver high product yields, and meet technical and economic feasibility requirements (Fig. 1.4).

Nowadays, most industrial microbial processes rely on heterotrophic organisms. However, future developments are expected to increasingly utilize autotrophic microorganisms, which do not require scarce organic substrates, as well as extremophiles—organisms that thrive under extreme conditions such as high temperatures (*thermophiles*), high alkalinity (*alkaliphiles*), or low pH (*acidophiles*).

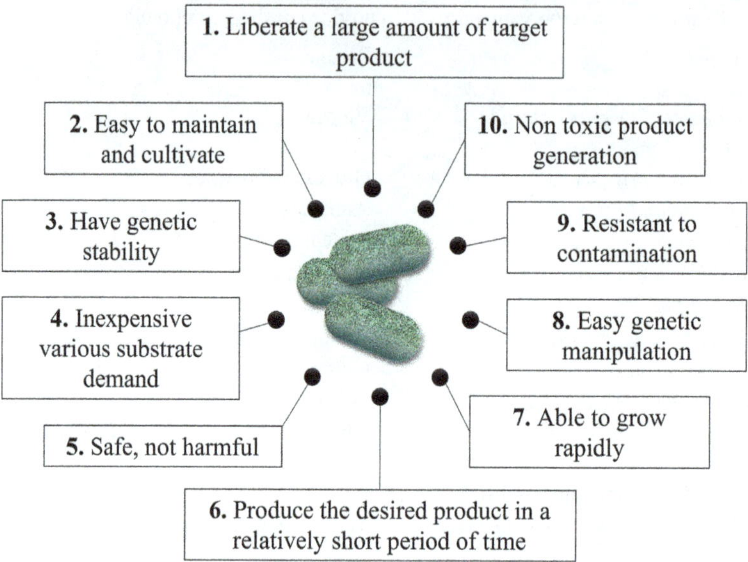

Fig. 1.4 Criteria for the selection of a biotechnological agent, particularly bacteria

Mixed microbial cultures and their natural associations are gaining international attention. Compared to monocultures, microbial consortia offer several advantages: they can degrade complex and heterogeneous substrates, demonstrate superior biotransformation capabilities, show greater resistance to toxic substances and environmental stress, and exhibit enhanced productivity. In some communities, microorganisms can also exchange genetic material, further improving adaptability. These consortia are particularly valuable in applications such as environmental protection, biodegradation, and the treatment of complex waste streams.

A distinct category in biotechnology involves *enzymes*, which are biological catalysts widely used in industrial processes. Despite their broad applicability, free enzymes face challenges such as instability, high production costs, and difficulty in recovery. Immobilized enzymes, however, offer several benefits: increased operational stability, retention within the reactor system, simplified product separation, and reusability in continuous fermentation. Their effectiveness depends on the proper selection of the enzyme, carrier material, and immobilization technique.

Immobilized enzymes now play a key role in the development of microscale biotechnological devices, particularly in analytical applications and bio-electrocatalysis. In parallel, non-traditional biological elements—such as hybridomas, animal tissues, and transplantable structures—are increasingly being incorporated into biotechnological systems. Advances in genetic engineering enable the creation of novel agents, including transgenic cells derived from diverse biological kingdoms. The development of artificial cells, constructed from synthetic and natural components, has also made significant progress. Furthermore, enzyme design technologies are producing catalysts with enhanced reactivity and stability.

In the field of bioactive compound production, eukaryotic cell cultures are indispensable. Plant cells serve as natural reservoirs for secondary metabolites such as alkaloids and saponins. Animal and human cell cultures are widely used to produce biologically active molecules, including hormones, lipotropins, and monoclonal antibodies, the latter being critical for medical diagnostics. Biotechnology also plays a growing role in veterinary science, supporting the development of advanced techniques such as cell and embryo culture, in vitro ovogenesis, and artificial insemination.

Thus, biological agents—from molecular to cellular scales—are key sources of biotechnological products. Throughout their functional lifecycle, these agents generate a wide array of metabolites with diverse physicochemical properties and biological activities.

These bio-products can be classified into four main groups (Fig. 1.5):

- Whole cells that directly serve as the product, e.g., bacteria or viruses for live or subunit vaccines; yeast as a feed protein or a source of hydrolysates for culture media
- Macromolecules synthesized during growth, such as enzymes, toxins, antigens, antibodies, and polysaccharides
- *Primary metabolites*, i.e., low molecular weight compounds essential for cell growth, including amino acids, vitamins, nucleotides, and organic acids
- *Secondary metabolites*, i.e., low molecular weight compounds not required for growth, such as antibiotics, alkaloids, hormones, and certain toxins

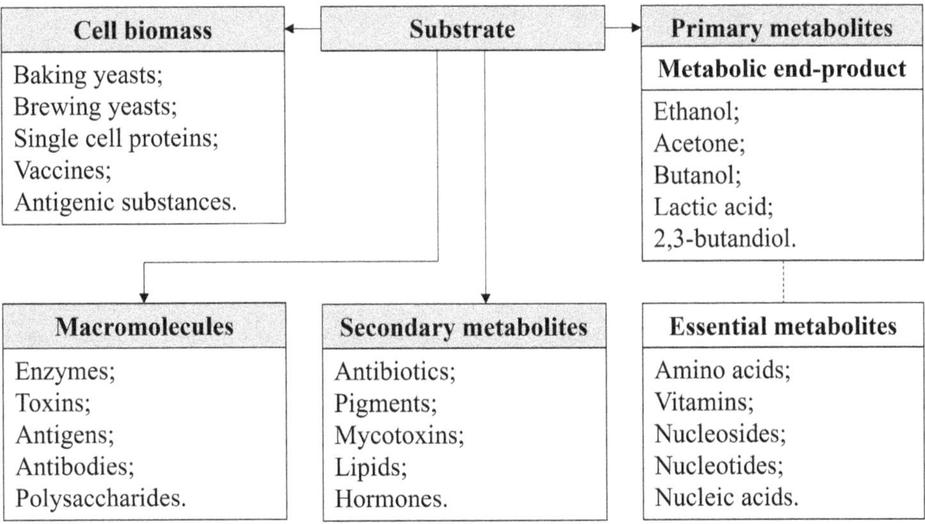

Fig. 1.5 The various groups of biological compounds synthesized by microorganisms

In industrial biotechnology, bio-agents can be employed through three main approaches:

- *Biomass production*: The most common method involves cultivating biomass, which serves as either an intermediate or a final product. Applications in this category include single-cell proteins (SCPs) for food and animal feed, yeast for brewing and baking, beneficial microbiota (normal flora), bio-pesticides, bio-fertilizers, selected vaccines, and diagnostic bacteriophages.
- *Metabolite utilization*: Another major approach focuses on harvesting valuable metabolic byproducts from bio-agents. These include amino acids, vitamins, enzymes, and antibiotics. A specialized form of this approach is biotransformation, where living cells or enzymes catalyze specific chemical reactions. Examples include the use of Acetobacter species for vitamin C synthesis (e.g., conversion of sorbitol to sorbose) and *Mycobacterium* species for steroid transformation (e.g., sitosterol to 17-ketoandrostan).
- *Immobilization techniques*: This approach enhances the efficiency and stability of bio-agents by anchoring whole cells or enzymes to solid supports. Immobilization enables easier product separation, reuse of the biological component, and greater process automation. Common matrices include gels, membranes, fibers, and inert micro-particles, with immobilization achieved via mechanical, physical, or chemical means.

In summary, biotechnological production utilizes bio-agents through biomass cultivation, the extraction and transformation of metabolites, and the application of immobilization technologies—each offering distinct advantages in terms of efficiency, scalability, and product specificity.

Brainstorming

1. What are the main components of a biotechnological process?
2. What role do biological agents play in biotechnological production?
3. Which organisms or systems can serve as biological agents in industrial biotechnology?
4. How are biological agents classified?
5. What are the different strategies for employing biological agents in biotechnological production?
6. What factors influence the selection of a microbial producer for industrial-scale production of specific products?
7. What are the advantages of using microbial agents compared to plant- or animal-based biological agents?
8. What are the key requirements for microorganisms used in industrial biotechnology?
9. What is meant by the term technological strain?

Take-Home Messages

- Biological agents are the core functional elements of biotechnological processes, ranging from whole cells and cell cultures to individual enzymes and macromolecules.
- Microorganisms are especially valuable in industrial biotechnology due to their high metabolic activity, rapid growth, adaptability, genetic variability, and ease of large-scale cultivation.
- The selection of a suitable bio-agent depends on multiple factors, including product yield, genetic stability, resistance to contamination, economic efficiency, and safety.
- Biotechnological production employs bio-agents in three main ways: (1) through biomass generation, (2) by extracting or transforming metabolic products, and (3) via immobilization techniques that enhance process stability and reusability.
- Mixed microbial cultures and extremophiles offer new biotechnological opportunities, particularly in the areas of environmental sustainability and complex substrate processing.
- Enzymes, especially in immobilized form, are powerful tools in modern biotechnology, enabling efficient, continuous, and cost-effective production processes in various industrial sectors.

1.3   Microorganisms as Biological Agents

Biotechnological processes depend extensively on microorganisms and the products they generate, which are crucial for a variety of industrial applications. Most commonly, these processes rely on *mesophilic organisms*—those that thrive under moderate conditions resembling those of human cells, including neutral pH, atmospheric pressure, moderate temperature, and the presence of oxygen, while excluding extreme salinity, xenobiotics, or volatility.

However, in many industrial settings, mesophilic conditions are suboptimal. Challenges such as poor substrate solubility, product instability, or high process viscosity necessitate alternative biological systems. For instance, hydrocarbon bioremediation in cold Arctic environments or starch hydrolysis at elevated temperatures requires specialized organisms. In these cases, *extremophiles*—organisms adapted to harsh environments— offer superior catalytic performance. Enzymes derived from such organisms, known as *extremozymes*, are particularly valued for their stability and specificity.

While large-scale cultivation of extremophiles can be cost-intensive, their enzymes are often expressed in mesophilic hosts *generally recognized as safe (GRAS)*, such as *Escherichia coli* or *Bacillus subtilis*. These hosts eliminate the need for extreme growth conditions while preserving enzymatic function.

A wide range of GRAS microorganisms are commonly used in industry. These include bacterial species such as *Bacillus subtilis*, *B. amyloliquefaciens*, *Lactobacillus* spp., and *Streptomyces*, as well as fungi like *Aspergillus*, *Penicillium*, *Mucor*, and *Rhizopus*, and yeasts like *Saccharomyces cerevisiae*. These strains are non-pathogenic, non-toxic, and do not produce antibiotics—making them ideal candidates for new biotechnological applications.

The microbiological industry currently employs thousands of microbial strains, initially isolated from natural environments and optimized through selection and engineering. However, the depth of characterization remains limited: few strains have been studied as thoroughly as *E. coli* or *B. subtilis*, due to the significant time and resource investment required.

This presents a strategic challenge—how to identify and harness the most promising features of newly isolated microorganisms with reasonable research effort. The classical approach remains effective: isolating potential producers from natural habitats, culturing them on selective media, and enriching for strains capable of synthesizing the desired product.

Selection criteria go beyond product yield. High growth rate, use of inexpensive substrates, and resilience to contamination are crucial for ensuring low-cost and efficient production. While many unicellular organisms grow rapidly, others—despite slow growth—can produce valuable compounds.

Photosynthetic microorganisms such as cyanobacteria (*Spirulina*, *Nostoc*) are notable for using solar energy and CO_2 as a carbon source, with some species also capable of nitrogen fixation. These microbes are promising producers of ammonia, hydrogen, protein, and other organic substances.

Thermophilic microorganisms, thriving in temperatures from 60–100 °C, include producers of alcohols, amino acids, enzymes, and molecular hydrogen. Their *growth and metabolic rates* are often 1.5–2 times higher than those of mesophiles. Their enzymes, known as *thermozymes*, are valued for heat stability—though often less catalytically active, which may be advantageous in certain processes. For example, *Taq polymerase*, an enzyme from *Thermus aquaticus*, has become a cornerstone in genetic engineering due to its thermal stability in PCR.

1.3.1   Overview of Extremophiles

The systematic exploration of extremophiles began in the mid-1960s when microbial ecologist Thomas Brock discovered thermophilic filamentous bacteria thriving in the hot springs of Yellowstone National Park at temperatures between 82 °C and 88 °C. This pivotal finding spurred extensive research into microbial life in geothermally heated environments such as geysers, volcanic fields, and deep-sea hydrothermal vents. The latter, discovered in 1977, revealed diverse microbial and invertebrate communities thriving under extreme conditions, including pressures exceeding 100 MPa and temperatures approaching 400 °C. Microorganisms that grow at elevated temperatures are classified as *thermophiles* (optimal growth between 45 °C and 80 °C) or *hyperthermophiles* (optimal growth above 80 °C).

Thermophilic microorganisms exhibit remarkable tolerance to high temperatures, often exceeding the boiling point of water. The current record-holder is *strain 121*, a hyperthermophilic archaeon related to *Pyrodictium occultum*, which can grow at 121 °C under high pressure. This surpasses the previous record set by *Pyrolobus fumarii*, with an upper growth limit of 113 °C. These organisms produce enzymes—known as *thermozymes*—that are structurally stable and catalytically active under extreme thermal conditions.

It is important to distinguish between *intracellular* and *extracellular enzyme stability*. In vivo, intracellular enzymes may be stabilized by molecular chaperones, compatible solutes, or specific protein–protein interactions. In contrast, extracellular enzymes must be inherently thermostable to function effectively in high-temperature environments. This property makes them particularly valuable in industrial applications such as polymer degradation, starch processing, and biofuel production.

Although the direct use of thermophiles in industrial processes remains limited due to factors such as low biomass yield and challenging cultivation conditions, their enzymes are extensively utilized. Advances in *recombinant expression* and *directed evolution* have enabled the production of thermostable enzymes in mesophilic hosts. Nevertheless, significant challenges remain, including low catalytic activity at moderate temperatures, poor secretion efficiency in native hosts, codon usage mismatches, improper protein folding, and the complexity of multienzyme assemblies or cofactors.

As a result, ongoing research focuses on developing genetic tools and optimized expression systems for producing thermozymes in thermophilic hosts. Despite these technical hurdles, thermostable enzymes continue to offer clear advantages over mesophilic counterparts for applications involving elevated temperatures and harsh processing conditions.

At the opposite end of the thermal spectrum, *psychrophiles* are adapted to permanently cold environments, including polar regions, glaciers, and permafrost. Some species remain metabolically active at temperatures as low as −20 °C. Although cold-adapted microbes were first isolated as early as 1887, significant advances in understanding their molecular adaptations—such as cold-active enzymes and membrane fluidity—occurred only in recent decades.

Halophiles are microorganisms that thrive in high-salt environments. While seawater has a salinity of approximately 35 g/L, hypersaline habitats can reach concentrations of 200–340 g/L, such as in the Dead Sea or solar salterns. Despite lower biodiversity, these environments often harbor dense microbial communities capable of producing unique bioproducts, including *bacteriorhodopsin* (for optical data storage), *exopolysaccharides*, and *polyhydroxyalkanoates (PHAs)*—the latter being biodegradable alternatives to conventional plastics.

Certain environments, like soda lakes characterized by high pH, are inhabited by *alkaliphiles*. Alkaliphiles are divided into alkali-tolerant bacteria, capable of growing at pH around 9, and obligate alkaliphiles, thriving at pH higher than 9. Alkaliphiles maintain a cytoplasmic pH much lower than the external pH, and their extracellular enzymes are particularly valuable in processes involving alkaline detergents, such as washing.

On Earth, various processes, whether biological or industrial, give rise to acidity, leading to the production of sulfuric acid. One common biological acceleration of this acid formation occurs in extremely acidic environments often linked to volcanic activities. In these environments, sulfur-containing compounds undergo oxidation by sulfur-oxidizing microorganisms, specifically bacteria and Archaea known as 'acidophiles.' Acidophiles thrive in highly acidic conditions, with an optimal pH below 3.0. For instance, *Ferroplasma acidarmanus* can even grow in extremely acidic conditions with a pH near 0, while moderate acidophiles prefer pH levels between 3 and 5. Additionally, the mining industry, dealing with mineral ores and coal, contributes to the creation of acidic environments by exposing sulfide mineral-rich rocks like pyrite (FeS2).

Acidophiles play a crucial role in mineral destruction under acidic conditions, contributing to the dissolution of various metals that give rise to significant global pollution.

Additionally, there is a group of extremophilic microorganisms known as *piezophiles*, residing in the deep sea, deep lakes, and subsurface regions. These organisms, also called *barophiles*, thrive at pressures higher than atmospheric pressure, with resistance extending up to 100 MPa (1100 atm), equivalent to the depths of approximately 10,000 meters in the Mariana Trench. The extreme conditions of deep-sea habitats—low temperatures, high pressures, and limited nutrient availability—support diverse microbial communities, often in association with invertebrate and vertebrate life.

Common piezophilic genera include *Colwellia*, *Moritella*, *Photobacterium*, *Psychromonas*, and *Shewanella*. Many of these microorganisms exhibit both pressure and cold adaptation, highlighting a strong physiological link between these two environmental stressors. High pressure affects biological systems at the molecular level, influencing processes such as protein unfolding, DNA-protein interactions, tubulin polymerization, and ribosomal stability.

Cultivating pressure-tolerant microorganisms remains technically challenging, which has limited our understanding of their full diversity. However, significant advances in deep-sea microbiology have been made using manned and unmanned submersibles like *Shinkai* and *Kaiko*, which have recovered both piezophilic and non-piezophilic species from the ocean's greatest depths.

Despite these cultivation challenges, piezophiles show considerable biotechnological potential. High-pressure processing (200–800 MPa) is already applied in food sterilization, preserving flavor and color while ensuring microbial safety. Pressure also influences cheese maturation and the synthesis of fine chemicals. Some deep-sea microorganisms produce omega-3 polyunsaturated fatty acids, which are of interest for nutritional and medical applications.

Additionally, pressure can regulate gene expression. Pressure-sensitive promoters have been identified in piezophilic bacteria, offering new tools for biotechnology. For example, the *lac* promoter in *Escherichia coli* has shown increased activity under high pressure, allowing for reversible control of heterologous protein production in mesophilic hosts without the need for additional regulatory elements.

1.3.2 Thermophilic Microorganisms and Biomass Conversion

Lignocellulose, derived from plant cell walls, is a vital renewable raw material. A key challenge in its sustainable use lies in converting it into fermentable sugars. Abundantly available as a byproduct of agriculture and forestry, lignocellulose exhibits considerable compositional variability depending on its origin. It primarily consists of cellulose, hemicellulose, pectin, and lignin, arranged in complex structural patterns. Upon appropriate processing, it can yield fermentable sugars such as glucose, xylose, arabinose, galactose, and mannose, which are essential substrates for the production of bioethanol. In addition, anaerobic microbial consortia can convert lignocellulose into methane.

However, enzymatic hydrolysis of lignocellulose is hindered by various physicochemical and structural factors, including lignin content, the degree of polymerization, limited surface area, cellulose crystallinity, and the tight binding of hemicellulose to cellulose microfibrils. To address these limitations, pretreatment of the raw material is essential. The main goal of pretreatment is to disrupt the lignin barrier—an amorphous, heterogeneous polymer—and reduce cellulose crystallinity, thus enhancing enzymatic accessibility. Pretreatment methods include mechanical size reduction, steam explosion, hot water treatment, or the use of acidic or alkaline chemicals, often tailored to the specific lignocellulose source. In some cases, extremophilic microorganisms are employed to facilitate high-temperature pretreatment in the presence of catalytic agents.

Despite its benefits, high-temperature pretreatment can also lead to undesirable effects such as the release of phenolic inhibitors, lignin solubilization, and the formation of organic acids like acetic acid—substances that can interfere with downstream microbial or enzymatic steps. Thus, achieving an optimal balance between pretreatment efficiency and its impact on further processing is critical.

Cellulose, which can constitute up to 50% of lignocellulosic waste, must be hydrolyzed into glucose via the action of three key enzymes: endoglucanases, cellobiohydrolases, and β-glucosidases. Given the high temperatures associated with efficient hydrolysis, thermostable enzymes and microbial systems are preferred. Recent advances include the cloning and expression of thermostable cellulases in *Trichoderma reesei*, enabling hydrolysis at 60 °C–15 °C higher than that achievable with standard commercial enzymes. Although numerous thermophilic cellulases have been identified, only a few, such as those from *Caldicellulosiruptor saccharolyticus* and *Anaerocellum thermophilum*, exhibit the multifunctional and multidomain characteristics required to degrade cellulose fully to glucose. However, these enzymes are not yet available in isolated form for industrial use.

The genome of *Caldicellulosiruptor saccharolyticus* encodes cellulases with domain structures optimized for efficient cellulose degradation. One of these enzymes consists of an endoglucanase catalytic domain from glycoside hydrolase family GH10, three cellulose-binding modules (CBMs), and an exoglucanase domain from family GH5. Another encoded cellulase contains an endoglucanase domain (GH9), three CBMs, and an exoglucanase domain (GH48). These modular enzymes likely underlie the organism's ability to utilize microcrystalline cellulose as a carbon source.

By contrast, while other thermophiles such as *Thermotoga maritima* and *Pyrococcus furiosus* encode endoglucanases, their enzymes lack cellulose-binding domains, limiting their ability to degrade microcrystalline cellulose—one of the main structural components of lignocellulosic biomass. To date, no thermophilic microorganism capable of growth near the boiling point of water has been found to produce cellulases with the complete set of functional domains required for efficient cellulose degradation, including endoglucanase, exoglucanase, and cellulose-binding activities.

The expression of highly thermostable cellulases in mesophilic hosts remains technically challenging and economically unfeasible for large-scale applications. Simultaneously, the genetic engineering of hyperthermophilic hosts is still in its early development. Current efforts aim to enhance their ability not only to degrade crystalline cellulose but also to tolerate inhibitory byproducts and high concentrations of glucose and ethanol—conditions typical in biofuel production.

Hemicellulose, accounting for approximately 25–35% of lignocellulosic biomass, is composed of heterogeneous polymers containing pentoses, hexoses, and uronic acids. Xylan, a major hemicellulose component, is a key source of fermentable xylose. Its enzymatic hydrolysis typically involves endo-1,4-β-xylanases and β-xylosidases. As with cellulases, thermostable xylanases are preferred in applications such as lignocellulose conversion, animal feed enhancement, and chlorine-free pulp bleaching.

A notable example is Xyn10A from *Rhodothermus marinus*, which exhibits an optimal operating temperature of 85 °C. This enzyme has demonstrated high efficiency in the bleaching of both hardwood and softwood pulps, supporting its use in sustainable and high-temperature bioprocesses.

1.3.3 Peculiarities of the Cold-Adapted Organisms

Organisms adapted to cold environments—also known as *psychrophiles*—exhibit unique physiological and molecular features that enable them to thrive in habitats where temperatures approach or even fall below the freezing point of water. Despite the inhibitory effect of low temperatures on biochemical reaction rates, these organisms maintain metabolic activity comparable to that of mesophiles and play critical roles in cold ecosystems, which make up more than two-thirds of the Earth's biosphere.

Although less extensively studied than thermophiles, psychrophiles are of increasing scientific and technological interest. Their importance lies in understanding fundamental aspects of enzyme kinetics, protein folding, structural flexibility, membrane dynamics, and resistance to freezing. They are also valuable for biotechnological applications such as low-temperature biocatalysis and bioremediation.

Recent research has identified several key features underlying cold adaptation. Psychrophilic enzymes consistently exhibit two defining characteristics:
(1) High catalytic activity at low and moderate temperatures, and
(2) Low thermal stability, making them easy to inactivate under mild conditions.

These properties are particularly desirable in industrial processes that require high enzymatic efficiency without the need for elevated temperatures.

The molecular basis of this adaptation lies in discrete structural changes that increase enzyme flexibility. These modifications reduce the number and strength of weak bonds stabilizing the protein's structure, thereby lowering the enthalpy of activation required for catalysis. This is often accompanied by a greater reduction in activation entropy, indicating a more disordered ground state. As a result, these enzymes are more flexible and dynamic, a trait measurable through techniques such as fluorescence quenching. This flexibility helps preserve function at low temperatures and explains why thermophilic enzymes generally show poor activity under such conditions.

Structural evidence supports these findings: crystallographic studies of approximately 20 cold-adapted bacterial enzymes consistently show fewer stabilizing interactions compared to their mesophilic counterparts.

In terms of biodiversity, psychrophilic ecosystems are far more diverse than initially assumed. For example, microbial cell densities of 10^5–10^6 cells/mL have been recorded in Antarctic seawater—values comparable to those in temperate zones. Even in frozen environments such as glacial ice, sea ice, or permafrost, microbial life persists, although diversity may be constrained by the limited volume of residual liquid phases. In such conditions, compact cell forms with low surface-to-volume ratios may be favored, helping minimize thermal energy loss.

Microbial activity has been observed at temperatures as low as $-20\,°C$. The remarkable physiological and structural adaptations that enable life under such extreme conditions also make cold-adapted organisms promising tools for the bioremediation of polluted low-temperature environments, including those contaminated with xenobiotics.

Enzymes produced by cold-adapted organisms possess two defining properties that are highly attractive for biotechnological applications: high specific activity at low and moderate temperatures and relatively low structural stability, allowing for easy inactivation under mild conditions, particularly when the influence of the substrate is minimal. Interest in these enzymes dates back to 1984, when Kobori and colleagues isolated a psychrophilic strain from Antarctica that produced a heat-sensitive alkaline phosphatase. This enzyme proved ideal for ^{32}P-labeling of oligonucleotides, as it could be rapidly inactivated before interfering with subsequent polynucleotide kinase reactions—unlike conventional phosphatases.

In molecular biology, cold-adapted strains capable of carrying and expressing mesophilic genes have been developed. Low-temperature expression systems are valuable for minimizing or avoiding inclusion body formation, as the hydrophobic interactions responsible for protein aggregation are weaker at reduced temperatures.

Cold-active hydrolases have great potential in detergent formulations, particularly in regions where manual washing is prevalent or where energy efficiency drives a shift toward lower washing temperatures. These enzymes maintain activity under such conditions and reduce the need for heating. The most widely used group is proteases, especially serine proteases such as subtilisin-like enzymes. Commercial examples include Kannase and

Polarzyme (Novozymes) and Purafect Prime and Properase (Genencor). Notably, a cold-adapted protease from Antarctic krill (*Euphausia superba*) has been patented and is commercially available.

In textile processing, cold-adapted amylases are used for desizing, while cellulases contribute to biofinishing (removal of surface fibrils and pills) and stone-washing for denim fading. Products like Celluzyme (Novozymes) offer low-temperature activity that helps preserve fabric quality.

Cold-adapted enzymes also enhance food processing through both their activity and ease of inactivation. In juice production, pectinases and xylanases facilitate extraction and reduce viscosity and turbidity. In the dairy sector, cold-active β-galactosidases improve lactose digestibility—relevant for lactose-intolerant populations—and enable environmentally responsible use of whey. Cold-active proteases are used in meat tenderization, while other hydrolases improve the digestibility of animal feeds. Many applications are protected by patents.

In the baking industry, cold-adapted xylanases (notably from glycoside hydrolase family 8) improve dough elasticity, machinability, and bread quality, including increased volume and improved crumb structure.

The cosmetics and pharmaceutical industries benefit from the use of cold-active enzymes in biotransformations involving volatile or temperature-sensitive compounds. In organic synthesis, these enzymes are preferred for producing enantiomers, peptides, lipids, and sugars, particularly via reverse hydrolysis in low-water environments. Their conformational flexibility gives them superior performance at low temperatures compared to mesophilic enzymes.

Finally, in the biofuel sector, cold-adapted enzymes offer energy-efficient alternatives to traditional enzymatic steps in low-temperature saccharification. Products such as Stargen (Genencor) and BPX technology (a collaboration between Novozymes and Broin Company) illustrate the growing role of cold-active enzymes in sustainable biofuel production.

1.3.4   Role of Extremophiles in Bioremediation

Currently, natural attenuation and induced biodegradation represent the predominant strategies for remediating soils and waters contaminated with petroleum products, xenobiotic compounds, and wastewater rich in organic substances such as proteins, lipids, and sugars. In cold climates, temperature plays a critical role, as the rate of microbial degradation follows the temperature dependence of chemical reactions, typically increasing exponentially with temperature in accordance with the Arrhenius equation.

The pivotal role of microorganisms in the natural breakdown of environmental pollutants is well documented. Historical events such as the Torrey Canyon oil spill (1967) and the Exxon Valdez disaster (1989)—in which approximately 40,000 tons of crude oil were released into the cold waters of Alaska's Prince William Sound—illustrate how indigenous microbial populations significantly reduced environmental damage. In both cases, bioremediation occurred at unexpectedly rapid rates, demonstrating the capacity of native psychrophilic and cold-tolerant microbes to mitigate pollution. As a result, bioremediation has become a core approach for restoring the ecological function of polluted environments.

In wastewater treatment, *bioaugmentation*—the introduction of specialized microbial degraders—is a widely practiced method. Psychrophilic microorganisms offer distinct advantages over mesophiles in colder regions, due to their ability to remain active across a broad temperature range. However, the success of introduced strains often depends on their ability to compete with native microbial communities and to tolerate environmental toxicity.

A broad array of microorganisms—including bacteria, fungi, cyanobacteria, and microalgae—have demonstrated the capacity to degrade hydrocarbons. In polar ecosystems, hydrocarbon-degrading microbes exhibit considerable phylogenetic diversity, suggesting multiple evolutionary solutions to this challenge. In such environments, bioaugmentation products (microbial consortia, inocula, or engineered strains) have been shown to enhance the degradation of oil-contaminated water and soil more effectively than nutrient addition alone.

The success of bioaugmentation, however, often depends on the composition and adaptability of the biostimulation consortium. In some cases, supplementing indigenous microbial communities with nutrient-rich additives such as Inipol EAP-22 has been more effective than introducing foreign strains. Moreover, cold-adapted microorganisms have shown promise in degrading persistent xenobiotic compounds, including chlorophenols, which are commonly found in legacy pollutants such as wood preservatives and pesticides.

1.3.5 Biological Agents in Low pH Environments

Microorganisms that thrive in consistently acidic environments are known as acidophiles. These include extreme acidophiles, which grow optimally at pH <3, and moderate acidophiles, which prefer a pH range of 3–5. Acidic conditions can arise through both natural and anthropogenic processes, such as fermentation, nitrification, sulfur oxidation, and sulfide mineral exposure, all of which contribute to sulfuric acid production that overwhelms the buffering capacity of soils and aquatic systems.

Natural acidic environments are common in volcanic regions, where reactions between hydrogen sulfide (H_2S) and sulfur dioxide (SO_2) lead to the formation of elemental sulfur, which is further oxidized by sulfur-oxidizing bacteria into sulfuric acid. This process results in extremely acidic habitats, such as the solfatara fields near Naples (Italy) and the hot springs of Yellowstone National Park (USA). Anthropogenic acidification, particularly from mineral ore mining, also creates highly acidic environments by exposing sulfide-rich rocks like pyrite (FeS_2) to oxygen and moisture, leading to acid mine drainage. These processes result in mineral dissolution and the release of toxic metal ions, which can be harmful to most organisms.

Acidophiles are widely distributed across the Archaea and Bacteria domains and frequently exhibit thermophilic traits due to the geothermal origins of many acidic habitats. Some researchers propose that these organisms reflect early life conditions on Earth, suggesting that key metabolic pathways may have originated on sulfide mineral surfaces. Acidophilic microbes, therefore, may be among the most ancient lineages of life, offering insights into prebiotic evolution.

Despite living in highly acidic surroundings, acidophiles maintain a near-neutral intracellular pH. They achieve this through several adaptations: developing proton-impermeable membranes, reversing the membrane potential, actively exporting protons (H^+) via ATP-dependent pumps, and enhancing cytoplasmic buffering capacity. These mechanisms confer high tolerance to metal cations, but also result in greater sensitivity to anions compared to neutrophilic organisms.

In terms of metabolism, acidophiles commonly utilize inorganic electron donors, such as ferrous ions (Fe^{2+}) and reduced sulfur compounds. While oxygen serves as the primary electron acceptor in most environments, the ferric ion (Fe^{3+}) can also act as an alternative in acidic conditions due to its high redox potential. Genomic analyses of model acidophiles—including *Acidithiobacillus ferrooxidans*, *Ferroplasma acidarmanus*, and archaeal species like *Thermoplasma acidophilum*, *Picrophilus torridus*, and *Sulfolobus tokodaii*—reveal the presence of genes involved in iron and sulfur oxidation, pH homeostasis, and metal resistance.

Members of the genus *Ferroplasma*, which thrive in extreme acidity, lack a cell wall and exhibit unique membrane structures composed of tetraether lipids, glycoproteins, lipoglycans, and lipopolysaccharide-like materials. Their ability to use iron as an electron donor is considered an evolutionary relic, supporting the hypothesis that early life may have originated on iron–sulfur-rich mineral surfaces, such as pyrite.

Genomic studies have also identified various transporters, including H^+/K^+-transporting ATPases, and multiple chaperone proteins that stabilize protein structure and function under acidic conditions. Furthermore, genomic expression libraries have revealed that some intracellular enzymes in *Ferroplasma acidiphilum* exhibit acidic pH optima, suggesting the possibility of subcellular compartmentalization or the influence of unknown selective pressures on enzyme expression.

1.3.6   Biological Agents in Highly Alkaline Environments

Microorganisms that thrive in environments with pH values above 9 are known as *alkaliphiles*. They fall into two major physiological categories:
- *True alkaliphiles*, which grow optimally at pH ≥ 9
- *Haloalkaliphiles*, which tolerate both high pH and high salinity (e.g., up to 5 M NaCl)

Alkaliphiles are found in diverse habitats, including those shaped by geological and climatic processes. Alkaline environments can result from the weathering of silicate minerals such as pyroxene ($MgCaFeSiO_3$) and olivine ($MgFeSiO_4$) in the presence of CO_2. In such settings, Mg^{2+} precipitates as serpentine ($Mg_3Si_2O_5(OH)_4$), producing highly reducing waters rich in $Ca(OH)_2$ and $Fe(OH)_2$, often with pH values exceeding 11.

Haloalkaliphiles are frequently isolated from extreme ecosystems such as soda lakes in East Africa and the western United States, characterized by high concentrations of sodium carbonate and bicarbonate. These waters are typically low in Ca^{2+} and Mg^{2+}, and the combination of CO_2 dissolution and rapid evaporation can raise pH levels to approximately 12. In contrast, calcium-rich environments—like the Dead Sea—lead to calcite ($CaCO_3$) and dolomite ($MgCa(CO_3)_2$) formation, resulting in near-neutral pH conditions.

Alkaliphiles maintain near-neutral cytoplasmic pH despite their external surroundings. They achieve this through multiple strategies:

- Cell wall acidification, via incorporation of teichuronic acids, glutamic acid, and aspartic acid.
- Na^+/H^+ antiport systems in the membrane, which drive nonstoichiometric exchange of one inward H^+ for more than one outward Na^+, thereby maintaining a negative membrane potential.
- In some hyperthermophiles, ATP synthesis is facilitated by Na^+-dependent ATP synthases rather than traditional proton-driven systems.

Additionally, metabolic adaptations include elevated production of organic acids and overexpression of enzymes such as amino acid deaminases, enhancing cytoplasmic buffering. Extracellularly, these organisms produce alkaline-adapted enzymes, especially hydrolases, to aid in carbon and nitrogen acquisition.

Alkaliphilic microorganisms span diverse genera, including *Bacillus*, *Pseudomonas*, *Streptomyces*, and *Synechocystis*, as well as numerous archaeal species. However, the most industrially relevant alkaline enzymes—such as proteases, lipases, amylases, cellulases, and xylanases—are predominantly sourced from Bacillus, Pseudomonas, and certain fungi. These enzymes have been extensively studied using rational design and directed evolution to enhance their activity under alkaline conditions.

One key factor in enzyme adaptation is the modulation of the pKa of catalytic residues, particularly in acidic side chains. Protein engineering techniques—such as modifying hydrogen bonds or altering solvent accessibility—can shift pKa values, optimizing enzyme activity at high pH. Comparative studies show distinct amino acid usage patterns between alkaline and non-alkaline enzymes.

In biotechnology, alkaline-active hydrolases have become essential components of laundry detergents. Since Otto Röhm's 1913 patent, the detergent enzyme market has grown significantly; alkaline enzymes now account for approximately 30% of global enzyme production. These enzymes function optimally at 20–60 °C and pH 7.5–10.5, aligning with modern energy-saving wash cycles.

Advances in enzyme formulation have accompanied environmental considerations. For instance, phosphate-based detergents have been replaced by zeolite- or silicate-based alternatives, in combination with tailor-made alkaline enzymes, to reduce ecological impact and energy consumption. Additional components may also include:

- Builders (e.g., sodium carbonate, zeolite) to soften water and destabilize particle adhesion
- Oxygen-based bleaching agents (e.g., sodium perborate, hydrogen peroxide) to enhance stain removal

1.3.7 Biotechnological Enhancement of Detergent Compositions: Proteases, Lipases, and Amylases

Since the discovery of the Carlsberg subtilisin protease from *Bacillus licheniformis* in 1960, numerous subtilisin-like serine proteases—members of the subtilase A superfamily—have been identified. These enzymes have become indispensable in industrial detergents due to their broad substrate specificity and remarkable stability under harsh washing conditions, including high temperatures and alkaline pH, which are necessary for effective removal of proteinaceous stains such as blood, grass, and sauces.

Since 1984, *subtilisins* have been subjected to extensive protein engineering to improve substrate specificity and enhance resistance to degradation by bleaching agents. For example, Subtilisin BPN' (*Bacillus protease Novo type*), a 275-amino-acid protein, has been modified via both directed evolution and rational design to yield oxidation-stable proteases (OSPs). Traditional subtilisins are susceptible to oxidation, especially in the presence of bleach, which results in rapid activity loss.

One successful variant, derived from *Bacillus* sp. strain KSM-KP43, demonstrated enhanced oxidation resistance and is now incorporated into Japanese laundry detergents. Given that subtilisins are calcium-dependent, their stability can be compromised by water-softening agents (e.g., EDTA), which displace Ca^{2+} ions. In subtilisin BPN', a deletion at a calcium-binding site (site A) enabled resistance to chelators but increased thermal sensitivity. Other mutations, however, enhanced thermostability by a factor of ten. Alternative stabilization strategies involve increasing free calcium concentrations, particularly important in formulations containing other Ca^{2+}-sensitive enzymes like amylases.

Lipases are another key class of detergent enzymes, hydrolyzing ester bonds in acylglycerols, and thus facilitating the removal of sebum and greasy food stains. Effective detergent lipases must function under alkaline, oxidative, and metal-rich conditions, and exhibit broad substrate specificity. However, lipid insolubility in aqueous systems necessitates the use of surfactants to improve enzyme access. The activity of lipases is also influenced by micelle structure, including size and distribution.

The first alkali- and heat-stable lipase introduced for commercial use was Lipolase (Novo Nordisk, 1988), derived from *Thermomyces lanuginosus* and expressed in *Aspergillus oryzae*. Although effective, it required multiple wash cycles for complete stain removal, with optimal activity occurring during drying. Subsequent protein engineering efforts, including site-directed mutagenesis, led to Lipolase Ultra (D96L mutant). This version, featuring a substitution in the lipid-binding region, reduced repulsion from fatty acids and anionic surfactants, increasing activity on mixed monolayers of diglycerides and surfactants.

Newer detergent lipases, effective in a single wash cycle, have been discovered via environmental screening. Examples include alkaline lipases from alkaliphilic *Pseudomonas mendocina* and *P. alcaligenes*, now marketed by Genencor as Lumafast and Lipomax, respectively.

Amylases are essential in detergents for starch degradation, preventing dirt from binding to starch residues on fabrics. Early detergent formulations used non-alkaline amylases from *B. amyloliquefaciens*, *B. stearothermophilus*, and *B. licheniformis*. The discovery of AmyK, an alkali-stable amylase from *Bacillus* sp. KSM-1378, marked a significant advance. AmyK exhibits optimal activity at pH 8–8.5 and 55 °C, and through mutation-driven enhancement of calcium binding, has been stabilized for use under oxidative and high-temperature conditions.

A homologous enzyme, AmyK38, was later isolated from *Bacillus* KSM-K38 and retains activity even in the presence of chelating agents, as it lacks a calcium-binding site. Chelator-resistant amylases, such as AmyK38, are now patented (US Patent 6916645) for use in formulations containing metal ion chelators.

In addition to proteases, lipases, and amylases, detergent formulations also benefit from *cellulases*, such as Celluzyme from *Humicola insolens* (Novozymes). These endo-1,4-β-glucanases (EC 3.2.1.4) assist in removing particulate soil by hydrolyzing cellulose residues on fabrics.

Overall, biotechnological applications in the detergent industry demand a diverse range of enzymes tailored for performance under extreme pH, temperature, and chemical conditions. Achieving optimal outcomes with minimal environmental impact and low production cost requires precise calibration of enzymatic properties. To meet these needs, manufacturers increasingly explore microorganisms from extreme environments, where novel enzyme candidates may be found.

However, cultivating extremophiles presents technical challenges due to their unique growth requirements. To overcome these limitations, recent advances in metagenomics and functional screening enable the discovery and exploitation of enzymes from uncultivable microorganisms. Combined with modern cloning and expression technologies, these tools now allow for rapid, large-scale production of extremophilic enzymes in standard industrial hosts—ushering in a new era of enzyme engineering for tailored biotechnological solutions.

Brainstorming
1. What are mesophiles, and why are they preferred for biotechnological processes?
2. Can you provide examples of environmental conditions that are considered ideal for biotechnological processes?
3. Why might the conditions mentioned in the text not be ideal for certain biotechnological applications?

4. How do extremophiles and their enzymes offer unique opportunities in biotechnology?
5. What are extremozymes, and how are they different from enzymes derived from mesophiles?
6. How can extremophile enzymes be integrated into industrial processes, and what advantages do they offer?
7. What role do recombinant enzymes play in biotechnology, and why are they considered suitable for industrial applications?
8. According to the text, what is the potential of biological tools in replacing traditional industrial processes, especially in the context of environmental concerns?
9. How does the search for extremophiles contribute to the extension of microorganism and product usage in unexpected transformations?
10. What are the challenges associated with the large-scale production of extremophiles for industrial applications, and how can these challenges be addressed?
11. How do subtilisin-like proteases contribute to the effectiveness of industrial detergents?
12. What modifications have subtilisins undergone through protein engineering, and how do these modifications enhance their properties?
13. Why is resistance to oxidation important for proteases in laundry detergents, and how has this been achieved in certain mutants?
14. How do lipases contribute to the removal of sebum and fatty food stains in detergents?
15. What challenges do lipases face in laundry detergents, and how have protein engineering techniques been applied to address these challenges?
16. Can you elaborate on the characteristics of Lipolase-Ultra and how it improves washing efficiency?
17. What role do amylases play in the detergent industry, and why is the discovery of high pH-stable amylases advantageous?
18. How has the enzyme AmyK been modified to enhance its stability and activity in different conditions?
19. What properties make amylase AmyK38 resistant to calcium-chelating agents, and how is it used in detergent formulations?
20. What is the role of cellulases, specifically endo-1, 4-β-glucanases, in detergent additives, and how do they contribute to soil particle removal from laundry?
21. How do manufacturers calibrate experimental conditions to achieve optimal results in biotechnological processes?
22. What challenges arise when working with extremophiles in biotechnological applications, and how have recent techniques like metagenomics addressed these challenges?
23. Can you explain the significance of metagenomics in the study of extremophiles for biotechnological applications?
24. How do new cloning techniques facilitate the expression of extremophilic proteins in easily cultivable microorganisms, and what impact does this have on biotechnological research?

Take-Home Messages

- Extremophiles are key to expanding the biotechnological toolbox-Microorganisms adapted to extreme conditions—such as high or low temperatures, pH extremes, salinity, or pressure—possess unique physiological and enzymatic traits that make them valuable for industrial applications, especially where conventional biocatalysts fail.

- Thermophilic microorganisms play a crucial role in biomass conversion. Thermostable enzymes from thermophiles (e.g., cellulases and xylanases) enable the efficient hydrolysis of lignocellulose under harsh pretreatment conditions. These enzymes are essential for converting agricultural waste into biofuels and bioproducts.

- Cold-adapted organisms offer advantages in low-temperature bioprocessing. Psychrophilic microbes and their enzymes exhibit high activity at low temperatures and rapid inactivation at moderate heat, making them suitable for energy-saving applications such as cold-water detergents, food processing, and molecular biology tools.

- Extremophiles contribute to the bioremediation of polluted environments. Psychrophilic, acidophilic, and halophilic microorganisms are capable of degrading hydrocarbons and xenobiotics under extreme conditions (e.g., cold, acidic, or saline), offering solutions for cleaning up contaminated soils and waters, especially in challenging climates.

- Acidophiles and alkaliphiles have evolved sophisticated pH homeostasis mechanisms
Acidophiles maintain internal pH neutrality in acidic environments through proton export and membrane impermeability, while alkaliphiles adapt through Na^+/H^+ antiport systems, cell wall acidification, and specialized metabolic adjustments.

- Extremophilic enzymes are integral to modern detergent formulations. Enzymes such as oxidation-stable proteases, alkaline lipases, chelating-agent-resistant amylases, and alkali-tolerant cellulases are engineered for high performance under detergent conditions (pH 8–11, 20–60 °C, surfactants, chelators), reducing environmental impact and improving stain removal.

- Metagenomics and protein engineering drive the discovery of next-generation enzymes. Due to the cultivation challenges associated with extremophiles, metagenomic approaches and recombinant expression systems have become essential for accessing and optimizing enzymes from uncultivable sources, accelerating innovation across biotechnology sectors.

1.4   Strategies to Improve the Efficiency of Bio-agents: Mutagenesis and Selection

The main objective of industrial biotechnology is to maximize the production yield of valuable compounds or biomass generated by biological agents. The key strategy involves increasing product yield per unit of microbial biomass. In most cases, wild-type organisms isolated from natural environments are not sufficiently efficient for commercial use; without further improvement, the production process may be economically unviable or technically impractical.

The essential goals for improving bio-producer strains include:

1. Enhancement of productivity to achieve high yields of target substances per unit biomass.
2. Adaptation to inexpensive and non-scarce media components.
3. Elimination of feedback inhibition in the biosynthesis of the final product.
4. Increased resistance to bacteriophages and viral infections.
5. Compatibility with basic or older bioreactor systems, ensuring biosynthesis performance is not reduced by suboptimal equipment (e.g., reducing foam formation).
6. Improved hygienic and organoleptic properties (e.g., absence of unpleasant odors).

Thousands of microbial strains are currently used in the biotechnology industry. The three most common industrial strains are shown in Fig. 1.6.

Most of these strains have been improved via induced mutagenesis followed by artificial selection, enabling large-scale synthesis of desired products (see Figs. 1.7 and 1.8).

Selection is the process of artificially isolating microbial variants that possess favorable traits from a population. It typically involves two steps: (1) identifying naturally occurring strains with potential and (2) enhancing their traits to develop industrial-grade strains. This process primarily aims to amplify microbial productivity.

Microbial populations are inherently heterogeneous, containing individuals with and without desirable traits—denoted as (+) and (−) variants, respectively. These differences arise from spontaneous mutations, which occur randomly and at a low frequency ($\sim 10^{-10}$ per gene per generation). However, in large microbial populations (10^{15}–10^{17} cells), a significant number of mutant cells will appear even in a single generation. This natural variability provides a foundation for selection and adaptive evolution under industrial conditions (Fig. 1.9).

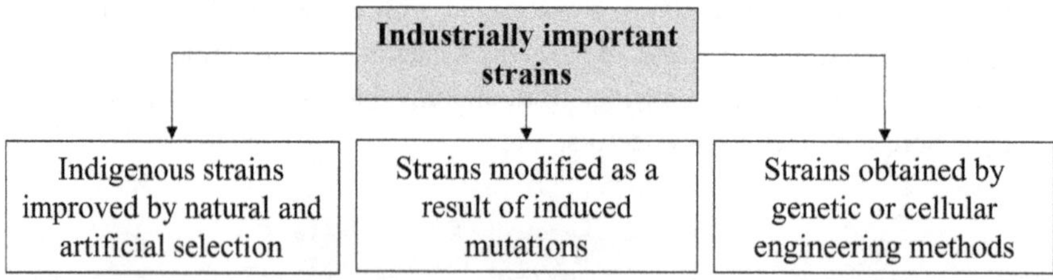

Fig. 1.6 Industrially important types of microbial strains

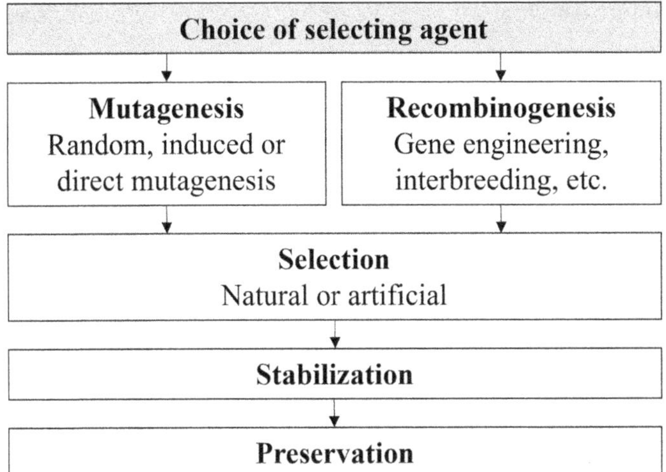

Fig. 1.7 General stages of microbial selection

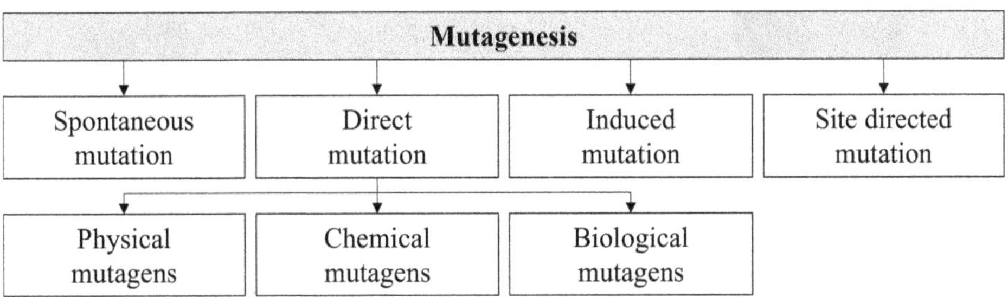

Fig. 1.8 Types of mutagenesis

Traditional strain improvement methods have relied on long-term selection. Examples include the development of *Saccharomyces cerevisiae*, *S. uvarum*, and various other yeasts, acetic acid bacteria, and propionic acid bacteria used in food and beverage production. In one notable case, ethanol-tolerant yeast strains were selected by cultivating *S. uvarum* in bioreactors under ethanol stress. Ethanol, both a metabolic product and an inhibitor, was introduced progressively, while CO_2 output served as a marker of metabolic activity. Through continuous cultivation, strains with ethanol tolerance up to 10% were obtained. Similar approaches can be used to select strains resistant to acidic or alkaline conditions, toxic metabolites, or heavy metal ions.

A major limitation of selection based on spontaneous mutations is the long time required due to the low frequency of mutation events. To overcome this, induced mutagenesis is employed as a more rapid and effective approach.

Induced mutagenesis is a method that intentionally increases the mutation rate by damaging DNA through physical, chemical, or biological mutagens. It enables rapid genera-

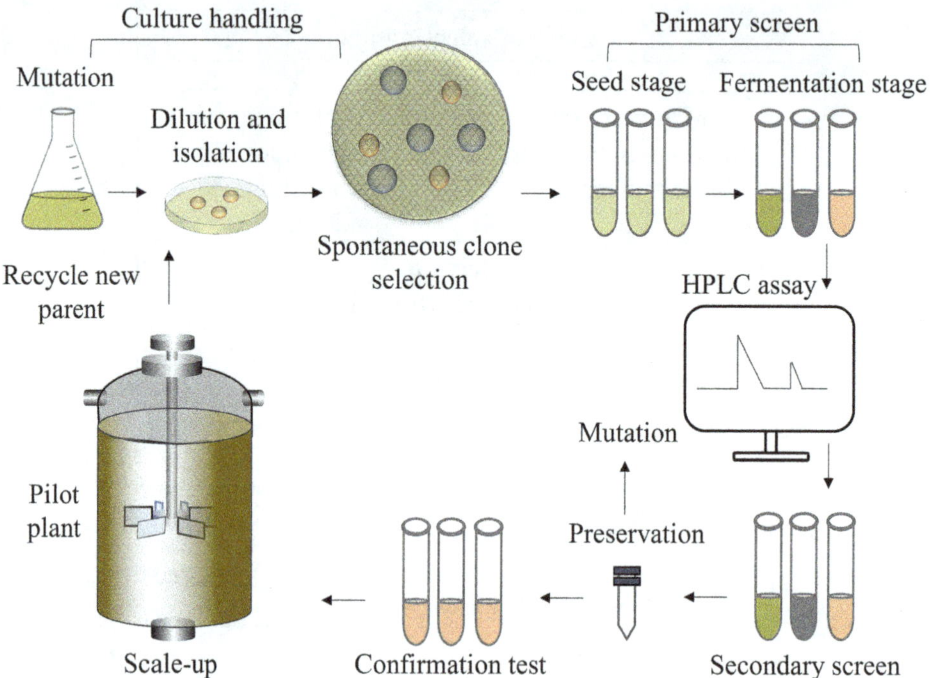

Fig. 1.9 The typical procedure in the mutation and spontaneous strain selection process

tion of microbial diversity for subsequent screening and selection. Common mutagenic agents include:

- *Physical mutagens*: ultraviolet (UV) radiation, X-rays, γ-rays, fast electrons, protons, neutrons, and positrons
- *Chemical mutagens*: nitrous acid, alkylating agents, acridine dyes, and bromouracil
- *Biological mutagens*: bacteriophages and plasmids

These agents alter the primary structure of DNA, resulting in nucleotide substitutions, deletions, or insertions that may yield beneficial phenotypic changes. A summary of common mutagens and their mechanisms is presented in Table 1.3.

The primary phenotypic trait used in microbial strain improvement is an increased ability to produce the target product. To identify such high-performing mutants, selection systems are designed to screen for desirable characteristics. Resistance to antibiotics, toxic compounds, or phage infection often serves as a selectable marker during mutant screening.

After mutagenic treatment, cells are typically plated on solid selective media, where only resistant colonies can grow. For the identification of auxotrophic mutants, replica plating is used on minimal media containing specific metabolites, allowing for the rapid analysis of hundreds of mutants per agar plate. A wide range of mutants can be obtained depending on the selective conditions and screening technique employed, including methods like replica plating, the penicillin enrichment technique, and others (see Fig. 1.10).

Table 1.3 Mutagens used for strain improvement

Mutagens	Mechanism	Applications
Physical agents		
Ionizing radiation (X-rays, γ-rays)	Leads to ss and dsDNA breakage	Major genetic alterations
Short wavelength (UV light)	Thymidine and cytidine form dimers	Point mutations
Chemical agents		
Nitrate	Deaminates adenine to hypoxanthine, cytidine to uridine	Point mutations
Alkylating agents (NTG, EMS)	Alkylate purines	Point mutations
Base analogs (5-Chlorouracil, 5-bromouracil)	Incorporated into replicated DNA	Major genetic alterations
Intercalating agents (acridine orange, ethidium bromide)	Intercalates into DNA	Major genetic alterations
Biological agents		
Phages, plasmids, DNA transposons	Transfer DNA elements within a chromosome	Gene markers

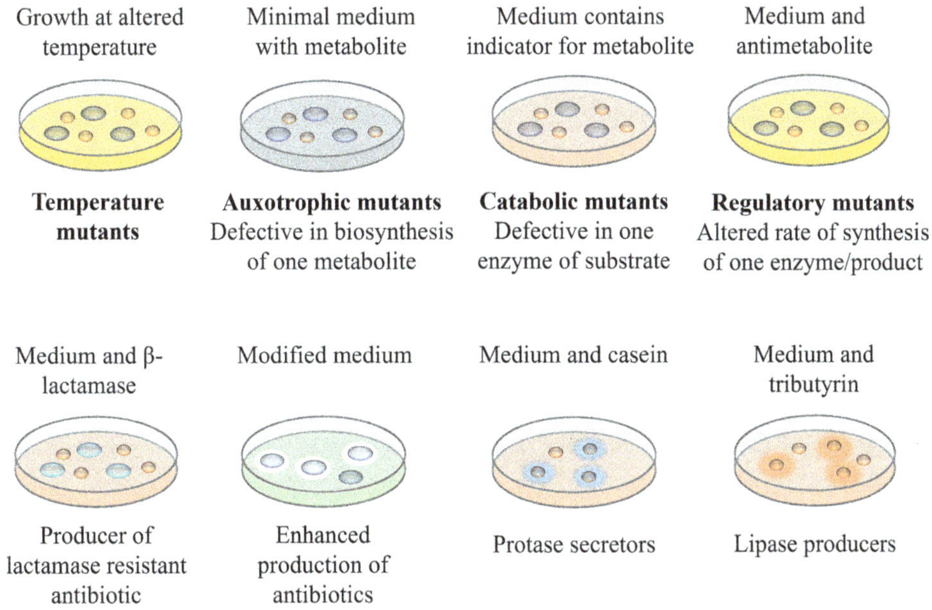

Growth at altered temperature	Minimal medium with metabolite	Medium contains indicator for metabolite	Medium and antimetabolite
Temperature mutants	**Auxotrophic mutants** Defective in biosynthesis of one metabolite	**Catabolic mutants** Defective in one enzyme of substrate	**Regulatory mutants** Altered rate of synthesis of one enzyme/product

Medium and β-lactamase	Modified medium	Medium and casein	Medium and tributyrin
Producer of lactamase resistant antibiotic	Enhanced production of antibiotics	Protase secretors	Lipase producers

Fig. 1.10 The selective media technique

Following mutagenesis, the number of both "positive" (productive) and "negative" (non-viable or undesired) mutants increases significantly. Higher doses of mutagens lead to a greater frequency of lethal and irrelevant mutations, but may also increase the fraction of desirable survivors. Therefore, balancing the mutagen dose is critical to ensure sufficient diversity while preserving viability.

Mutant selection can also be carried out in continuous culture systems, such as chemostats, where microbial populations are exposed to selective pressure in real time. For example, replacing a preferred carbon source with a suboptimal one can favor the survival of mutants capable of utilizing the alternative substrate. However, such systems are less suited for selecting overproducing strains, as metabolite production is not necessarily linked to survival under selective pressure. Thus, it is essential to evaluate both the production capacity of selected mutants and their ability to retain key traits of the wild-type strain. Candidate strains with superior attributes are typically cultivated in shaking flasks, mimicking industrial process conditions. To reduce the accumulation of deleterious mutations from repeated mutagenesis cycles, promising strains may be crossed with wild-type or minimally modified strains.

The metabolic networks of microbial cells are governed by complex regulatory systems that enforce stability and metabolic balance. To push cells beyond their natural production limits, these systems must be strategically disrupted. Through mutations affecting both regulatory and structural genes, the production of target compounds can be significantly enhanced. Such modified microorganisms are sometimes called *overproducers*.

For instance, mutagen-induced alterations such as deletions, inversions, transpositions, or point mutations (see Fig. 1.11) can lead to the emergence of novel traits or enhanced expression of existing pathways. Structural genes relocated by transposition or inversion may come under the control of new promoters, resulting in unusual hybrid proteins or

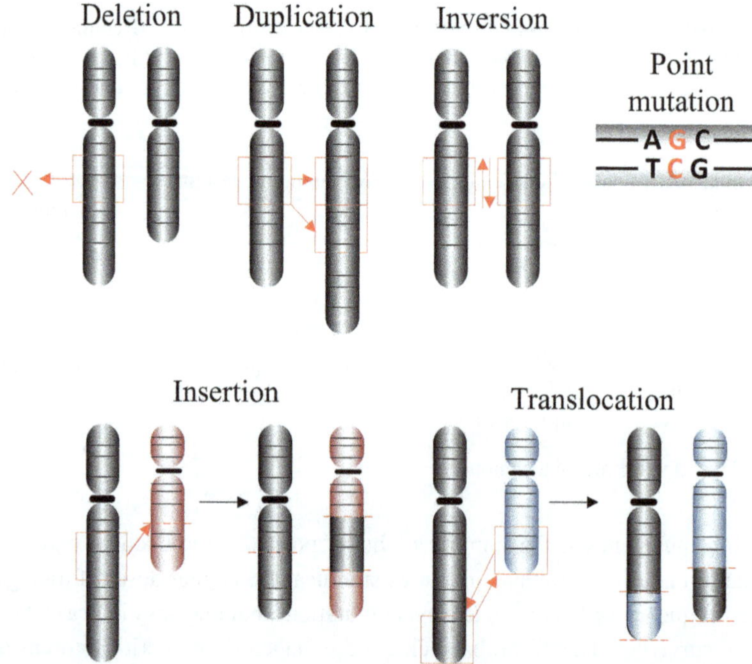

Fig. 1.11 Types of mutations

overexpression of biosynthetic enzymes. Likewise, repressor genes that normally inhibit production can be inactivated, and feedback inhibition loops can be disrupted, leading to unregulated metabolite synthesis.

Mutations may also modify the transport systems responsible for precursor uptake, increasing substrate availability for the desired biosynthetic pathway. In addition, enhancing the strain's resistance to its own metabolic products prevents self-toxicity—a critical requirement for the industrial production of antibiotics and other bioactive compounds.

A common mechanism underlying microbial strain improvement is the occurrence of point mutations, such as single-nucleotide substitutions. These changes may result in the replacement of one amino acid in a protein during translation, potentially altering the protein's structure, function, or stability—particularly when the mutation occurs within active or allosteric sites.

Another powerful strategy is the increase of gene dosage through duplication or amplification. Gene amplification involves a significant rise in the number of copies of a target gene, leading to a substantial increase in the biosynthesis of the encoded product. This is typically achieved via plasmid-based amplification, where plasmids—normally present in 1–30 copies per cell—can be increased to up to 3000 copies under optimized conditions. These plasmids may carry 2–250 genes, including structural and regulatory sequences.

Amplification has been widely used in *Escherichia coli* and other industrial microorganisms. It is now feasible to transfer any chromosomal gene or operon into a plasmid, which can then be introduced into recipient strains for amplification. Intergeneric plasmid transfer has also been achieved—for example, *Bacillus* plasmids can be transferred between cells using polyethylene glycol, while plasmids from *Pseudomonas* can be introduced into other Gram-negative bacteria (*Acinetobacter, Agrobacterium, Klebsiella, Proteus, Rhizobium, Salmonella*). Plasmids from *Staphylococcus aureus* have also been successfully replicated in *Bacillus subtilis*.

Many antibiotic-producing strains naturally carry plasmids encoding the biosynthetic genes or regulatory elements involved in antibiotic production. Amplifying these plasmids or their specific genes has proven highly effective in increasing antibiotic yields.

Microorganisms are ideal candidates for mutagenesis-selection cycles due to their rapid reproduction, genetic variability, and ease of handling. Repeated exposure to mutagenic factors combined with targeted selection allows for stepwise enhancement of strain productivity. For example:

- Penicillin production has been improved 100,000-fold by selecting mutants with increased LLD-tripeptide synthesis, reduced feedback inhibition, and enhanced cell wall permeability.
- The yield of streptomycin-producing strains has been increased 20,000-fold through sequential treatments with UV and X-ray irradiation.
- Mutants of *Streptomyces* strains capable of synthesizing chlortetracycline and oxytetracycline have also been found to produce novel derivatives, such as 6-dimethyltetracycline, 7-chloro-6-dimethyltetracycline, and others.

- Similar success has been achieved with strains producing vitamins and amino acids. For instance, improved *Bacillus subtilis* strains can yield up to 75 kg of vitamin B_2 per ton of nutrient medium.

In essence, mutagenesis-based strategies yield strains with genomic modifications that enhance biotechnological traits—most notably the overproduction of desired metabolites.

Advantages of the mutagenesis–selection approach:

- No prior knowledge is required about the gene or protein function linked to the phenotype.
- Effective for genetically uncharacterized organisms and for metabolites with unknown biosynthetic pathways.
- Generates novel biological insights relevant to molecular biology and biochemistry.

Limitations of the mutagenesis–selection approach:

- Labor-intensive, requiring extensive screening.
- Lack of precise information about the nature or location of mutations.
- Requires mutation mapping and phenotype validation to rule out off-target effects.
- The selection rate of productive mutants decreases with successive rounds.

These limitations can be mitigated by combining mutagenesis with genetic exchange techniques, enabling a transition from random screening to targeted genome construction. One promising approach is in vivo cell engineering, which integrates classical strain improvement with modern genetic tools to create designer bio-agents tailored for specific industrial applications.

Brainstorming

1. The purposes of bio-producer improvement.
2. What properties of the biological agents can be used to improve the efficiency and safe production of targeted products?
3. Describe the directions of bio-producer selection.
4. Describe the methods of bio-producer selection.
5. Enumerate the general stages of microbial selection.
6. What methods of mutagenesis are used to improve producers?
7. What mechanisms provide an opportunity to increase the yield of the target product in mutants?
8. What are structural and functional genes?
9. What are the functions of regulatory genes?
10. What mutations can lead to a change in the biosynthetic activity of microorganisms?
11. Describe the conventional procedure in the mutation and spontaneous strain selection process.
12. What is the selective media technique?
13. What are the advantages and disadvantages of spontaneous and induced mutagenesis in bio-producer selection?

Take-Home Messages

- Biotechnological productivity often requires strain improvement. Wild-type organisms typically lack the efficiency required for industrial-scale production. Enhancing their performance through genetic and physiological modification is essential to make processes technically feasible and economically viable. Key goals include increased product yield, resistance to stress and viral infection, reduced nutritional demands, compatibility with simple equipment, and improved hygienic or environmental characteristics.
- Mutagenesis combined with selection is a classical yet powerful tool. By selecting for high-performance variants—especially after exposure to mutagens—strains can be developed that exhibit enhanced biosynthetic capacity. Induced mutagenesis dramatically increases genetic diversity. Exposure to physical (UV, gamma rays), chemical (alkylating agents, nitrous acid), or biological (phages, plasmids) mutagens raises mutation rates, enabling the isolation of productive mutants at a higher frequency.
- Gene amplification is a critical strategy to boost product yield. Plasmid-based gene amplification can increase gene copy number up to thousands of times per cell. This method is widely used to enhance the production of antibiotics, vitamins, and enzymes. Regulatory and structural gene modifications unlock overproduction.
- Remarkable improvements have been achieved through mutagenesis. Productivity gains of 10,000- to 100,000-fold have been reported for antibiotics (e.g., penicillin, streptomycin), vitamins (e.g., B_2), and amino acids, confirming the effectiveness of sequential mutagenesis-selection cycles.
- Modern strategies integrate classical and targeted approaches. The limitations of blind selection can be overcome through integration with genetic exchange, recombinant DNA technology, and in vivo genome engineering, allowing rational strain design and accelerated development.

1.5 Cell Engineering

In cell engineering, the entire microbial or fungal cell is treated as the unit of manipulation to generate strains with desired industrial traits. One of the most powerful tools in this process is DNA recombination, which involves the rearrangement and fusion of genetic material from different organisms.

Recombination may occur via several biological mechanisms, including:

- Sexual and parasexual reproduction in eukaryotes
- Transformation, conjugation, and transduction in prokaryotes
- Protoplast fusion, a universal technique applicable across domains

1.5.1 Sexual and Parasexual Methods in Eukaryotes

In *eukaryotic systems*, sexual crossing remains the most straightforward approach for combining specific traits. Hybridization typically involves genetically marked strains (e.g., those resistant to specific growth inhibitors), allowing for the selection of successful recombinants. Hybrids formed in yeasts, fungi, and algae often exhibit either haploid or diploid chromosome sets, with diploids capable of undergoing meiosis and genetic recombination via chromosomal crossover.

Hybridization is especially useful for combining economically important traits within a single strain. Although underutilized in microbiology due to the rarity of natural hybridization events, many microorganisms—including industrial yeasts and fungi—are capable of engaging in natural or artificial hybridization.

Sexual and parasexual approaches are widely used in the genetic improvement of industrially important fungi. For instance, hybrid yeast strains resistant to molasses have been developed for baking applications, and others have been tailored to efficiently ferment sugars such as glucose and raffinose.

In fungi that do not undergo conventional sexual reproduction, vegetative hybridization can occur through a parasexual cycle. This involves the formation of cytoplasmic bridges between adjacent hyphae, enabling nuclear exchange and the eventual formation of diploid nuclei. These diploids may undergo mitotic recombination and haploidization, producing both haploid and diploid recombinants with enhanced metabolic performance.

The main goal of parasexual hybridization is to generate recombinant strains that combine beneficial traits from different parent organisms. This method has been particularly effective in industrial fungal genera such as *Aspergillus*, *Penicillium*, *Cephalosporium*, and *Fusarium*. Notable outcomes include a Penicillium hybrid that produces oxytetracycline with improved growth characteristics and reduced foaming in the culture medium—an important advantage in bioreactor operation. Similar strategies have yielded strains with enhanced enzyme production and organic acid biosynthesis.

1.5.2 Recombination in Prokaryotes

Unlike eukaryotes, true sexual reproduction is absent in bacteria. Bacterial cells lack specialized sexual structures, do not form gametes, and undergo neither karyogamy (nuclear fusion) nor meiosis. Furthermore, bacteria are typically haploid, meaning that they carry only one copy of each gene (Fig. 1.12).

Nevertheless, bacteria engage in horizontal gene transfer, which fulfills a similar evolutionary function. Recombination in prokaryotes can occur via:
- Transformation—uptake of free DNA from the environment
- Conjugation—direct cell-to-cell DNA transfer via plasmids
- Transduction—gene transfer mediated by bacteriophages

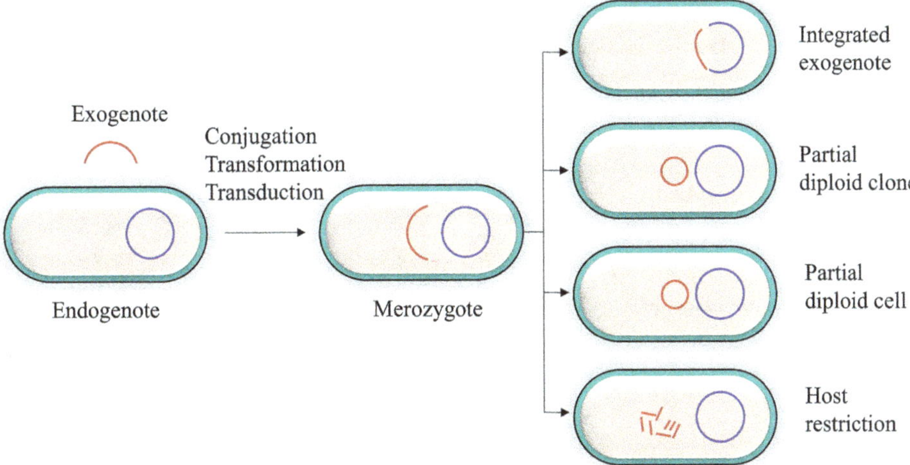

Fig. 1.12 The process of merozyogote formation

These processes are central to modern prokaryotic cell engineering, especially for the development of strains optimized for metabolite production, resistance traits, or complex biosynthetic pathways.

Although prokaryotes do not undergo true sexual reproduction, they are capable of gene exchange and genetic recombination. In addition to homologous recombination, which involves the exchange of similar chromosomal regions, other forms of recombination introduce entirely new genetic elements into the bacterial genome.

In bacterial recombination, gene transfer does not lead to zygote formation, as in eukaryotes, but instead results in the formation of partial diploids, known as merozygotes. In these cases, the recipient cell's original genome is called the endogenote, while the incoming DNA from the donor is termed the exogenote.

In nature, bacteria exchange genetic material through three fundamental mechanisms (see Fig. 1.13):

1.5.3   Transformation

Transformation involves the uptake of free DNA from the environment, typically released by lysed bacterial cells. To initiate this process, high molecular weight DNA must first bind to the recipient cell's surface, after which it is taken up across the membrane. The acquired DNA may then be:

- Integrated into the host chromosome via homologous recombination
- Maintained independently as a plasmid

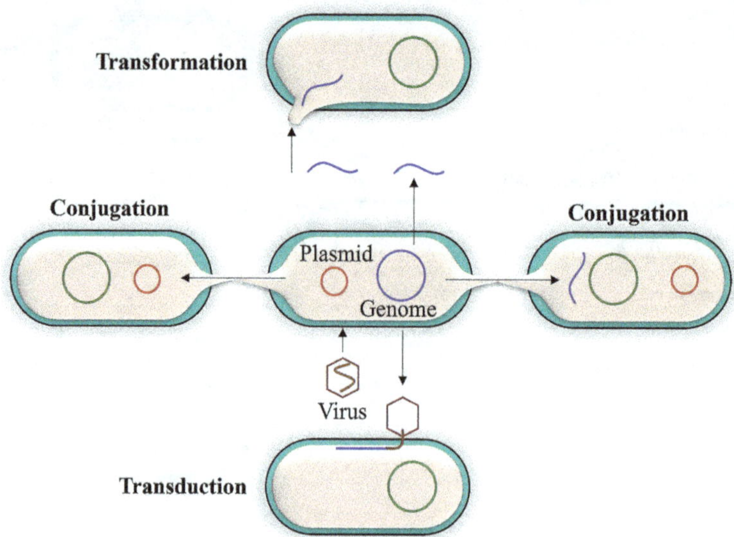

Fig. 1.13 Gene transfer types in bacteria

Natural transformation occurs in a number of bacterial species and can contribute to traits such as increased virulence or antibiotic resistance. It also forms the basis for many applications in recombinant DNA technology.

1.5.4 Conjugation

Conjugation is a process of direct DNA transfer between two bacterial cells, requiring physical contact. It is often considered the most prevalent form of gene transfer in bacteria. During conjugation, a donor cell connects to a recipient cell (usually of the same or a closely related species) via a conjugation pilus or tube, through which plasmid DNA is transferred.

Conjugative plasmids may carry a variety of functional genes, including:

- Genes conferring antibiotic resistance
- Genes involved in toxin production
- Other accessory or adaptive traits

In many cases, plasmids can integrate into the host chromosome, allowing stable inheritance of the introduced genetic information.

Among Gram-negative bacteria, conjugation is a major mechanism for horizontal gene transfer and can occur between different bacterial species. This process plays a central role in the rapid spread of antibiotic resistance, particularly when resistance genes are carried on conjugative plasmids. Once a plasmid is transferred, the recipient cell becomes a new donor, enabling the fast conversion of an entire sensitive population into a resistant one.

Gram-positive bacteria also harbor plasmids carrying multiple antibiotic resistance genes. These can be transferred through conjugation or, in some cases, through transduction, depending on the species and genetic context.

1.5.5 Transduction

Transduction is a form of genetic exchange mediated by bacteriophages (viruses that infect bacteria), during which DNA is transferred from a donor cell to a recipient cell. There are two main types of transduction:

1. Generalized Transduction

    In generalized transduction, bacteriophages accidentally package random fragments of host bacterial DNA during the lytic cycle. The process begins when a phage infects a bacterial cell, degrades the host chromosome, and uses the cellular machinery to synthesize phage components. Occasionally, bacterial DNA fragments are mistakenly enclosed within new phage particles.

    These recombinant phages, upon infecting new bacterial cells, inject the foreign bacterial DNA into the recipient, which may then be integrated into the host genome through homologous recombination. Since any gene from the donor cell has the potential to be transferred, this process is termed "generalized."

2. Specialized Transduction

    Specialized transduction was first described by André Lwoff, who observed that certain bacteriophage-infected bacterial strains could persist without immediate lysis. In such cases, the phage genome integrates into the bacterial chromosome, forming a prophage, and the host cell enters the lysogenic cycle. These bacteria are termed lysogenic.

    A special repressor protein, synthesized during lysogeny, inhibits phage particle production, allowing the cell to survive for multiple generations. When the repression is lifted, the phage enters the lytic phase, initiating synthesis of new viral components.

    During this transition, adjacent bacterial genes near the prophage integration site may be co-excised with phage DNA and packaged into new phage particles. When these modified phages infect new cells, they introduce specific donor bacterial genes into the recipient genome. Therefore, only genes located near the prophage insertion site can be transferred, giving the process its name—"specialized" transduction.

1.5.6 Artificial Hybridization and In Vivo Cell Engineering

Beyond natural mechanisms, hybridization can also be achieved artificially using laboratory-based approaches. One such strategy is in vivo cellular engineering, a key area in modern biotechnology. The technique involves somatic cell hybridization—the fusion of non-sex (somatic) cells from different organisms into a single, hybrid cell.

Somatic hybridization allows for the combination of genetic material from phylogenetically distant organisms, often beyond the reach of classical sexual crossing. The fusion can be complete—resulting in a single, genetically mixed cell—or partial, where only specific components of the donor cell (e.g., cytoplasm, mitochondria, chloroplasts, or nuclear DNA) are introduced into the recipient cell.

These techniques form the basis for a wide range of biotechnological applications, including the development of novel bio-agents, synthetic strains with enhanced production traits, and cells with combined metabolic capacities.

Cellular engineering is a powerful technique that enables the exchange of DNA fragments or entire chromosomal segments in prokaryotes, and even complete chromosomes in eukaryotes, irrespective of evolutionary distance between the organisms involved. Unlike traditional sexual reproduction, which is constrained by species and genus-level barriers, cellular engineering removes these restrictions, allowing genetic recombination between otherwise incompatible organisms. This opens up unprecedented opportunities for developing novel bio-agents with tailored properties.

The core methods of cellular engineering include:

- Cultivation and cloning of cells in specialized media
- Somatic cell hybridization
- Nuclear transplantation
- Various microsurgical manipulations that enable the "disassembly" and "reconstruction" of viable cells from selected subcellular components

One of the most important tools in this field is protoplast fusion, which allows for the forced hybridization of cells, including those that are genetically isolated. The removal of the cell wall to generate protoplasts permits direct membrane contact and subsequent recombination of nuclear and cytoplasmic material.

1.5.7   Protoplast Formation and Fusion Workflow

To initiate the process, cells are first converted into protoplasts—membrane-bound cells lacking cell walls. These protoplasts can then be fused together under controlled conditions to form hybrid diploid cells. After fusion, cells are incubated to allow DNA rearrangement and recombination, followed by culturing on selective media. Colonies that form are screened, and clones with desired traits are isolated for further study (Fig. 1.14).

The standard steps in protoplast fusion include:

1. *Cell wall removal*

   Enzymatic digestion is used to break down the cell wall polymers:
   - For bacteria: *lysozyme*
   - For fungi: *zymolyase* (e.g., derived from snails)
   - For plant cells: a combination of *cellulase, hemicellulase,* and *pectinase*

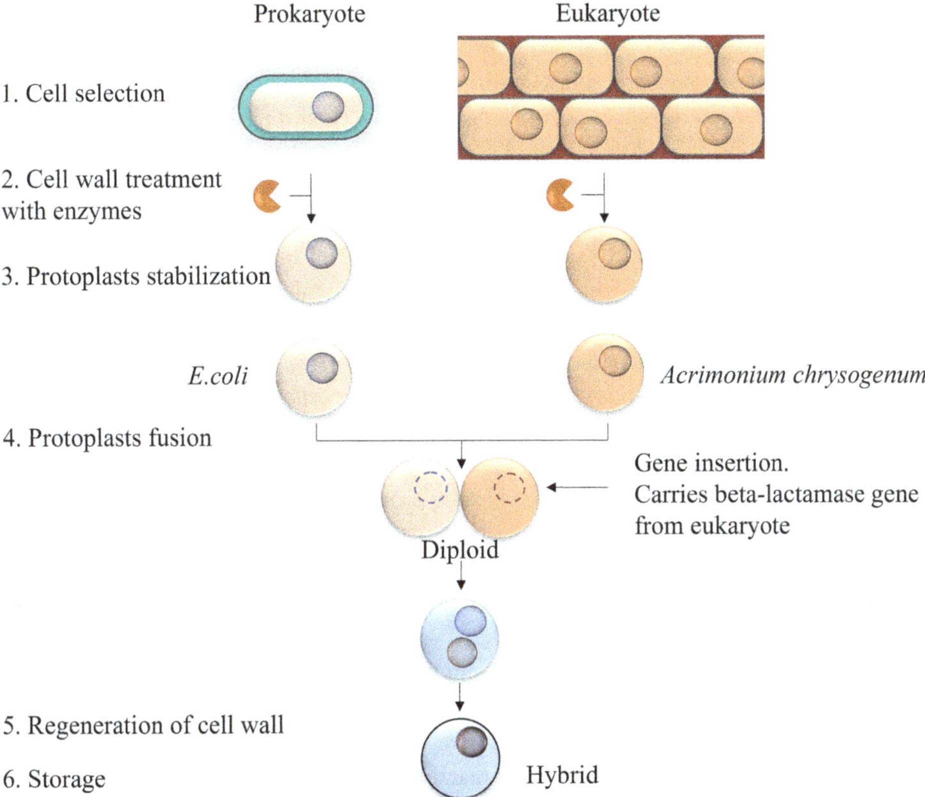

Fig. 1.14 Protoplast fusion technique

2. *Protoplast stabilization*

   To preserve membrane integrity, osmotic pressure between the cytoplasm and external medium must be equalized. This is achieved using osmotic stabilizers, such as:
   - 20% mannitol or sucrose
   - 10% NaCl

   Additional parameters—pH, temperature, and light—are optimized. The formation of protoplasts is typically monitored via phase-contrast microscopy.

3. *Protoplast fusion*

   Fusion requires close membrane contact, which is hindered by the natural negative charges on membrane proteins and lipids. To overcome this, fusion is induced by:
   - Polyethylene glycol (PEG)
   - Divalent calcium ions (Ca^{2+})
   - Electrofusion (alternating electric fields)
   - Magnetically induced fusion in some advanced systems
   - In animal cells, the Sendai virus is sometimes used to hydrolyze cytoplasmic membrane proteins and promote fusion

Additionally, preconditioning treatments—such as enzyme digestion, freeze-thaw cycles, or exposure to heavy metals—can render cells more competent for fusion.

The formation of a diploid protoplast confirms that DNA recombination has occurred during the hybridization process. These diploids can be further cultured to regenerate whole organisms or used in industrial applications to produce enhanced bio-agents.

4. *Regeneration and Screening of Hybrid Cells*

    Regeneration refers to the restoration of the cell wall after protoplast fusion. Once regenerated, the protoplast behaves like an independent cell, capable of division and colony formation. Some regenerated cells contain recombined chromosomes, and the resulting cell lines may exhibit new and valuable phenotypic traits.

Screening of the resulting hybrid cells typically follows two main strategies:

1. Evaluation of phenotypic characteristics, such as morphology, growth patterns, or metabolic product profiles
2. Selective cultivation, using media in which only hybrid cells—those integrating genetic material from both parents—can survive and proliferate.

1.5.8   Examples: Hybrid Antibiotic Production

One notable application of cellular engineering is the creation of "hybrid antibiotics." For instance, among *Streptomyces* species, various antibiotics differ in glycosidic structure, including variations in aglycone cores and sugar moieties:

- Erythromycin features a 14-membered macrolide aglycone and two sugar units
- Anthracycline antibiotics contain a tetracyclic aromatic aglycone linked to an amino sugar

    Through protoplast fusion, researchers successfully created recombinant strains with novel hybrid structures:

- Erythromycin's macrolide aglycone was fused with the sugar moiety of anthracyclines
- Conversely, anthracycline aglycones were modified with erythromycin-like sugars

    Similar experiments in *Nocardia mediterranei*, a rifamycin-producing organism, yielded new rifamycin analogs after fusion with mutant strains. Comparable approaches have been applied to develop hybrid β-lactam antibiotics.

1.5.9   Improvement of Industrial Strains

Protoplast fusion can also restore and enhance industrial strains that have accumulated undesirable mutations through prolonged selection. For example, two strains of *Nocardia lactamdurans*, both cephamycin producers, were fused. The resulting recombinants produced 10–15% more antibiotic than the best parental strain.

An additional benefit of fusion is the potential for gene amplification. In some hybrids, biosynthetic gene clusters may be duplicated, resulting in an enhanced expression of the desired metabolic pathway. These "+ variants," with amplified gene copies, demonstrate improved biosynthetic performance (e.g., increased production of vitamins, hormones, or antibiotics).

However, DNA repair systems in cells may gradually revert such recombinants to the wild-type genotype, leading to a loss of beneficial traits. To overcome this:
- "+" variants may be treated with mutagens to disrupt the repair systems
- Alternatively, the strains may be immobilized to stabilize their genetic configuration

1.5.10 Significance of the Hybridization Method

- Enables the combination of desired traits from different species or strains into a single organism
- Allows the selection of second-generation recombinants with novel properties absent in parent cells
- Facilitates genomic enrichment via independent mutations from multiple parental strains
- Makes it possible to introduce foreign genes and amplify endogenous gene copies, strengthening associated metabolic capabilities

In vivo *cell engineering* methods are versatile tools for introducing genetic material into cells of diverse biological origin. Their simplicity and broad applicability make them highly valuable for the selection and improvement of industrially relevant bio-producers.

This approach enables the generation of interspecific and intergeneric hybrids, facilitating the combination of traits across phylogenetically distant organisms. Notable achievements include:
- Interspecific hybrids of tobacco, potato, and petunia
- Sterile intergeneric potato–tomato hybrids
- Interspecific and intergeneric yeast hybrids
- Hybrids formed between fungi and bacteria

In addition to symmetrical hybrids, asymmetric hybrids—cells carrying a complete genome from one parent and a partial genome from the other—can be generated. These often result from fusion events between evolutionarily distant species. Asymmetric hybrids are generally more stable, fertile, and viable than symmetrical counterparts, making them especially attractive for industrial use.

Another promising development is the fusion of cells carrying distinct functional properties, such as:
- Fusion of cells from different tissues or organs
- Fusion of normal cells with those exhibiting neoplastic transformation

The latter leads to the creation of hybridomas—hybrid cells that combine the biosynthetic capability of normal cells with the immortalized growth potential of tumor cells. This strategy is widely used, for instance, in the production of monoclonal antibodies.

In conclusion, strain selection has achieved significant success, further enhanced by advances in molecular and cellular biology. Many microbial strains now routinely produce high yields of valuable substances under industrial conditions. Nevertheless, the potential of strain improvement can be greatly expanded through the integration of cellular engineering (in vivo) with genetic engineering (in vitro)—paving the way for more precise, stable, and productive bio-agents in biotechnology.

Brainstorming

1. Define cellular engineering.
2. What is genetic recombination?
3. List the types of recombination patterns.
4. What is the significance of hybridization in selecting industrial bio-producers?
5. What ways of sharing genetic information exist in prokaryotes?
6. What is a protoplast?
7. What techniques of protoplast separation are used in biotechnology?
8. How are protoplasts cultured?
9. Describe the process of obtaining protoplasts from prokaryotes and their stages.
10. Discuss the protoplast fusion technique in selecting high-productive, valuable strains of microorganisms.
11. What are the advantages of the protoplast fusion technique over other breeding methods?
12. What are the promising areas of development of cellular engineering in biotechnology?
13. What are the areas of the practical application of cellular engineering achievements?

Take-Home Messages

- Cellular engineering enables genetic recombination beyond natural species boundaries, allowing the creation of hybrid organisms by fusing protoplasts or somatic cells from phylogenetically distant species.
- Protoplast fusion is a central technique in cell engineering, involving the enzymatic removal of the cell wall, osmotic stabilization, and induced membrane fusion, followed by regeneration and screening of hybrid cells with desired traits.
- Somatic hybridization expands breeding potential, making it possible to produce both symmetrical and asymmetric hybrids with improved stability, viability, and productivity.
- Cell engineering has been successfully applied to industrial microorganisms, including *Streptomyces*, *Nocardia*, and *Bacillus*, resulting in enhanced production of antibiotics, vitamins, and other bioactive compounds.
- Hybridoma technology combines the biosynthetic function of normal cells with the proliferative capacity of tumor cells, making it a cornerstone of monoclonal antibody production and therapeutic biotechnology.
- Cell engineering bridges classical strain selection with modern molecular tools, and when combined with in vitro genetic engineering, it offers powerful strategies for the development of next-generation microbial producers.

1.6   Genetic Engineering

Advances in the field of genetics and molecular biology have enabled biotechnologists, since the 1970s, to move from blind selection of mutant strains to conscious genome construction — the basis of modern genetic engineering using recombinant DNA technology. By recombinant means, DNA is formed by combining in vitro two or more DNA fragments isolated from various biological sources.

Recombinant DNA is constructed in two ways. The first is the preliminary isolation of the necessary gene from the donor organism or its synthesis, followed by integration of the gene into the vector molecule. The second way is to create collections of recombinant DNA and search within the collection for recombinants containing the necessary built-in gene.

The essence of genetic engineering is the combination of DNA fragments in vitro, followed by the introduction of the isolated DNA into a living cell.

The scope of genetic engineering involves combining isolated DNA fragments of natural or synthetic origin, or a combination of both, in vitro. Further, it is necessary to introduce the resulting recombinant structures into a living cell. Finally, the inserted DNA fragment is either incorporated into the chromosome and replicated or expressed autonomously. Consequently, the introduced genetic material becomes part of the cell's genome.

Hence, the necessary conditions for the implementation of genetic engineering are the following:
1. There must be a biological agent capable of synthesizing a heterologous protein and perceiving and transmitting genetic information.
2. The host organism should not reject the product synthesized by the producer.
3. The cell must be capable of division; the genes producing the desired product must be expressed in the cells formed after division.
4. A transport tool is required for introducing DNA into the producer's cell—a vector in the form of plasmids, cosmids, or phages.

Natural methods of transferring genetic information are usually used to introduce a vector into a prokaryotic cell:
1. *Conjugation*—Upon approaching, the genetic material passes from one cell to another in the form of a plasmid.
2. *Transduction*—Transfer of genetic material through a virus or phage into a cell.
3. *Transformation*—Transfer of genetic material into a cell using isolated DNA, resulting in alteration of the genome.

Finally, the cells containing the recombinant DNA are identified and selected.

This is mainly conducted on a selective medium using marker genes. The most widely used markers are antibiotic resistance genes. For example, suppose the vector molecule contains a gene for resistance to an antibiotic. In that case, only the correspondingly transformed cells will resist this antibiotic. On this basis, they are selected.

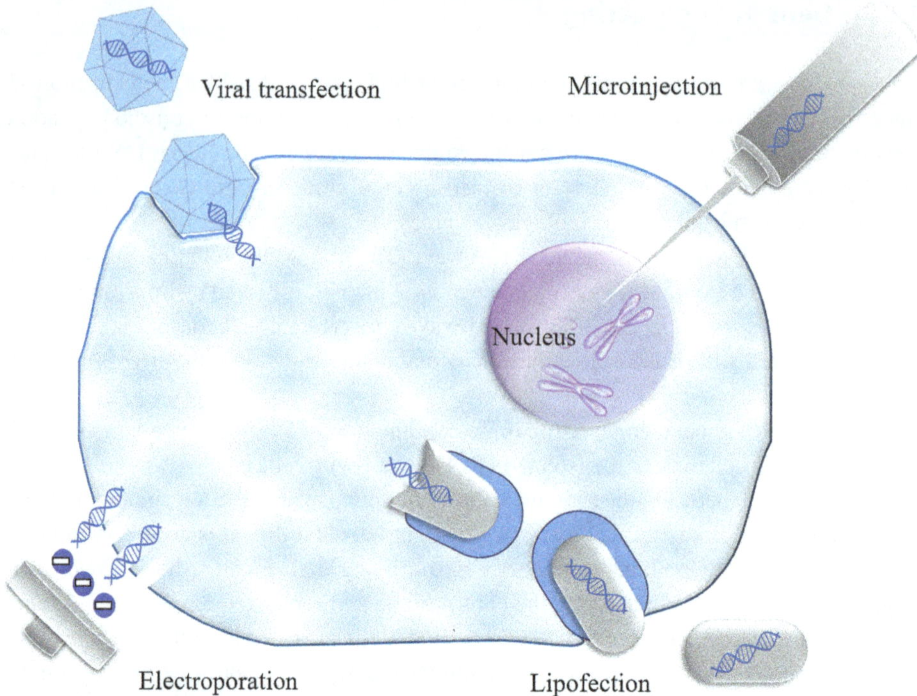

Fig. 1.15 Methods of the vector delivery into eukaryotic cells

Various methods with different efficacies are used to introduce recombinant DNA (vector) into a host eukaryotic cell. All of them are based on *pore formation in the cytoplasmic membrane*, through which DNA can penetrate into the cell (Fig. 1.15).

1. *Microinjection.* By means of the finest glass tube and a micromanipulator, vector DNA can be inserted into the cell's nucleus with the transgene included. The number of DNA molecules inserted per injection can range from 100 to 300,000.

2. *Electroporation.* The permeability of membranes is reversibly increased under the influence of high-voltage electric pulses. As a result, through the micro-pores, which are formed for a short time in the membrane, DNA from the surrounding medium penetrates the cell.

3. *Transfection.* The vector is treated with calcium ions. The resulting nanocomplexes of ions and vectors penetrate the cell by pinocytosis. The method is used to introduce transgenes into eukaryotic cells.

4. *Packing in liposomes (lipofection).* Liposomes are spherical formations coated with phospholipids and containing a vector inside that can penetrate the cell due to their dissolution in the lipids of the plasma membrane.

5. *Microparticle bombardment (gene gun, ballistic transformation, and agrolistic method).* This technique is highly effective for plant transformation. Immature seed embryos serve as the target for bombardment by gold or tungsten particles coated with

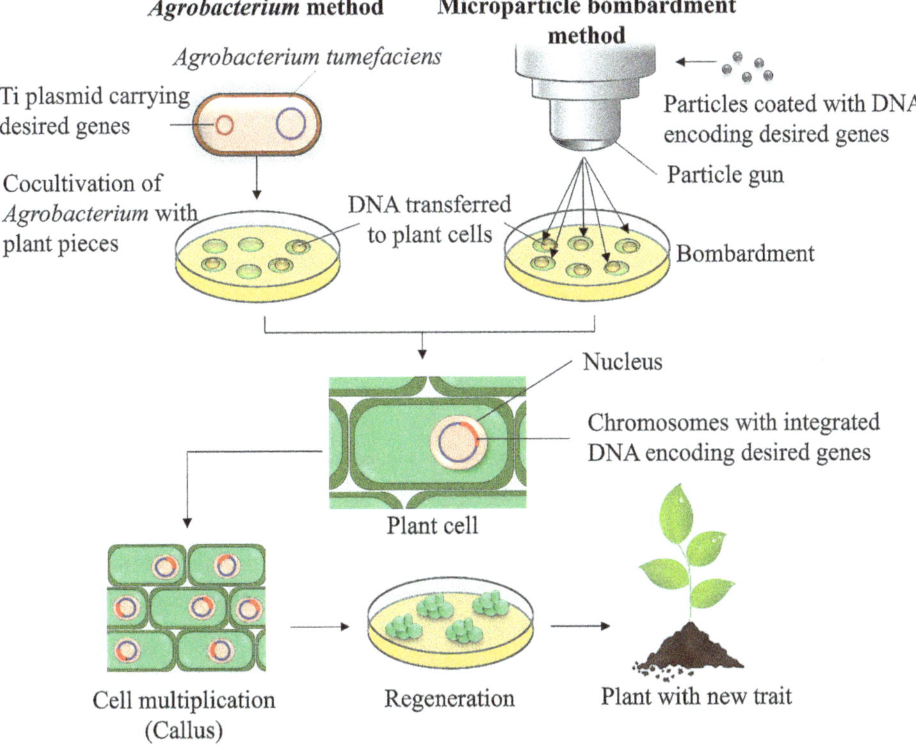

Fig. 1.16 Agrobacterium and microparticle bombardment methods

vectors. These particles are loaded into "gene guns" and then shot into the cells. Cells at the impact site often die, while those within a 0.6–1 cm central zone typically undergo successful transformation. The particles can penetrate up to 2–3 cell layers deep and may directly enter the nucleus, enhancing transformation efficiency. This method also allows for the transformation of DNA in cellular organelles like chloroplasts and mitochondria (Fig. 1.16).

6. *Natural vectors for horizontal gene transfer in dicotyledons* exist, known as agrobacterial plasmids. During transformation, bacterial enzymes damage the plant cell wall, facilitating close contact with the plant cell membrane. This enables the transfer of bacterial DNA into the plant cell. Two prominent types of agrobacteria are widely studied and used. The first is *Agrobacterium tumefaciens*, which carries a Ti-plasmid that integrates into plant cell chromosomes. The second is *Agrobacterium rhizogenes*, responsible for excessive root generation in damaged root zones due to its Ri-plasmid.

7. *Infection with a virus.* The most promising is the white cabbage mosaic virus, which mainly affects plants of the cabbage family. The small genome size (circular DNA of 8000 nm length) allows for manipulation of viral DNA in vitro as a bacterial plasmid and its subsequent introduction into plants by rubbing the leaves. In this case, all cells

become infected because the virus multiplies rapidly and is transmitted from cell to cell.

The advantage of virus-based vector systems is the small size of the genome, the high concentration of viral DNA in plant cells (up to 50,000 copies per cell), and the presence of robust promoters that ensure the effective expression of foreign genes.

8. *In recent years, a novel approach using "antisense RNA" has been proposed* to create transgenic plants. This technique involves rotating a copy of the inserted gene's DNA (cDNA) by 180 degrees when constructing the vector. This action produces an inverted mRNA molecule alongside the regular mRNA. Due to its complementarity with the normal mRNA, a complex forms that inhibits protein synthesis. The antisense construct strategy has broad applications for modifying gene expression. It is used both for developing plants with new traits and for foundational research in plant genetics (Fig. 1.17).

Following plant tissue transformation, the tissue is cultured in vitro using a medium enhanced with phytohormones to stimulate cell multiplication. This medium typically includes a selective agent to which only transgenic cells exhibit resistance. Regeneration typically involves a callus phase, followed by organogenesis or shoot formation. The resulting shoots are then placed on a rooting medium with a selective agent, ensuring stringent selection of transgenic specimens.

In addition, the main object of cellular engineering—*protoplasts*, used extensively in genetic engineering—is central to the concept of *cell genetic engineering*. This is because molecules and DNA fragments are exchanged in protoplasts more easily than in cells with their complex cell walls.

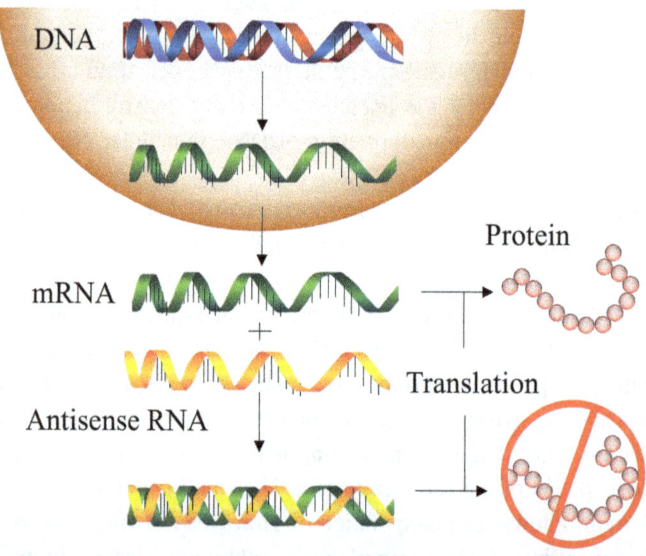

Fig. 1.17 Antisense RNA strategy

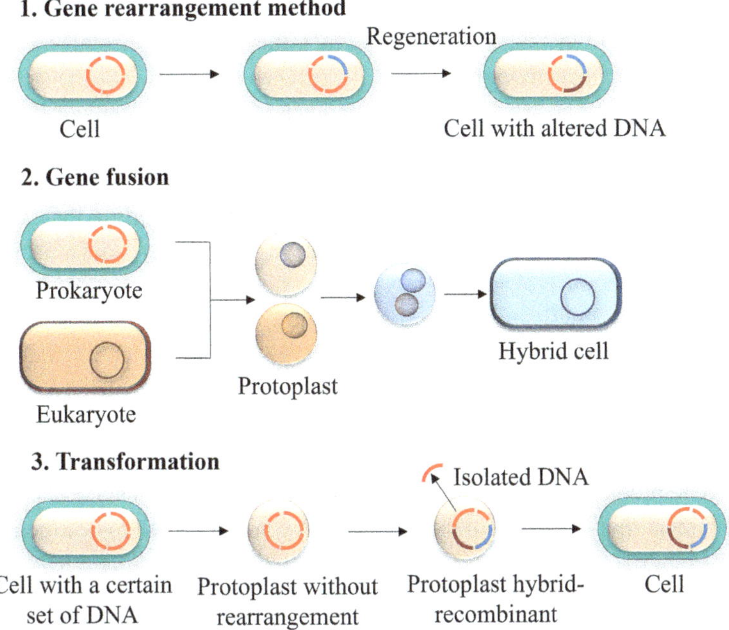

Fig. 1.18 Techniques of cellular genetic engineering

There are three principal ways of cell genetic engineering (Fig. 1.18):

1. *The method of gene rearrangement.* There is a certain set and sequence of genes in the cell. The genes are rearranged during protoplast fusion. There is also the activation of silent genes.

2. *Fusion of genes.* Prokaryotes are fused with eukaryotes; a recombinant contains both original and foreign DNA.

3. *Transformation.*
    (a) A cell with a certain set of genes in DNA → living protoplast without rearranging the genes → protoplast hybrid-recombinant → recombinant regenerating cell.
    (b) Isolation of DNA, i.e., a *vector* (a piece of isolated DNA from another microorganism). For example, suppose there is a cell and an isolated DNA fragment with the desired gene. The vector can be a *phage, plasmid, virus,* or *cosmid* (plasmid + phage). It is possible to integrate the vector into an isolated protoplast.

For transformation, it is necessary to make the cell *competent*, i.e., to increase its permeability. This is achieved as follows:

1. By exposing the cell to heavy metal ions (zinc, cobalt, lithium, magnesium)

2. By treating the cell with enzymes (e.g., lysozyme or the complex enzyme of the grape snail)

3. By quick freezing and thawing

Competent cells more easily absorb the vector. Their cytoplasmic membrane, in some places, forms a pore to the surface, through which the vector—as isolated DNA, phage, virus, or cosmid—can penetrate the cell.

The necessary gene construction by gene and cell engineering methods allows for the management of heredity and the vital activity of animals, plants, and microorganisms, enabling the creation of organisms with desired traits not previously observed in nature. Therefore, the principles and methods of genetic engineering are primarily established on microorganisms, mainly bacteria and actinomycetes (prokaryotes), and yeasts and mold fungi (eukaryotes).

With the help of genetic engineering techniques, according to a specific plan, new forms of microorganisms can be designed to synthesize various products (including those of plant and animal origin). In addition, microorganisms have a high growth rate and productivity and can recycle multiple raw materials.

However, while constructing new microbial producers, specific problems arise:

Firstly, gene products of plant, animal, and human origin are secreted into an intracellular environment that is foreign to them, where they are subjected to destruction by proteases. For example, peptides like somatostatin are hydrolyzed within a few minutes. To protect genetically engineered proteins in a microbial cell, the following approaches are used:

- Applying protease inhibitors.
- Obtaining the peptide of interest as part of a hybrid protein molecule; for this purpose, the peptide gene is cross-linked to a natural gene of the recipient organism.
- Amplification (increase in the number of copies) of genes.

Secondly, in most cases, the product of the transplanted gene is not released into the culture medium. It accumulates inside the cell, making it difficult to isolate. For example, insulin production using E. coli involves the destruction of cells and subsequent purification of insulin. It should be noted that E. coli excretes relatively few proteins. In addition, its cell wall contains a toxic substance that must be carefully separated from products used for pharmacological purposes. Currently, more promising objects of genetic engineering are gram-positive bacteria of the genera Bacillus, Staphylococcus, Zymomonas, and Streptomyces, as well as eukaryotic microorganisms such as the yeasts Saccharomyces (wine, baker's, brewer's yeast), Candida, and Pichia (single-cell protein).

Thirdly, most hereditary characteristics are encoded by several genes; thus, genetic engineering must include the consecutive transplantation of each gene. In some cases, simultaneous transplantation of whole blocks of genes using a single plasmid is possible.

Microorganisms that have been genetically modified can produce substances not naturally found in them. This is achieved by incorporating genes into their genetic makeup. Bacteria, yeasts, and viruses can be manipulated to create hundreds of these modified strains that produce various biologically active compounds such as antigens and hormones. Genetic engineering has played a significant role in the development of numerous medications that are widely used in medical practice today. These medications include hormones like insulin, anticoagulants, vaccines, immunomodulators, and diagnostic drugs, among others.

Genetic engineering in biotechnology is particularly useful when specific substances are unattainable by other means and are safe for both humans and the environment. For instance, antigens for vaccines against non-cultivable pathogens like Plasmodium (malaria) can only be sourced through genetic engineering. Genetically engineered interferon is both more active and cost-effective than interferon derived from blood leukocytes. Additionally, the production of drugs for high-risk infections like plague can be carried out using recombinant strains of non-pathogenic bacteria, replacing traditional methods.

The method of genetic engineering is being increasingly used in biology and medicine. This method will allow the development of new, effective drugs, particularly new polyvalent live (vector) vaccines and regulatory proteins, to be implemented in gene diagnostics and gene therapy.

Genetic engineering can also enhance the effectiveness of microorganisms used in production. A standard method of increasing the yield of a useful product is amplification, i.e., an increase in the number of gene copies. By amplifying genes within vectors, highly efficient producers of threonine and proline are created.

Antibiotics and vitamins are two examples of biotechnologically useful substances that often involve complex biosynthetic processes controlled by multiple genes. Isolating and amplifying these genes can be challenging. For instance, if antibiotic synthesis relies on a multi-enzyme complex encoded by a single operon, this operon can be readily inserted into a vector for cloning. When genes are dispersed across the genome, product yield is enhanced by cloning the genes associated with these complex biosynthetic pathways.

Strains can be constructed by adapting to the specific conditions of a technological process, called *technological engineering*. For example, antibiotic-producing operons can be transferred from slowly growing streptomycetes to fast-growing bacteria, such as *E. coli* or *B. subtilis*. This allows a much higher yield to be achieved through better control of cultivation conditions. In addition, the structural genes for the biosynthesis of certain antibiotics are usually concentrated in chromosomes. They can be incorporated into plasmids and transferred to other actinomycetes or *E. coli*. Such plasmids were constructed based on plasmids SLP1.2 of *S. lividans* and SCP2 of *S. coelicolor*.

Conventional types of fermentation can be improved by transferring the genes encoding amylase from *Aspergillus* or *Rhizopus*, glucose isomerase from *Streptomyces*, or rennin from *Mucor* into bacterial cells. As a result, the production of enzymes becomes more economical, and the fermentation processes more effective. In addition, the introduction of amylase and glucoamylase genes into *Saccharomyces cerevisiae* cells ensures the production of alcohol from starch.

Many bacteria have a problem not with synthesizing the product but with its excretion from the cell. Increasing the efficiency of transporting the product from the cell to the culture liquid accelerates the synthesis of new product portions, for example, amino acids.

Intensifying the effectiveness of traditional methods is achieved with the help of localized (site-specific) mutagenesis in vitro. For example, instead of treating all genomes in the cells, chemical mutagens are applied to gene fragments of interest. A skilled biotechnologist uses genetic engineering to transition from gene to product, modifying DNA nucleotide sequences to control corresponding protein changes. This serves as the basis for protein engineering, enabling the optimization of enzymes, hormones, and vaccine antigens.

The insertion of heterologous genes and regulatory elements allows the construction of new metabolic pathways with production-friendly characteristics. This approach is aimed at optimizing cellular activity. The change in the enzymatic, transport, and regulatory properties of bacteria using recombinant DNA technology is called metabolic engineering.

One of the first successes in this field was the production of a recombinant strain with altered nitrogen metabolism, used as fodder microbial protein. Assimilation of ammonia in methylotrophic bacteria *Methylophilus methylotrophus* occurs with the participation of aminotransferase and glutamine synthetase. This process runs with 1 mole of ATP per 1 mole of assimilated ammonia. Simultaneously, in *E. coli*, only glutamate dehydrogenase (GDH) participates in ammonia assimilation, which does not require ATP. The *E. coli* GDH gene was cloned and introduced into glutamine synthetase-free mutant *M. methylotrophus* cells. When this gene was expressed, the yield was 5% higher than in the parent strain.

Cloning and subsequent introduction into cells and expression of heterologous genes can be used to enhance the catabolic capabilities of the microorganism and, accordingly, the spectrum of substrates used. For example, the lac operon of *E. coli* was chosen to create lactose-utilizing strains of *Alcaligenes eutrophus*, *Corynebacterium glutamicus*, and *Xanthomonas campestris*.

Optimization of strains can reduce the cost of production by enabling growth on cheap substrates or by improving specific properties. Furthermore, this can change the producer's nutritional needs, expanding the industry's raw material base and increasing the efficiency of substrate conversion. For example, after introducing appropriate genes into *E. coli* cells producing threonine, the strain was able to utilize sucrose.

Thus, genetic engineering opens excellent prospects for biotechnologists, including the creation of fundamentally new producers of valuable compounds for humans and the increased efficiency of producers already used in production.

Currently, genetic engineering has mastered all realms of living organisms. The main directions of the development of plant genetic engineering include:

- Enrichment of cultivated plants with additional reserve substances using genes taken from other plants
- Increasing plant photosynthesis efficiency based on ribulose-1,5-bisphosphate carboxylase genes, chlorophyll a/b-binding proteins, etc.
- Changing nitrogen metabolism
- Providing resistance to herbicides, salinization of soils, high and low temperatures, and other unfavorable environmental factors

The current stage in the development of plant genetic engineering is called metabolic engineering. Mainly, this is because the goal is not only to improve certain existing qualities of the plant, as in traditional breeding, but to manipulate the plant to produce entirely new compounds important in medicine, chemical production, and other fields. These compounds can be, for example, special fatty acids, valuable proteins with a high content of essential amino acids, modified polysaccharides, edible vaccines, antibodies, interferons, and other "medicinal" proteins, as well as new polymers that are safe for the environment.

Genetic engineering methods allow solving several critical tasks to advance the resistance of agricultural plants' new forms, lines, varieties, and hybrids to various pathogens.

However, genetic engineering manipulations with plants can lead not only to the expected results. For example, resistance to herbicides, caused by the transplantation of a single gene, can cause serious problems in crop rotations; the herbicide-tolerant plant, cultivated on a given crop area, will act in the following year as a weed against its successor crop, against which herbicides are ineffective. Another threat is that biochemical changes caused by genetic modifications can lead to plants losing their food or fodder value—or even acquiring toxicity. This problem is inherent not only in genetic engineering but also in traditional selection methods.

With the help of genetic engineering manipulations in animals, the genes of β-globin in mice, tyrosine tRNA of *E. coli*, thymidine kinase, and human β-interferon have been cloned into insect and mammalian cells.

Genetic engineering methods in livestock farming are promising for altering several organism properties, such as increasing productivity, resistance to diseases, growth rate, and improving product quality.

Thus, the creation and development of genetic and cellular engineering methods make it possible to artificially create new highly productive forms of organisms suitable for industrial use.

The incorporation of genetic engineering has elevated biotechnology, enabling deliberate control over complex cellular processes. Firstly, it has significantly enhanced the productivity of industrial microbial producers by adding or activating extra genes. Secondly, new genes have altered the nutritional requirements of these microbes. Thirdly, microbes have been engineered to produce substances previously foreign to them, thereby diversifying biotechnological products. Lastly, the approach to selecting microbial producers has shifted. Instead of searching for an inherently active strain, a strain can be adapted to production conditions and genetically modified for efficient target product synthesis.

There are now more isolated plant and animal tissues and cells in the pool of microbial producers. This development has led to new biotechnological methods and innovative eukaryotic selection techniques. Consequently, significant strides have been made in microclonal plant reproduction and the production and application of transgenic plants and animals.

1.6.1 Brainstorming

1. What is the fundamental difference between cellular and genetic engineering methods?
2. What are the necessary conditions for the implementation of genetic engineering?
3. What is recombinant DNA?
4. What are the main methods of creating recombinant DNA?
5. List the main types of enzymes used in genetic engineering technology and briefly describe their functions.
6. What is a vector? What can be used as vectors?
7. List the main stages of genetic engineering techniques.
8. What natural mechanisms of genetic information transfer are commonly used to introduce vectors into prokaryotic cells?
9. What problems arise when designing new microbial producers using genetic engineering methods?
10. How are genetically transformed microorganisms selected?
11. What methods are used to insert recombinant DNA into recipient eukaryotic cells?
12. What are the main directions in the development of plant genetic engineering?
13. What is the purpose of applying genetic engineering methods in animal husbandry?
14. What is cellular engineering? What are the main approaches to its implementation?
15. What is "metabolic engineering"?
16. List the key prospects for applying genetic engineering methodology in science and practice.

Take-Home Messages
- Genetic engineering enables the precise construction of new microbial, plant, and animal strains by transferring and modifying genes, allowing for the production of valuable compounds and traits not naturally present in the organism.
- The creation of recombinant DNA involves combining natural or synthetic DNA fragments in vitro and introducing them into host cells using vectors such as plasmids, phages, or viruses.
- Transformation methods vary depending on the organism, including electroporation, microinjection, lipofection, and natural mechanisms like conjugation, transduction, and agrobacterial transfer in plants.
- Genetic engineering allows for the expression of foreign proteins, but challenges such as intracellular degradation, accumulation of products, and multi-gene regulation must be addressed through strategies like gene amplification, fusion proteins, or the use of alternative host organisms.
- Metabolic and protein engineering are advanced applications of genetic manipulation, enabling the optimization of biosynthetic pathways, enzyme activity, and the expansion of substrate utilization for industrial and medical purposes.
- The integration of genetic engineering across microorganisms, plants, and animals opens new avenues in agriculture, medicine, and environmental protection—while also requiring careful consideration of biosafety and long-term ecological effects.

2.1   Main Stages of the Biotechnological Process

Biotechnology focuses on producing valuable substances and services using the biochemical processes of biological entities. This field is often viewed as an eco-friendly technology.

- *Benefits of biotechnological methods include:*Production of unique natural substances that traditional methods cannot achieve.
- Operations at milder temperatures and pressures.
- Rapid growth and high biomass accumulation of bio-agents.
- Efficient one-step synthesis of intricate compounds.
- Capability to utilize low-cost agricultural and industrial by-products.
- Eco-friendly nature with minimal hazardous waste generation.
- Products are often biodegradable.
- Simple and cost-effective equipment and technology.

Biotechnological green processes differ from chemical synthesis methods. They operate under mild conditions—normal pressure and low ambient temperatures—assuring active reactions. Additionally, these processes generate less environmental waste and fewer by-products. They are also less affected by weather and do not require vast land areas or xenobiotics. Consequently, biotechnology ranks high among scientific and technological advancement priorities. It epitomizes sustainable technologies and holds promise for numerous industrial applications. While biological technologies are rapidly evolving, a nation's scientific and technical capacity significantly influences their development.

The primary goal of biotechnological processes is to devise and refine technologies and equipment based on scientific principles. Unlike chemical methods, biotechnology

I. Digel et al., *Introduction to Industrial Biotechnology*, Learning Materials in Biosciences,
https://doi.org/10.1007/978-3-032-07918-3_2

employs more intricately organized substances—namely, biological systems. Each biological entity functions as an independent, self-regulating system.

Biotechnological production varies in complexity depending on its components. The specific biotechnological process dictates its challenges, which are influenced by the biological agents used and the desired product. When producing biomass like single-cell protein (SCP), the process is more straightforward. For high-purity pharmaceuticals, it becomes more intricate and involves a longer processing line. Cultivating microorganisms as product sources requires aseptic conditions, specialized equipment, and a proper processing plan.

- *In general, any biotechnological process involves three main stages:*Upstream processing (pre-fermentation)
- Midstream processing (fermentation)
- Downstream processing (post-fermentation)

The basic scheme for the realization of biotechnological processes in a general form can be represented by a flowchart (Fig. 2.1).

In upstream processing, the inoculum is stored, activated, and resources such as nutrient substrates, equipment, and sterile air are prepared. Maintaining a pure culture of the biological agent is vital, as it significantly influences the success of the biotechnological process. Inoculum cultivation is performed in stages, starting with small containers and scaling up to fermenters. This cultivated inoculum is then introduced into the bioreactor for fermentation.

Special reactors with agitators are used to prepare the nutrient media. The type of reactor depends on the characteristics of the media components. Media containing insoluble components complicates preparation. Different biotechnological processes require specific substrate preparations. Each facility individually selects and doses the nutritional components.

Large-scale dosing is carried out using weight and volume-based measuring devices. Conveying substances is done with pumps and conveyors, and vacuum pumps are used to transfer bulk components into fermenters. Given the diversity of biotechnological procedures, the specifics are closely tied to the particular production process.

The fermentation stage plays a central role in biotechnology, as it is where the bioproducer interacts with the substrate to create the desired outcomes, such as biomass or extracellular components. This stage takes place in a biochemical reactor specifically designed to meet the requirements of the bio-producer and the intended final product (Fig. 2.2). Fermentation can be conducted strictly under aseptic conditions.

Downstream processing produces final products and handles waste. Equipment and methods vary based on the product's cellular location and nature. Post-fermentation begins by fractionating the culture and isolating the biomass. Separation, done in separators, is a prevalent method. For enhanced separation, cultures undergo treatments like pH adjustments or heating. The product's nature determines its formulation. Additives like fillers and lubricants stabilize biotechnological products.

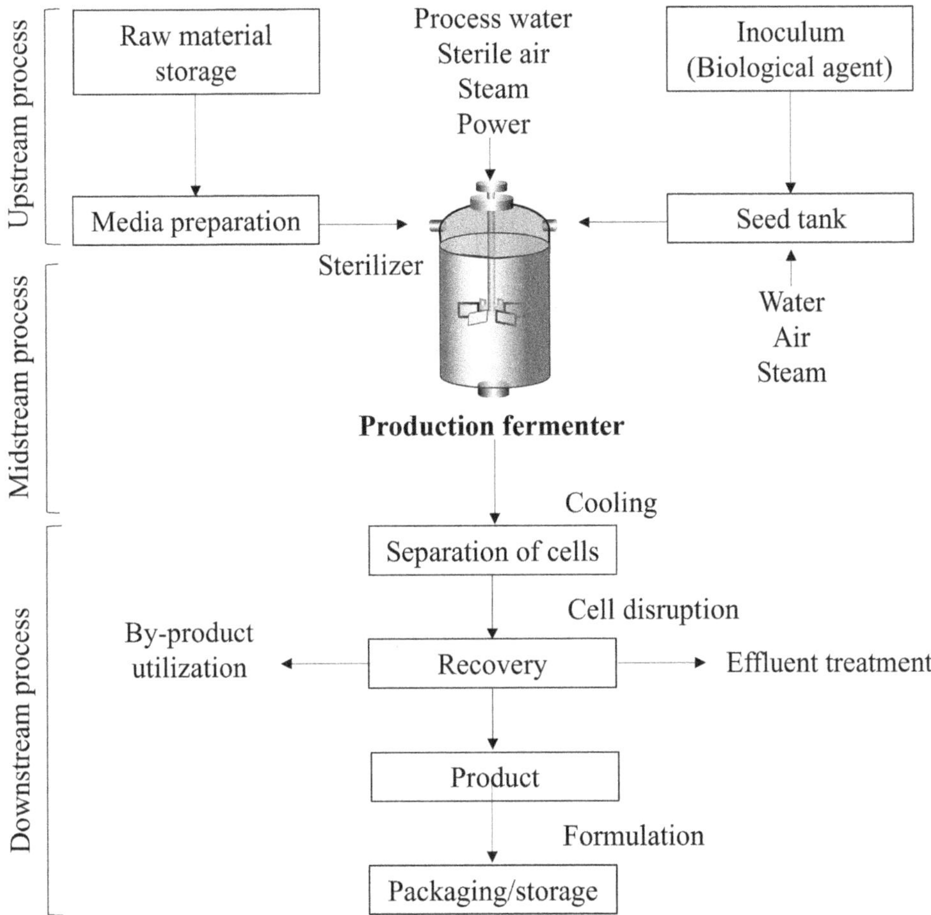

Fig. 2.1   The general process scheme for biotechnological production

The main elements that make up biotechnological processes are shown in Fig. 2.3:

Recombinant bio-producer-based biotechnological production has unique features. It requires rigorous control of product stability and strict measures to prevent environmental contamination by microorganisms. This approach also demands specific equipment and adherence to particular technological protocols.

Modern biotechnological production commonly uses microorganism strains cultivated in different types of fermenters or bioreactors.

Maintaining a controlled environment is essential for the successful execution of bioconversion or fermentation. However, ensuring sterility can be quite challenging due to factors such as the large volume of the fermenter, the complex composition of the medium, the significant volume of air involved, and the intricate equipment design. The presence of contamination can significantly affect the composition, pH, and rheological properties of the culture medium, ultimately leading to a reduction in product yield.

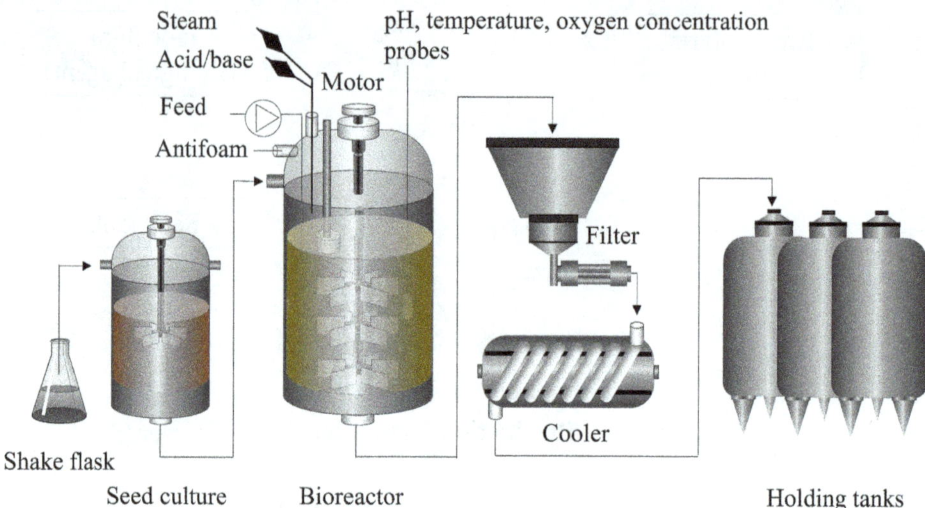

Fig. 2.2 Flow diagram of a typical industrial biotechnological process

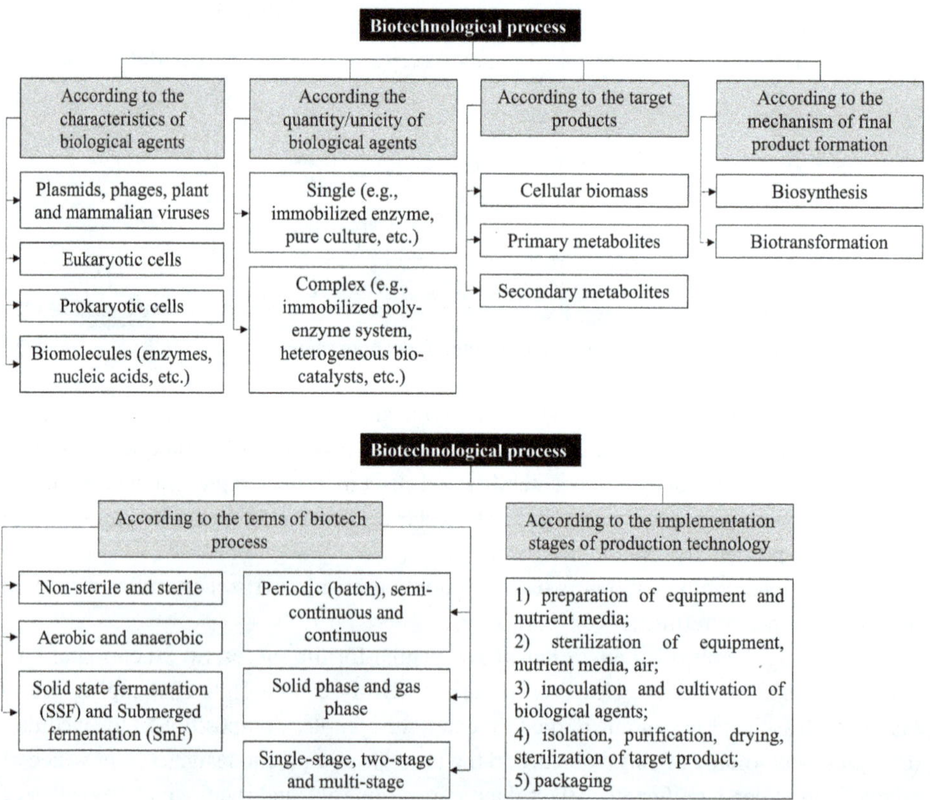

Fig. 2.3 The core elements of biotechnological processes

Such contamination may occur due to impurities or compounds introduced as a result of microbial activity.

2.1.1   Brainstorming

- Describe the modern concepts behind the organization of industrial biotechnological production.
- What is the structural organization of a biotechnological production facility?
- What are the distinctive features of biotechnological production compared to traditional technologies?
- List the advantages and disadvantages of biotechnological production relative to conventional technologies.
- Present a typical flowchart or scheme of a biotechnological production process.
- Describe the main components that make up a biotechnological process.

Take-Home Messages
- Biotechnological production is based on the use of living organisms or their components to produce valuable products in an eco-friendly, efficient, and often more sustainable way than traditional chemical methods.
- A typical biotechnological process includes three main stages:
  - Upstream processing (preparation of inoculum and nutrient media),
  - Midstream processing (fermentation or bioconversion), and
  - Downstream processing (product isolation and purification).
- The success of the process depends on maintaining sterility, proper equipment design, and accurate preparation of substrates and environmental conditions, especially in large-scale fermentations.
- Recombinant microorganism-based production requires strict biosafety measures, including containment and stability monitoring, due to the potential risks of environmental contamination.
- Biotechnological production offers multiple advantages, such as mild operational conditions, the use of inexpensive raw materials, biodegradability of products, and lower energy consumption.
- Process design and reactor configuration must be tailored to the type of bioproducer and product, which may range from simple biomass to complex pharmaceuticals.

2.2   Sterilization of Technological Flows and Equipment

The term "sterilization" is understood as the destruction of all viable organisms. Terminology is essential when controlling microorganisms is considered because some terms are often used mistakenly.

- *Sterilization* (Latin *sterilis,* unable to produce offspring) is the complete inactivation (distraction or removal) of all forms of microbial life (viable cells, spores, infectious agents) and acellular entities (viruses, viroids, and prions) about the ability to reproduce.
- The chemical compound used for sterilization is referred to as *sterilant.* This concept should be distinguished from the idea of "disinfection."
- *Disinfection* means only reducing the number of contaminating microorganisms to the "safe" level, not causing infection, but not necessarily their complete destruction.
- *Disinfectants* are chemical compounds used to conduct disinfection and are typically applied only to not-alive objects. Some spores and viable microorganisms remain indestructible after treatment with disinfectants.
- *Sanitization* is closely associated with disinfection. Here, the viable microorganisms are decreased to levels believed safe by public health standards.
- Sanitizer is designed to kill germs and are usually applied to food-processing machinery and equipment. It is also often essential to control foreign cells on or in living tissue with chemical agents.
- *Antisepsis* (Greek *anti,* against, and *sepsis,* decay) prevents infection, putrefaction, or sepsis. Antiseptics are chemical agents applied to living tissues to prevent infection by inhibiting growth of microorganisms. Since they are used on host tissues, antiseptics are generally less harmful than disinfectants.
- *Chemotherapy* is the application of chemical compounds to kill or inactivate microbial growth within the host tissue.
- Various agents that inactivate organisms often have the suffix *cide* (Latin *cida,* to destroy); thus, *Germicide* deprives pathogen or nonpathogen existence (and many nonpathogens) but not certainly endospores. Other categories include *bactericides, fungicides, viricides, and algaecides.*
- Some chemical agents do not inactivate but rather inhibit growth. If these chemicals are removed, growth will continue and have the suffix *static* (Greek *statikos,* causing to stop), examples, *bacteriostatic, fungistatic*, and *algistatic.*

2.2.1   Aseptic Requirements in Biotechnological Processes

Aseptic or sterilization is a network of efforts aimed at preventing contamination. It must be considered that any input or material flow is a potential source of microorganisms.

Table 2.1 Recommended limits for microbial contamination in cleanroom design

Cleanroom class	Grade	Microbial cells, CFU/m^3	Mechanical impurities (0.5 μm)
1	A	<1	<3500
2	B	10	<350,000
3	C	100	<3,5 million
4	D	200	

Air dust or droplets of moisture in the air contain a layer of attached microorganisms or spores on their surface. The dust particles have different sizes, and they are divided into three fractions:

- Larger particles (>100 μm diameter)
- Smaller particles (<100 μm diameter)
- A phase of bacterial dust (with a particle diameter from 1 to 100 microns)

In the absence of air flows, particles of larger fractions settle on the surface for several seconds. However, particles belonging to smaller fractions can stay in the air long enough to form a stable colloidal system.

Industrial facilities for the degree of air contamination by microorganisms and mechanical particles per 1 m^3 are classified into four cleanliness classes (Table 2.1).

The first class of cleanliness is accessible under the laminar flow of sterile air. There are no microorganisms, and mechanical particles up to 0,5 microns in size should not be more than 3500. Sterile medicines and drugs are prepared on such premises.

The second class of cleanliness is a room that contains no more than ten microbial cells, and mechanical impurities of 5 microns in size do not have more than 2500 particles and particles of 0,5 μm in size to 350,000.

The third class of purity is up to 100 microbial cells and mechanical particles with five μm in size to 25,000 and 0,5 μm to 3,5 million in 1 m^3.

For the fourth class of cleanliness, it is sufficient to fulfill the conditions by State standards (no more than 200 CFU/m^3).

Employees working in clean rooms should wear unique technological clothing or lab coats (according to the requirements of GLP). So, put on a sterile suit on the first class of cleanliness premises. This headdress covers the hair, including the beard, and is wrapped under the costume collar. Additionally, a face mask is worn, sterile gloves are put on the hands, and shoe covers are placed on the feet. Such protective clothing is either disposable or intended for single-day use. The doors of these rooms are equipped with airlocks. In working areas, sterile air is supplied under positive pressure and exchanged 600 to 200 times per hour.

2.2.2   Aseptic Techniques and Agents

A sufficiently large number of agents and techniques ensure the inactivation of microorganisms. The general methods of microbial control and their application are shown in Fig. 2.4.

2.2.3   Steam Sterilization Under Pressure

Sterilization with saturated steam under pressure in autoclaves is highly effective and reliable. This method is universal, given that the materials can withstand high temperatures and humidity. Saturated steam deactivates all microorganisms, including extremophiles

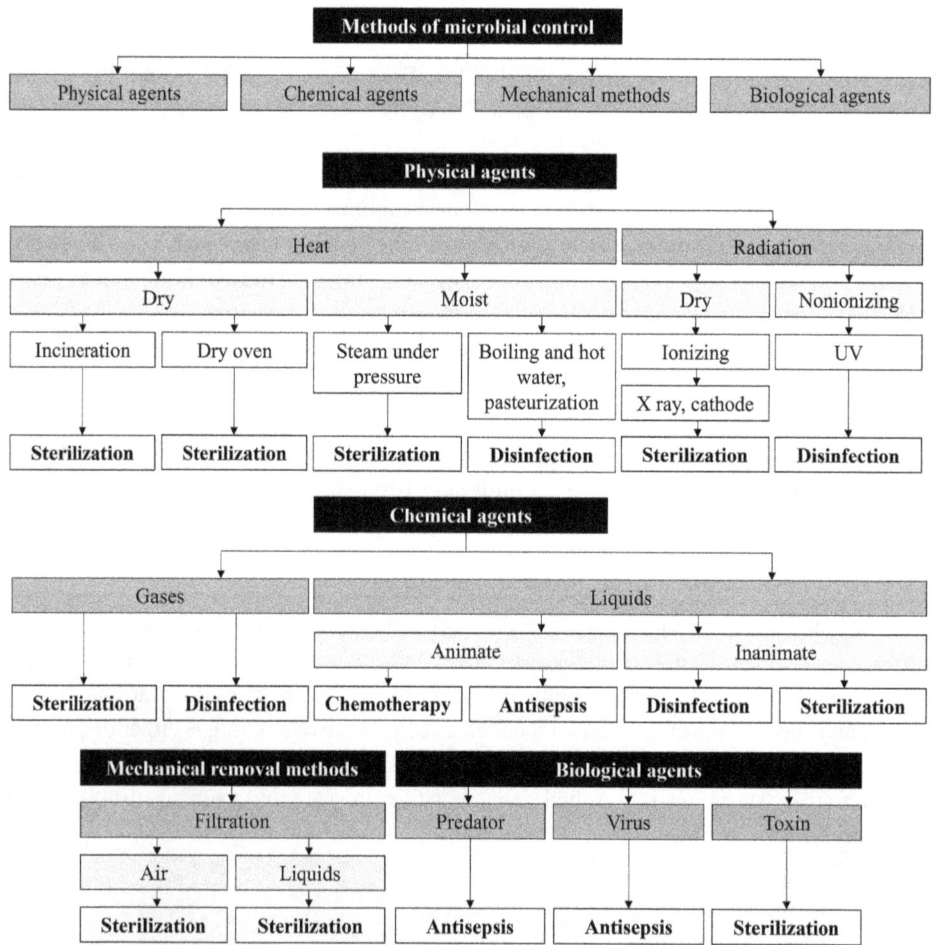

Fig. 2.4 Microbial control methods

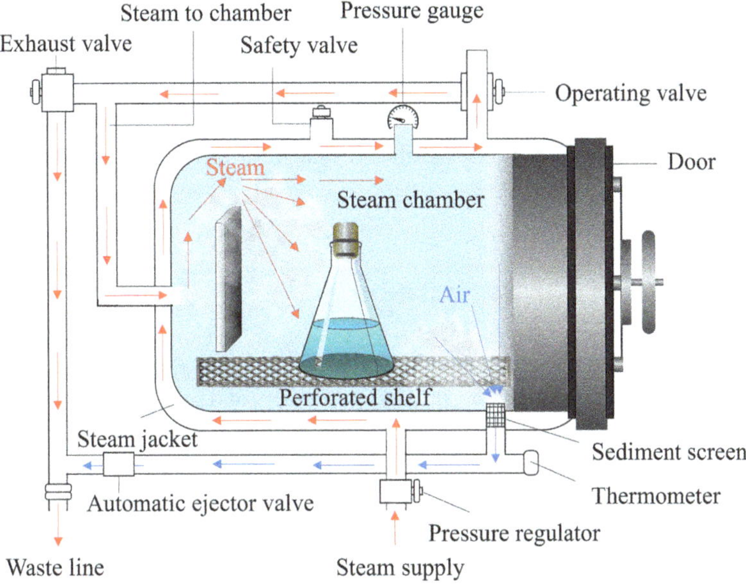

Fig. 2.5 The autoclave sterilization process

and bacterial spores, due to its high heat capacity. At 100 °C, steam's heat capacity is seven times that of water. Sterilization efficiency hinges on the steam's temperature, typically at 121 °C for 20–30 min. The duration depends on the material's volume and composition, as well as the autoclave chamber size..

Sterilization by steam is carried out in autoclaves with a different design solution (Fig. 2.5).

A primary challenge in steam sterilization is the differing properties of air and steam. Air, being denser and cooler than steam, settles at the chamber's bottom. Since air and steam don't mix well, a temperature gradient forms, leading to uneven sterilization. To address this, air is pre-emptively removed from the sterilization chamber.

2.2.4 Sterilization with Dry Heat

Hot air sterilization is suitable for tools like metal and glass, and for pharmaceutical items like powders and ointments that can't endure steam. This method uses drying cabinets and ovens with adjustable temperatures and durations. Its limited use stems from design flaws in heat-producing devices. The process operates between 100 °C and 180 °C, tailored to the items being sterilized. Pathogens like Bacillus anthracis and Clostridium deactivate at 170 °C in 60 min.

2.2.5 Sterilization with Chemical Agents

Delicate and fragile tools or equipment that are easily damaged by the effect of physical agents are sterilized by various chemicals (Table 2.2). It was found that sterilization is affected by air humidity. The most significant impact of sterilization is detected with a relative humidity of more than 25%.

2.2.6 Sterilization with UV Radiation

Sterile air is crucial for various technological processes, especially in fermentation and downstream processing. Clean air ensures product purity and safeguards both workers and the environment. Atmospheric air contains numerous fine particles, including microbial cells and spores. Absolute sterility is virtually unachievable, but air can be purified to minimize microbial contamination. Advanced air purification systems have demonstrated the capacity to significantly diminish microbial content to below detectable levels.

Table 2.2 Chemicals used for microbial sterilization

Chemical agent		Mechanism of action	Application
Halogens		Restriction of protein function, alteration of cellular components	Disinfection of water, dairy equipment, household things, etc.
Alcohols		Protein denaturation	Various instruments, working surfaces
Heavy metals and their compounds		Enzyme and protein denaturation	Treatment of wounds, burns, algicide
Surface- active agents	Detergents	Removal of biological agents through scrubbing	Skin disinfection
	Acid-anionic sanitizers	Enzyme inactivation	Dairy and food-processing plants
Chemical food preservatives	Organic acids	Metabolic inhibition	Personal care products: cosmetics, shampoos, etc.
	Nitrates and nitrites	Inhibition of iron-containing enzymes of anaerobic microorganisms	Meat and poultry products
Aldehydes		Protein denaturation	Disinfection of medical equipment
Chemical sterilization	Ethylene oxide	Inhibition of cellular functions	Disinfection of many materials
	Plasma sterilization	Inhibition of cellular functions	Disinfection of medical equipment
	Supercritical fluids	Inhibition of cellular functions	Sterilization of organic medical implants
Peroxygens		Oxidation	Contaminated surfaces, deep wounds

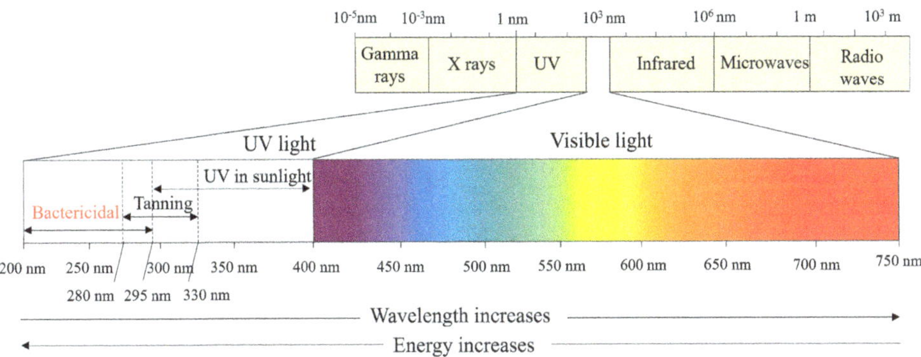

Fig. 2.6 The radiant energy spectrum

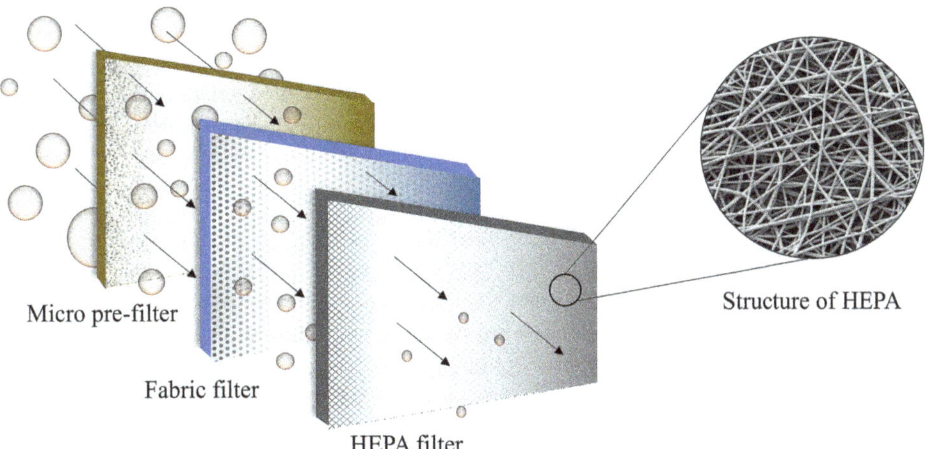

Fig. 2.7 Clogging of different-sized particles on various membrane filters

Several methods are used for air purification, including radiation, heat treatment, and air filtration.

Ultraviolet radiation with a wavelength of 260 nm is capable of killing bacteria and viruses contained in the air (Fig. 2.6). The lethal effect of UV light depends on both the temperature and the relative humidity of the air.

UV radiation does not provide high air sterility; this technique can only be combined with other sterilization methods.

2.2.7   Filtration

Ceramics, sintered glass, and various fibers/wools are used as filter materials. Due to modern engineering, high-efficiency filters—HEPA (High-efficiency particulate air) have been created (Fig. 2.7). Filtration using HEPA filters is the most preferred method when it is

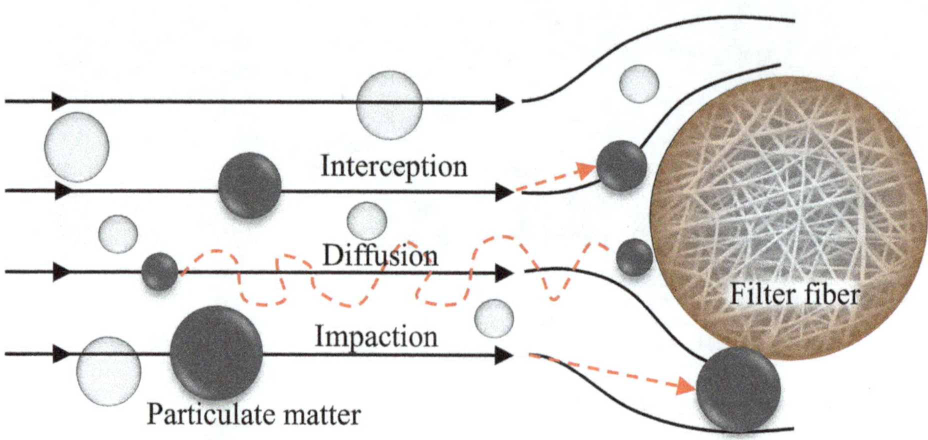

Fig. 2.8 Capture mechanisms of filters

necessary to extract air from the vessels, where cells are cultivated, creating safe zones and others. Particles suspended in the air are retained as they pass through fibrous material by inertial or diffuse deposition mechanisms.

During the filtration through the fibrous filters, three mechanisms function: impaction, interception, and diffusion of particles (Fig. 2.8).

- *Inertial impaction*: Airborne particles deviate from the airflow due to inertia and settle on fibers. This is especially effective for particles >5 μm and at higher airflow rates.
- *Interception*: Affects particles 0.5–5 μm in size. These particles follow the airflow and are trapped upon contact with fibers.
- *Diffusion*: Particles <0.3 μm follow Brownian motion, increasing the probability of colliding with fibers and being captured.
  Selecting an appropriate filter structure is essential for effective air purification. Primary filters use coarse fibers to capture nearly all microorganisms, while subsequent filtration stages use superfine fibers or membranes to eliminate residual contaminants.

2.2.8   Exhaust Air Cleaning

During fermentation, exhaust air may contain compounds like amines, ketones, and alcohols. It may also carry producer cells from the culture medium, making air purification necessary. Common methods include:

- *Catalytic conversion*: Exhaust air passes through a catalyst containing pyrolusite and palladium at ~335 °C. This energy-intensive method neutralizes up to 98% of contaminants.

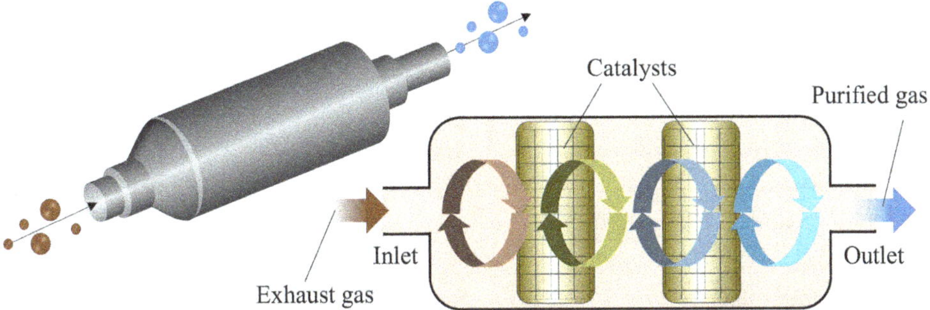

Fig. 2.9 Catalytic converter

- *Liquid-phase oxidation*: Potassium permanganate or sodium hypochlorite solutions are used as oxidants in a closed loop, replaced periodically. This achieves ~93% purification.
- *Screen filters*: Cylindrical filters with metal knitted netting purify exhaust air up to 99% (Fig. 2.9).

2.2.9   Sterilization of Nutrient Media

In upstream processing, nutrient media preparation is essential, as product yield depends on its quality.

Thermal sterilization of media is done in two ways:

1. *Batch sterilization*: Heating, holding at high temperature, and cooling occur in one device. While simple, it requires more energy and is harder to automate.
2. *Continuous sterilization*: Each phase—heating, holding, and cooling—is handled by separate equipment (heaters, thermostats, coolers), allowing fast heat-up and shorter exposure to high temperatures (Fig. 2.10).

Advantages of continuous sterilization:

- Reduces nutrient degradation due to brief high-temperature exposure
- Easier to control and automate
- Allows partial heat recovery
- Uses indirect or direct (steam-based) heat exchangers

Ultrafiltration is also used for sterilizing media. Filters with 0.15–0.25 µm pore size retain large molecules and let smaller ones pass. Ultrafiltration can also separate proteins or cells (Fig. 2.11).

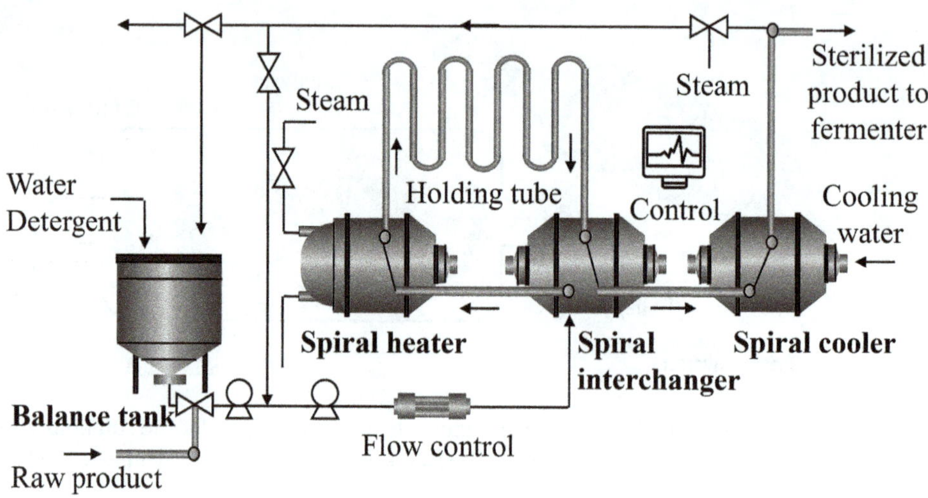

Fig. 2.10 Flow diagram of a typical continuous sterilization system

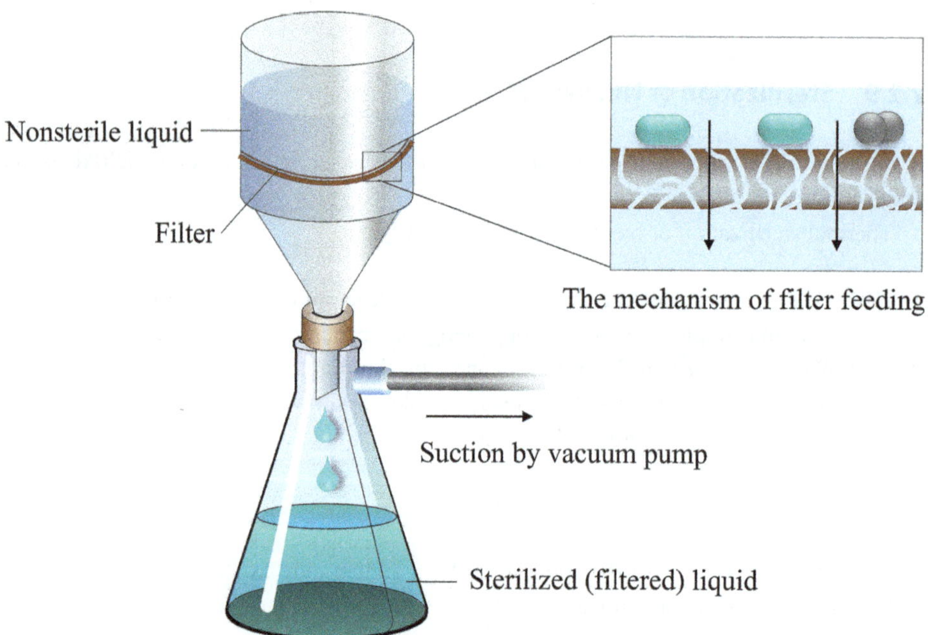

Fig. 2.11 Typical membrane filter for the sterilization of media and organic liquids

Membrane filters—made from cellulose esters or plastic polymers—are increasingly used in industrial and lab settings. They are prepared by:

- Etching nuclear bombardment sites
- Washing a solution to form a thin film on a moving tape

Sterility Control

Maintaining sterility requires strict technological protocols. Sterility is tested microbiologically by sampling, dilution, and culturing on selective or differential agars, followed by incubation. If no bacterial growth occurs, the sample is considered sterile.

However, conventional media may not detect organisms like mycoplasmas. Their identification requires special methods. Viruses are detected using PCR, microarrays, or specialized equipment. Therefore, inspecting equipment and systems at least once a year is recommended.

Brainstorming

1. Define the following terms: sterilization, disinfection, aseptic, and antiseptic.
2. What is the significance of aseptic cultivation in the production process?
3. Describe the principles of creating and maintaining aseptic conditions in biotechnological production.
4. What are the recommended limits for microbial contamination in cleanroom design?
5. Describe microbial control methods.
6. Describe the methods of sterilization by steam under pressure.
7. Describe the methods of sterilization using chemical agents.
8. Explain sterilization with UV radiation.
9. Name the mechanisms and materials used in sterilizing filtration.
10. How is air decontamination carried out in production facilities?
11. How is the sterilization of nutrient media performed?
12. List several advantages of continuous sterilization.

Take-Home Messages
1. Sterile air is essential in biotechnological production, especially during fermentation and downstream processing, to prevent contamination and ensure product quality.
2. UV radiation at 260 nm effectively inactivates airborne bacteria and viruses. Its efficacy depends on environmental factors like temperature and relative humidity.
3. HEPA filters are the industry standard for air filtration, employing impaction, interception, and diffusion mechanisms to trap particles across a wide size range, including microorganisms.
4. Exhaust air purification is critical due to the presence of volatile compounds and microbial contaminants. Effective methods include catalytic conversion, chemical oxidation, and screen filters, achieving up to 99% purification.
5. Sterilization of nutrient media is a crucial step in upstream processing. Continuous sterilization is favored over batch methods due to faster heating, better control, nutrient preservation, and heat recovery potential.

6. Ultrafiltration using membrane filters with pore sizes of 0.15–0.25 μm is a reliable alternative to thermal sterilization for media and is also useful in protein and cell separation.
7. Sterility validation requires regular microbiological testing. While conventional media detect common bacteria and fungi, specialized techniques are needed to identify mycoplasmas and viruses, underscoring the need for comprehensive sterility assurance.

Basic Methods and Approaches in Industrial Biotechnology   3

Biotechnological processes have vast applications, from food production to green energy. These processes are compact yet large-scale, technologically advanced, and efficient. They can be easily controlled, regulated, and automated. Unique technologies and agents are used to produce biotechnological products. Each industry has variations, but a standard model exists for all.

The determining factors of the biotechnological process include:

- Type of biotechnological process used
- A substrate with biochemical and biophysical characteristics
- Equipment, including a control system and technological regime

Industrial biotechnology typically involves three main stages when using microorganisms or cell cultures. These stages produce valuable metabolites for commercial use. (Fig. 3.1):

1. Pre-fermentation: Involves preparing nutrients and the inoculum.
2. Fermentation: Targets microorganisms that grow in a bioreactor, producing desired compounds.
3. Post-fermentation: The final product is purified from culture medium or cell components.

3.1   Upstream Processing

Upstream processing is necessary for the preparation of raw materials used in the biotechnological process (Fig. 3.2). Depending on the target product, the process includes:

- Preparation of the medium that includes the necessary components of nutrition for biological agents and its sterilization
- Preparation and sterilization of inlet gases by purification from impurities, moisture and microorganisms and spores found in the air

I. Digel et al., *Introduction to Industrial Biotechnology*, Learning Materials in Biosciences,
https://doi.org/10.1007/978-3-032-07918-3_3

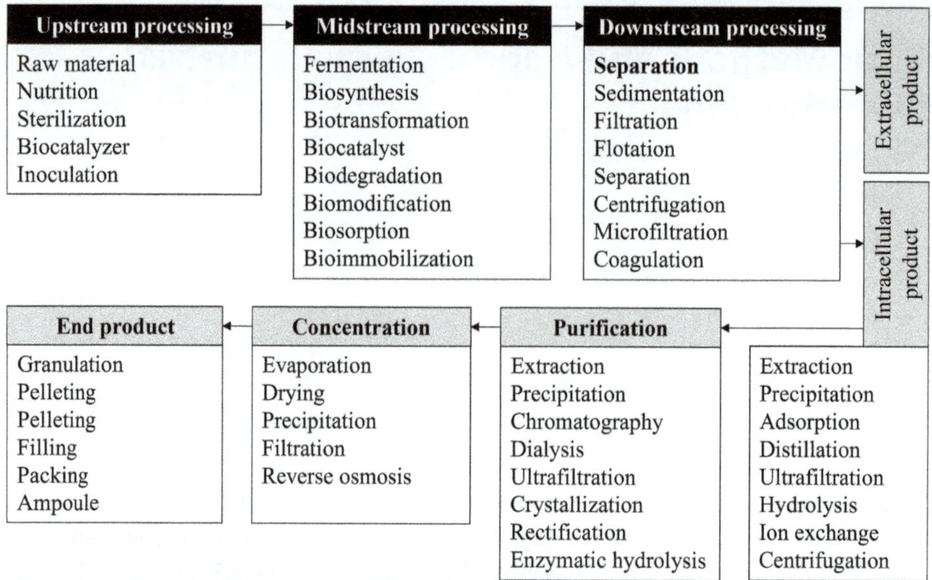

Fig. 3.1 Generalized flowchart of biotechnological production

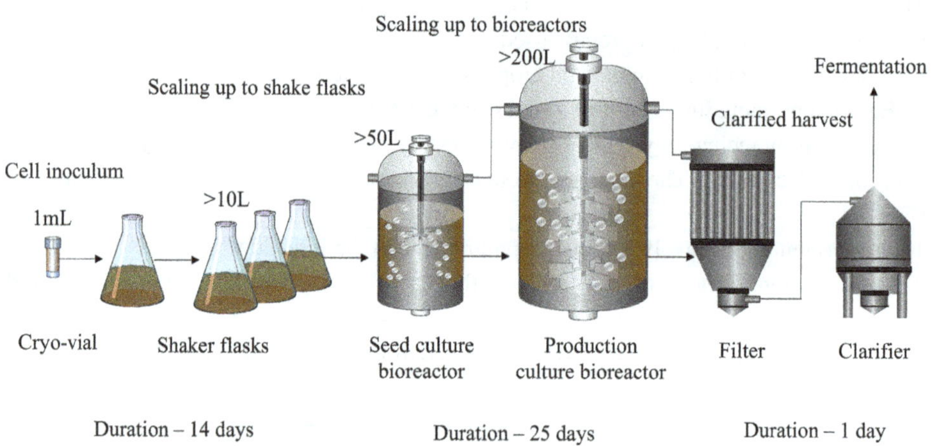

Fig. 3.2 Cell culture process design in upstream processing

- Preparation and cultivation of inoculum, such as microorganisms and cell lines of plants or animals
- Preparation of biocatalysts, either an enzyme in a free or immobilized form or microbial biomass, which possess high physiological activity
- Preliminary treatment of raw materials if they come in a form unsuitable for direct use

3.1.1 Raw Materials Base

Biotechnological processes become effective through the following steps:
1. Isolating highly efficient biological agents

    Biotechnological processes begin with the identification and isolation of biological agents that exhibit high efficiency in performing specific tasks. This involves selecting and isolating microorganisms, enzymes, or other biological entities that possess desirable characteristics for the intended bioprocess.
2. Implementing technological systems for optimal performance

    Once efficient biological agents are identified, the next step is to develop and implement technological systems that ensure their optimal performance. This includes creating suitable growth media, optimizing environmental conditions (such as temperature, pH, etc.), and establishing controlled settings for the cultivation of the biological agents.
3. Selecting the substrate or raw material to support the process

    The choice of substrate or raw material is a critical step in biotechnological processes. It involves selecting appropriate organic or inorganic materials that serve as feedstock for the biological agents. The substrate should provide essential nutrients and support the growth and activity of the biological components.
4. Designing specialized equipment tailored to the specific process requirements

    To facilitate biotechnological processes, specialized equipment must be designed to meet the specific requirements of the chosen biological agents and the overall process. This may include bioreactors, fermenters, or other custom-designed apparatuses that ensure efficient growth, metabolism, and product formation.
5. Incorporating automation for control of the entire process

    Automation is a crucial aspect of modern biotechnological processes, enabling precise control and monitoring throughout the entire production cycle. Automated systems regulate variables such as temperature, pH levels, nutrient supply, and agitation, ensuring consistency and reproducibility in the bioprocess.
6. Developing purification techniques to obtain the desired end product

    After the bioprocess is completed, the next step involves the separation and purification of the desired end product. Various techniques—such as chromatography, filtration, and centrifugation—are employed to isolate and purify the target biomolecules or compounds produced by the biological agents.

In microbial media, microorganisms rely on substrates as their primary source of carbon and energy for synthesizing microbial proteins. Other media components serve as auxiliary nutrients and functional additives.

The microbiological industry heavily depends on raw material sources, with approximately 90% allocated to ethanol production, 5% for bakery yeast, and 2% for organic acids, among other applications.

In general, biotechnologically important substrates can be classified as follows (Fig. 3.3):

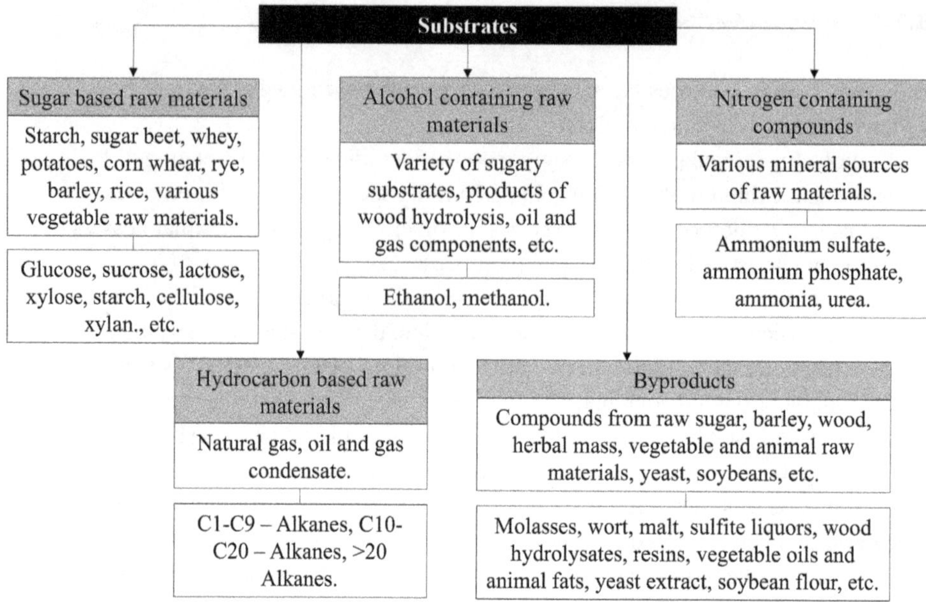

Fig. 3.3 Classification of substrates used in biotechnological production

3.1.2　Overview of Biotechnologically Important Substrates

- Beet molasses: A byproduct of sugar extraction, rich in organic and mineral contents, widely used for producing citric acid, ethanol, and other fermentation products.
- Distillers grains: Residues from molasses and alcohol production; composition depends on the initial molasses type. These are important for fodder yeast production due to their high vitamin content. Variants include WDG (wet distillers grains) and DDGS (dried distillers grains with solubles).
- Brewing byproducts and unmalted barley: Modest sources of carbohydrates, useful for producing microbial proteins.
- Wheat bran: A milling byproduct used as a base in solid-phase cultivation, often enhanced with low-cost additives like wood chips.
- Milk whey: A dairy byproduct containing nearly half the nutritional value of milk and a rich source of biologically active elements.
- Corn extracts: Nutrient content varies by maize variety and cultivation method. These are high in B vitamins and, after enzymatic treatment, also rich in amino acids.
- Wheat extracts: Obtained from soaked grains, similar in use and composition to corn extracts.

3.1.3 Nitrogen Sources

- Yeasts primarily metabolize nitrogenous compounds.
- Fungi can be optimized by adding organic nitrogen sources.
- Media formulations must include major macroelements and trace minerals.

3.1.4 Water and Media Additives

- Pure water, free of impurities and sediments, is essential for preparing growth media.
- A combination of natural and synthetic antifoaming agents is used to reduce foaming.
- Chalk and ammonia help maintain pH balance and nitrogen content, while specific acids regulate acidity.

3.1.5 Additional Components

- Rapid microbial growth requires vitamins and other growth factors.
- Complex media often include peptones and plant/animal extracts:
- For animal cells, placenta extract is frequently used.
- For plant cells, pumpkin extract is a common additive.

Brainstorming

1. What are the critical factors influencing a biotechnological process?
2. List the main stages of biotechnological production.
3. Provide a generalized flowchart or schematic of a biotechnological production cycle.
4. What are the main differences between biotechnological and chemical production methods?
5. Give a brief description of the preparatory (upstream) stage in the biotechnological process.
6. What technological operations are typically carried out during the preparatory stage?
7. Describe the types of substrates that are biotechnologically important.
8. What components are commonly included in nutrient media used in biotechnological production?
9. What does the concept of preliminary preparation in fermentation refer to?
10. How is sterility ensured throughout the entire biotechnological production process?
11. What are the key steps in applying a systematic approach to biotechnological processes?
12. How does each process step contribute to the overall success of the bioprocessing cycle?
13. Why is a carefully planned and executed process crucial for obtaining a high-quality final product?

14. What is the final stage in biotechnological production after the main bioprocess is complete?
15. What are some typical techniques used for the separation and purification of the end product?
16. Why is the isolation and purification of target biomolecules critical in the bioprocessing industry?

Take-Home Messages
- Biotechnological processes begin with the selection of high-performing biological agents, such as specific microorganisms or enzymes, that can efficiently carry out targeted reactions under controlled conditions.
- Optimizing environmental parameters (e.g., temperature, pH, oxygen levels) and designing appropriate growth media and cultivation systems are crucial for ensuring the effective performance of biological agents.
- Substrate selection is a critical factor, as it directly influences microbial metabolism, growth rate, and product yield. Substrates should be nutritionally suitable, economically viable, and compatible with the process.
- A wide variety of agricultural and industrial byproducts (e.g., beet molasses, whey, wheat bran, distillers grains) serve as cost-effective substrates in microbial cultivation media, enhancing sustainability and reducing waste.
- Automation in biotechnological production ensures consistency, reproducibility, and efficient process control, minimizing human error and optimizing resource use.
- Specialized equipment, including bioreactors and fermenters, must be tailored to meet the specific biological and engineering requirements of each process.
- Downstream purification of the desired product is a vital stage that often employs techniques like filtration, centrifugation, and chromatography, especially in pharmaceutical and food biotechnology.
- Microorganisms require complex nutrient formulations, including carbon and nitrogen sources, minerals, vitamins, and water of high purity, to support high-yield, high-quality bioprocesses.
- The quality and preparation of the nutrient media—including pH adjustment, sterilization, and antifoaming strategies—directly affect fermentation efficiency and final product quality.
- Modern industrial biotechnology relies on an integrated, stepwise approach: isolating bio-agents → optimizing conditions and media → selecting substrates → engineering equipment → automation and monitoring → product recovery and purification.

3.2 Midstream Processing. Technological Equipment for Industrial Use

The biotransformation stage is the crucial step in biotechnological production, where a particular biological agent (microorganisms, cell cultures, enzymes, or cellular organelles) is used, and the conversion process occurs—raw materials are transformed into a specific target product.

However, this stage usually includes not only the synthesis of new organic compounds but also several other biotechnological processes, such as:

- *Bioconversion*—the transformation of organic materials into usable bio-based products or energy sources
- *Fermentation*—a process carried out by enzymes in cultivated microorganisms
- *Biotransformation*—the modification of a chemical structure under the enzymatic activity of cells or isolated enzymes
- *Biocatalysis*—the chemical transformation of substances using biocatalysts (enzymes)
- *Bio-oxidation*—the aerobic oxidation of pollutants using microorganisms or microbial consortia
- *Bio-composting*—the reduction of harmful organic substances in solid waste
- *Biosorption*—the binding and removal of harmful impurities from gases or liquids by microorganisms attached to carriers or matrices
- *Bacterial leaching*—the microbial transfer of water-insoluble metal compounds into a dissolved state
- *Biodegradation*—the microbial breakdown of harmful compounds

The typical midstream processing stage results in the production of one or more liquid or gas products, and in some cases, processed solid products.

3.2.1 Industrial Technological Equipment

Different types of equipment are used to cultivate tailored biological agents depending on the scale and purpose. In laboratory settings, equipment such as flasks, tubes, rollers, seeding reactors, and incubators is used for small-scale cultivation. In contrast, bioreactors and fermenters are essential for large-scale industrial production.

In laboratories, agitating vessels are often used to prevent cell sedimentation and to maintain adequate concentrations of dissolved oxygen.

A *bioreactor* or *fermenter* is a device designed to optimize the conditions necessary for the growth of biological agents and the synthesis of desired products (see Fig. 3.4).

Fermenters are integral to many biotechnological processes, including the production of valuable compounds, substrate transformation into metabolites, organic waste degradation, and biogas generation.

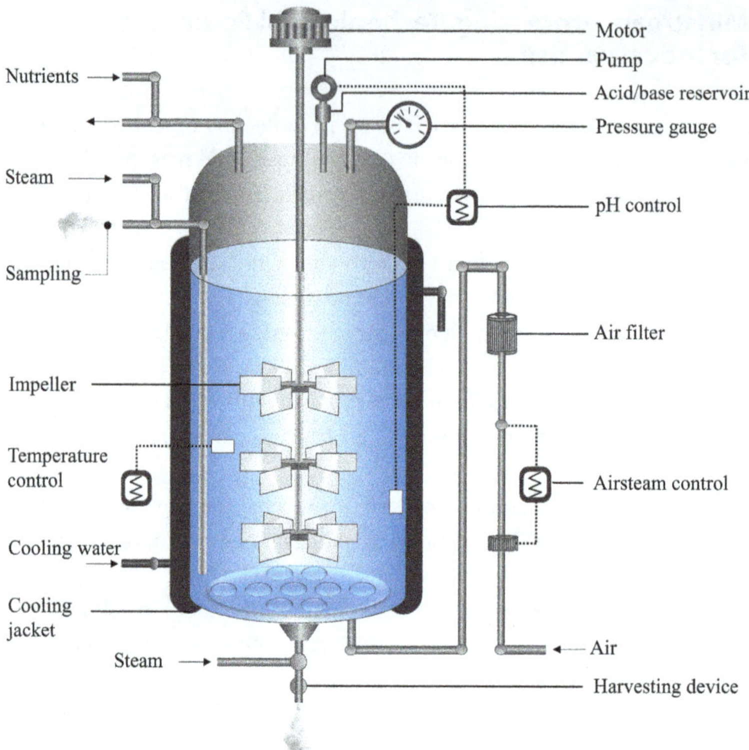

Fig. 3.4 Generalized view of a typical bioreactor

Within a bioreactor, various biological agents and biocatalysts such as microorganisms, mammalian cells, plant and insect cells, enzymes, and subcellular fractions drive the reactions.

Typically, product generation occurs in bioreactors ranging in volume from 15 liters (pilot scale) to 300 m^3 (industrial scale). Even larger units are required for processes such as wastewater treatment and methane production.

3.2.2 Key Operational Parameters for Bioreactor Function

To ensure efficient bioreactor function, the following parameters must be carefully controlled:

1. Maintain biomass concentration and prevent contamination—cells must remain in the reactor without being lost through the outlet.
2. Achieve proper mixing through agitation to keep the biological catalyst uniformly suspended.
3. Prevent the entry of foreign microbes (sterile conditions).

4. Manage oxygen supply and carbon dioxide removal effectively.
5. Provide an adequate supply of nutrients to support optimal biological activity.
6. Ensure effective waste removal, gas and liquid dispersion, and suspension of particles.
7. Control shear stress to avoid damaging sensitive biological agents.
8. Regulate temperature by removing excess metabolic heat.

The complexity of biological systems necessitates multifunctional control systems in modern bioreactors (Table 3.1).

The evolution of bioreactors began with simple earth pits used for organic waste disposal (Fig. 3.5). Over time, closed vessels made from various materials were used to produce alcoholic beverages such as wine, beer, and vinegar. A major advancement came during the time of Louis Pasteur, when the principles of sterile media, pure cultures, and refined production processes were established and integrated into industrial biotechnology.

Table 3.1 An indicative list of controlled parameters for the growth of biological agents in bioreactors

Parameter	Type of control
Recruitment factor	Physical (light-optical)
Agitator power	Mechanical (external-torsional dynamometry, internal strain gauge)
Rotary speed of mixer	Mechanical (tachometry)
Redox potential	Physical/electrical conductivity (determination of oxidation-reduction potential)
Amount of dissolved $ORR2$	Physicochemical (electrochemical, etc.)
Amount of dissolved $CORR2$	Physicochemical (IR spectrometry, etc.)
Foam detection	Physical (electrical conductivity)
Foam regulation	Mechanical (mechanical device), physicochemical (e.g. silicone emulsion defoamer)
Glucose consumption	Chemical, physicochemical
N consumption	Chemical, physicochemical
Amount of biomass	Physical (spectrophotometry, nephelometry)
Temperature	Physical (thermometry)
Gas flow rate	Mechanical (rotameters, flowmeters)
The nutrient rate, addition to solutions	Mechanical/flowmeters (drip counter, dynamometric sensors, etc.)
Pressure	Mechanical (manometry)
Viscosity	Physical (viscometry)
pH	Chemical (pH-metry)
Sampling of culture fluid	Mechanical (level control, dynamometer)
Enzymatic activity	Biochemical, chemical, physicochemical
Antibiotic activity	Biological, chemical, physicochemical
Determination of nucleic acids	Chemical, physicochemical
Determination of ATP	Chemical, physicochemical

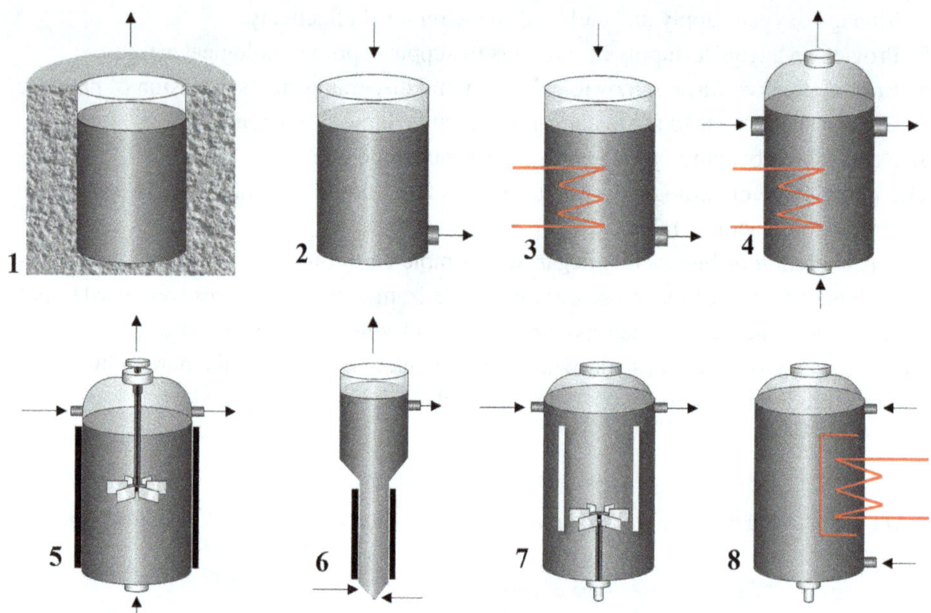

Fig. 3.5 Bioreactors from a historical perspective. 1—covered hole in the earth for biogas production; 2—classic container (wood, metal, leather, plastic) to produce alcoholic beverages and sour milk; 3—brewery (open reactor) with temperature control; 4—aseptic close reactor for controlled fermentation; 5—stirred reactor for antibiotic production; 6—tubular tower reactors for alcoholic drink production; 7—airlift reactor with internal recirculation; 8—airlift reactor with external recirculation

Classification of Bioreactors

Different types of bioreactors can be categorized in various ways.

According to volume, they are differentiated as:

- Laboratory bioreactors (up to 50 L)
- Experimental and pilot-plant bioreactors (50–5000 L)
- Industrial or commercial bioreactors (from 50 L up to 1,500,000 L)

  According to the height/diameter ratio, reactors are distinguished:

- Tank reactors (H/D = 3)
- Reactor columns (H/D >3)

  According to their energy input, they are categorized as:

- Mechanically stirred bioreactors are the most diverse (e.g., usually for producing secondary metabolites)
- Hydrodynamic reactors equipped with a pump (e.g., for producing single-cell proteins and yeasts)
- Pneumatic reactors with energy input (e.g., for producing important metabolites, wastewater purification, etc.)

Currently, there are several types of bioreactor configurations according to gas exchange and mixing design (Figs. 3.6, 3.7 and 3.8), but extensively used ones include:

- Stirred tank reactors
- Bubble columns
- Airlift reactors
- Packed and fluidized beds

In general, bioreactors employed in industrial production belong mainly to one of the three classes:

1. Without agitation, non-aerated (kinetically controlled anaerobic process)
2. Without agitation, aerated (batch mode process)
3. With stirring and aeration (versatile types of reactors)

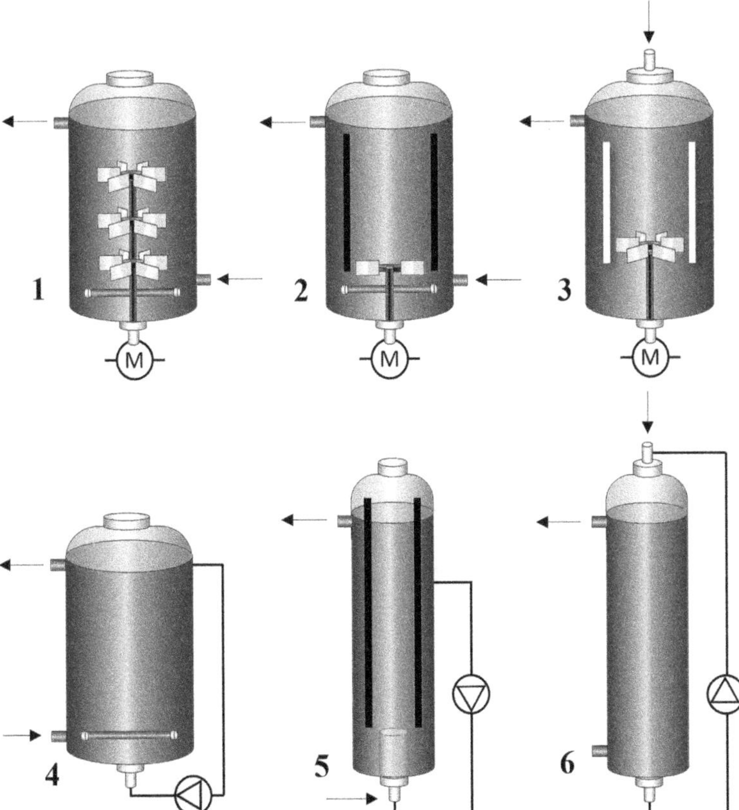

Fig. 3.6 Bioreactor types based on the gas distribution by mixing: 1—installation with rotating stirrers; 2—rotating vessel with a deep tube; 3—mixer with rotating stirrer and engulfing air tube. Bioreactor types based on the distribution of gas using pumps: 4—fused disc with recycling; 5—distribution of gas by the principle of ejection; 6—water jet aerator

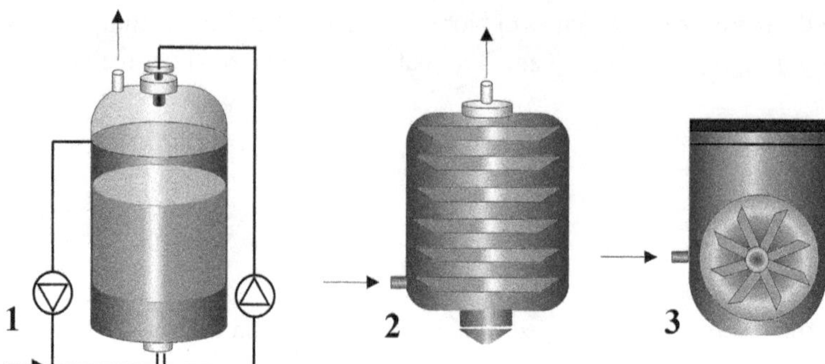

Fig. 3.7 Bioreactors with continuous gas phase: 1—jet film reactor; 2—surface film reactor; 3—wheel-blade reactor

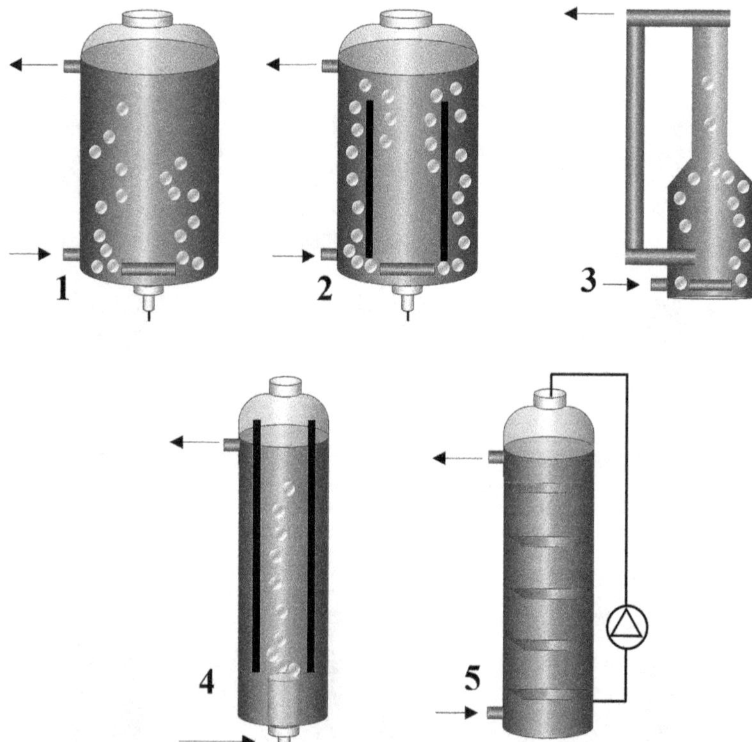

Fig. 3.8 Bioreactors based on the gas distribution due to excess gas pressure: 1—fused disk system; 2—airlift system; 3—reactor operating under pressure and recycling; 4—giant tubular reactor; 5—the system of cascade net cymbals

Industrial cultivation of plant cell suspension culture is carried out in various bioreactors, and generally, they are divided according to the design and the principle of agitation culture fluid.

As recorded by design categorization, the bioreactors stir the cells by aerating them with air. This is a *bubble column reactor*, in which the process of mixing the suspension is accomplished by rising air bubbles. Here, mainly good growth characteristics are retained for many cell cultures. However, the complexity of maintaining the suspension in a homogeneous state at high concentrations of the biomass of cells restricts the scope of their use.

Packed bed reactors belong to the non-agitated bioreactor group and are extremely versatile and advantageous in that a vast amount of cell suspensions can be immobilized per unit volume to synthesize or catalyze biologically active substances (Fig. 3.9).

The maximum concentration of cellular biomass can be achieved using *airlift bioreactors*, in which directional circulation flows are created. In airlift bioreactors, the suspension is stirred using a unique design that creates a density gradient (Fig. 3.10).

The second group of bioreactors is based on mechanical agitators. Bioreactors of this type make it possible to study plant cell populations in a wide range of cell biomass concentrations (Fig. 3.11).

Three main scientific and technical issues are involved in bioreactor design and process operation for plant cell cultures:

- Plant cell growth and product formation evaluation
- Modeling of the plant cell culture dynamics
- Technical characteristics (flow, agitation, and mass transfer) between the phases concerning bioreactor design and scale-up

Various strategies are available for expanding volumetric productivity in animal cell cultures (Fig. 3.12).

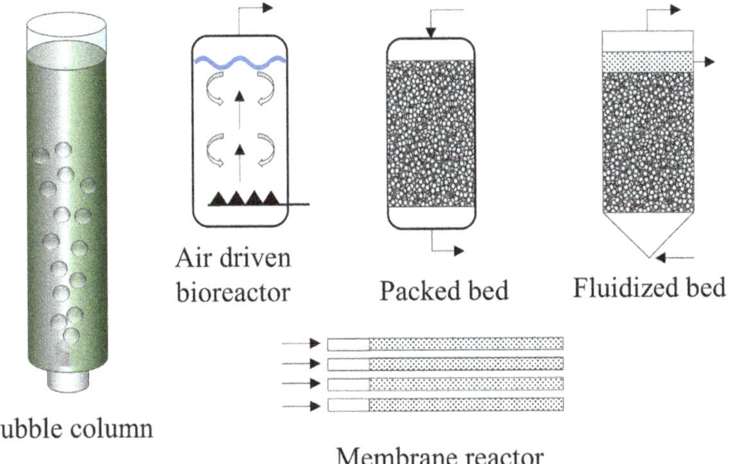

Fig. 3.9 Bubble column and non-agitated bioreactor types for suspended plant cell cultivation

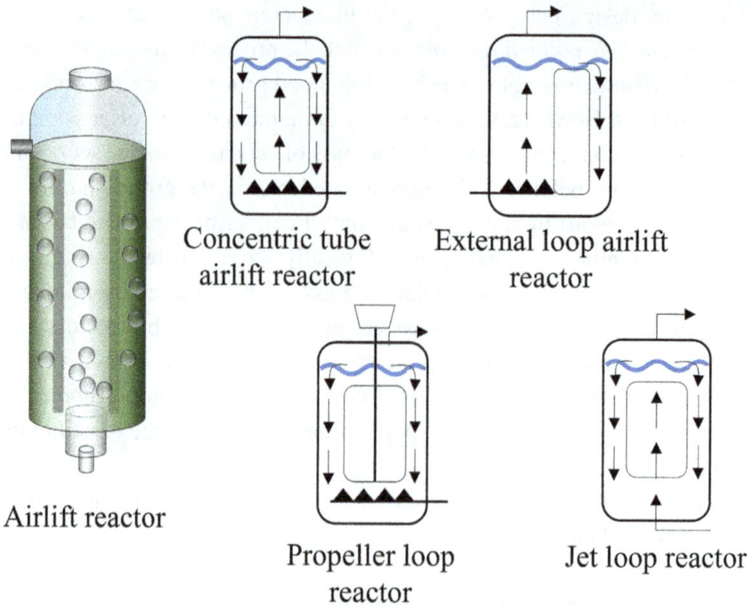

Fig. 3.10 Airlift reactor types for suspended plant cell cultivation

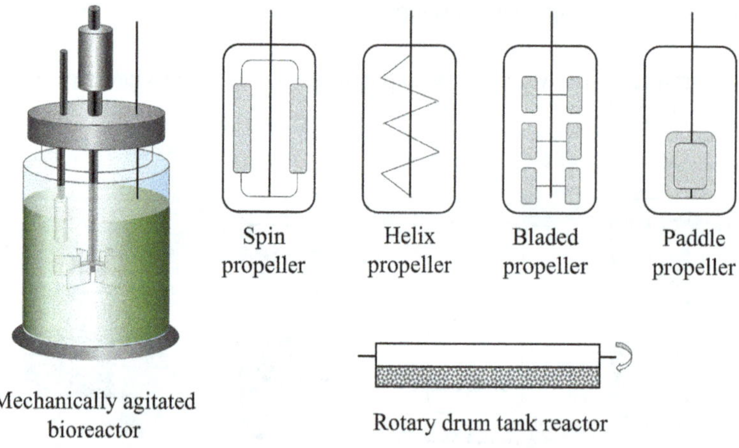

Fig. 3.11 Mechanically agitated bioreactor types for suspended plant cell cultivation

1. Batch cultures are in which cells are introduced into a fixed medium volume. As the cells grow, nutrients and metabolites accumulate, so the medium must be updated periodically, which leads to a change in the cellular metabolism, also called physiological differentiation. Over time, because of the depletion of the medium, cell proliferation ceases.

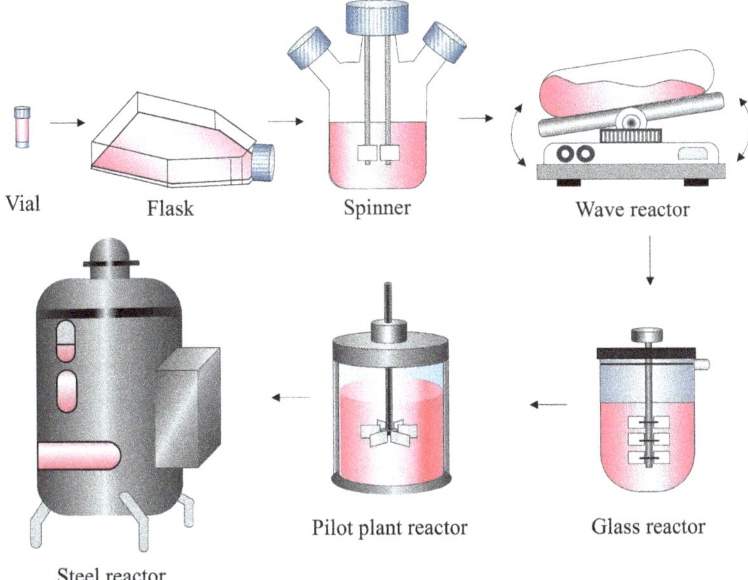

Fig. 3.12 Various bioreactor systems used in mammalian cell culture biomanufacturing (from 2 ml vial up to 5000 L bioreactor)

2. Continuous cultures provide true homeostatic conditions without changing the concentration of nutrients and metabolites, as well as the number of cells. Homeostasis is due to the constant supply of the medium into the culture and the simultaneous removal of an equal volume of the medium with cells. Such systems are suitable for suspension cultures and monolayer cultures on microcarriers.

A typical mammalian cell culture process starts with thawing a cryopreserved cell-bank vial, followed by subsequent expansions into larger flasks, shake vessels, spinners, wave reactors or bags, and stirred glass reactors. When cell density and culture volume satisfy predestinated criteria, the culture is transferred to a pilot-plant reactor or production bioreactor in which cell culture continues to propagate and produce the product.

Brainstorming

1. What are the stages of the industrial biotechnological process?
2. What biotechnological processes are carried out at this stage?
3. Describe and characterize a typical bioreactor.
4. What kind of operational parameters must a bioreactor have?
5. Enumerate the controlled parameters for the growth of biological agents in bioreactors.
6. What principles are used to classify bioreactors?
7. There are three main types of bioreactors. Define.
8. Give a classification of bioreactors depending on the type of cultivated biological agent.

9. How can we ensure active mass transfer in the fermenter, considering the specificity of cultured bio-agents?
10. What types of bioreactors are categorized according to the energy consumption methods, mixing, and energy input?

Take-Home Messages
- Midstream processing is the central stage in biotechnological production, where raw materials are biologically converted into target products via processes such as fermentation, biotransformation, or biocatalysis.
- Bioreactors are specialized devices that provide optimal conditions (e.g., pH, temperature, oxygenation) for the growth of bio-agents and synthesis of desired compounds. Their design and operation directly influence process efficiency.
- Biotechnological conversions include not only synthesis but also biodegradation, bioleaching, biosorption, bio-composting, and bio-oxidation, demonstrating the versatility of midstream processing.
- Bioreactors are classified by volume, geometry, energy input, and gas-liquid mixing principles. Widely used types include stirred-tank reactors, bubble columns, airlift reactors, and packed-bed reactors.
- Bioreactor scale ranges from laboratory models (<50 L) to industrial fermenters (>1 million L), depending on the production stage and process requirements.
- Effective bioreactor operation demands strict control of sterility, nutrient delivery, waste removal, oxygen supply, and shear force to ensure reproducibility and protect sensitive biological systems.
- Animal and plant cell cultures use batch or continuous cultivation strategies. Packed-bed and airlift bioreactors are particularly suitable for shear-sensitive cells and immobilized cultures.
- The historical evolution of bioreactors—from open pits to sterile, computer-controlled systems—reflects the growing complexity and control requirements of modern biotechnology.

3.3   Fermentation Processes in Biotechnological Production. Kinetics of Microbial Growth

Fermentation is an energy-yielding metabolic process in which biological agents convert nutrients into desired target products. In industrial contexts, the term is broadly used to refer to any method of microorganism cultivation, whether aerobic or anaerobic.

Aerobic cultivation is applied when aerobic microorganism-producers are involved in the process. The medium is oxygenated by supplying air or other gases through gas-supply tubes, nozzles, etc. Anaerobic processes, on the other hand, take place in sealed containers or by saturating the cultured medium with inert gases. The construction of bioreactors for

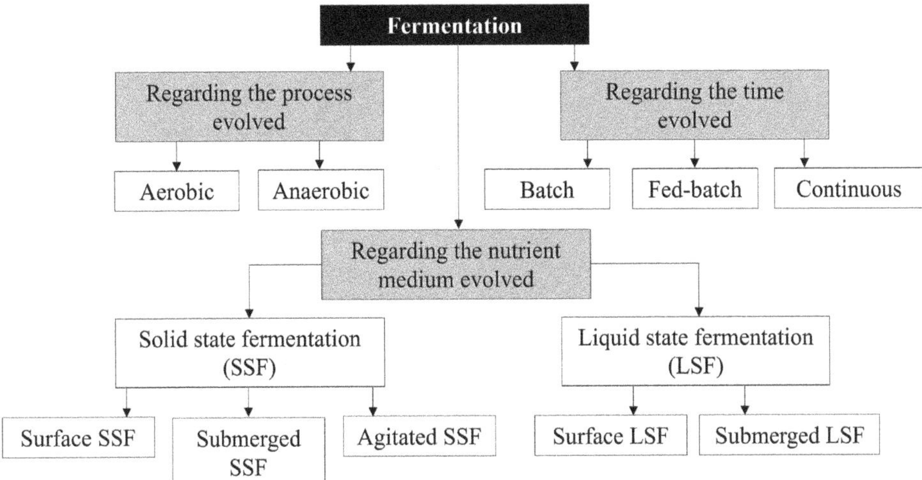

Fig. 3.13 Classification of fermentation process

anaerobic fermentation is simpler than for aerobic ones. The growth of plant cell cultures and microorganisms in bioreactors can be carried out in batch (periodic) or continuous cultivation regimes (Fig. 3.13).

Fermentation can be performed in both *liquid* and *solid phases*:

- Liquid-phase cultivation can be divided into:
  - *Surface fermentation*, implemented in shallow trays or dishes filled with medium. These dishes are placed in air-ventilated chambers, where biomass forms on the surface as a film or solid layer.
  - *Submerged fermentation*, which occurs throughout the volume of the liquid medium. This type of fermentation can be conducted using both batch and continuous methods.
- *Solid-phase fermentation*, using solid, friable, or pasty media with a moisture content of 30–80%, can be achieved in three ways:
  1. The substrate for surface processes is spread in thin layers (3–7 mm) on trays.
  2. Submerged solid-phase fermentation occurs in deep, open vessels without substrate mixing.
  3. Solid-phase fermentation with aerated substrate mixing ensures better mass transfer.

Batch cultivation is analogous to growing cell cultures in shake flasks. It is considered a *closed system*, as only oxygen, antifoam agents, and pH regulators (acid or base) are added during the fermentation. All other components are introduced at the beginning. As the cells grow and metabolize, the composition of the medium, as well as the concentrations of biomass and metabolites, change over time.

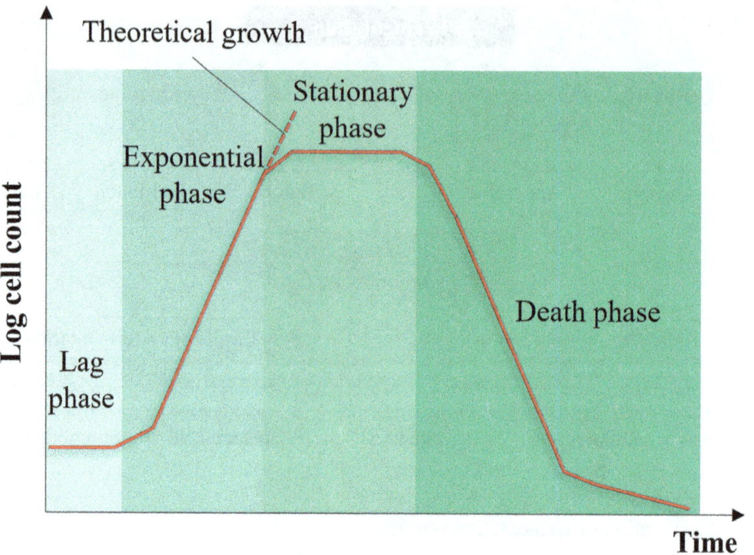

Fig. 3.14 Growth curve of microorganisms

To describe the *growth curve* during cultivation, four main phases are distinguished (Fig. 3.14).

- Lag phase
- Log (exponential) phase
- Stationary phase
- Death phase

The *lag phase* may be absent if the inoculum is already in the log phase.

If microbial culture grows at a constant and maximum rate during the log phase, the increase in cell mass can be described quantitatively as the time of doubling the number of cells or biomass. The growth rate of biomass is proportional to its quantity and the time of continuity of this process.

In a *stationary phase*, the amount of biomass can grow or remain constant, although the chemical composition of cells may change. Products that have biotechnological value are often synthesized in this phase.

The *death phase* is defined by any medium components being exhausted. In industry, fermentation is most often completed at the end of the log phase or before the onset of the death phase.

The conditions in a periodic culture are changeable; the density of the culture increases, the concentration of the substrate decreases, etc. However, the exponential growth phase for a long time at a constant substrate concentration is reliable. This can be achieved if a new nutrient solution is continuously loaded into the vessel containing the cell culture and simultaneously removes the corresponding amount of the cell suspension.

Quantitative characteristics, including cell culture growth and productivity, are particularly important for biotechnological production.

The growth rate is often characterized by the specific growth rate μ (h^{-1}):

$$\mu = \frac{1}{X}\cdot\frac{dX}{dt}\ \mu = \mu_{max} = \text{const.}$$

where

X—biomass (g/L)
dX/dt—the rate of population growth

For strains of microorganisms, the critical characteristic is the maximum possible rate of growth—μ_{max} (h^{-1}). This value depends on the specificity of the organism and on the conditions of fermentation. It is determined from the growth curve of a homogenously growing periodic culture.

The biomass doubling time (t_D) is as follows:

$$t_D = \frac{\ln 2}{\mu}$$

The generation time (t_G) is as follows:

$$t_G = \frac{\ln 2}{\upsilon}$$

where
υ (h^{-1})—the division rate, and it is calculated as:

$$\upsilon = \frac{1}{N}\cdot\frac{dN}{dt}$$

where N—cell count.

At present, high-capacity biomass build-up systems are required. Their implementation is related to the continuous fermentation process. Continuous cultivation is characterized by the ongoing supply of fresh nutrient medium to the bioreactor. A continuous culture is an open system that seeks to establish a dynamic equilibrium and maintains biological agents under unchanged medium conditions. This type of cultivation is structurally more complex and amenable to automatic regulation since it is associated with introducing additional devices into the overall process.

There are three types of open continuous culture systems that are commonly used (Fig. 3.15):

1. *The chemostat,* as illustrated in Fig. 3.16, is a bioreactor that operates by continuously supplying nutrient medium while removing cell suspension at an equal rate. This maintains a constant volume of the growing culture. The chemostat includes a propagator vessel that receives a steady flow of nutrient solution from a reservoir. Through aeration

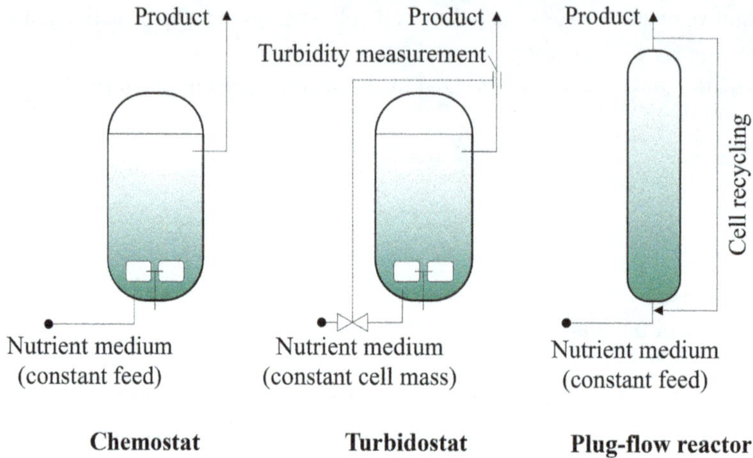

Fig. 3.15 Closed culture system types

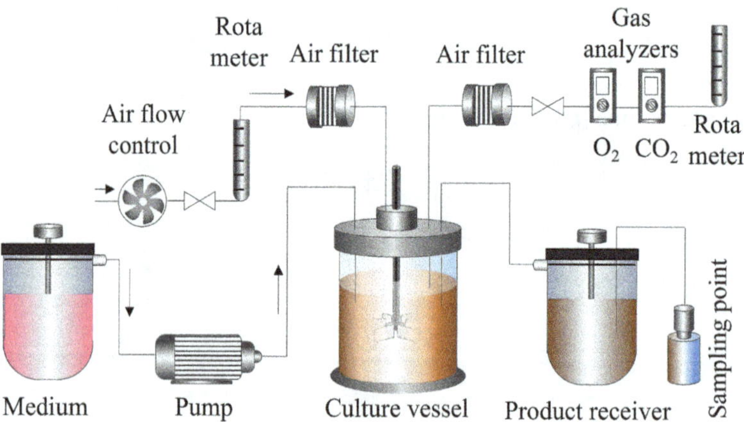

Fig. 3.16 A typical bioreactor operating system on a continuous cultivation basis

and agitation, the propagator creates optimal conditions for oxygen transfer and uniform nutrient distribution to the inoculum. The growth of the culture in a chemostat is regulated by the *concentration of a limiting substrate.*

2. *The turbidostat* operates by continuously monitoring the concentration of cellular biomass within the bioreactor and automatically adjusting it to a constant level by modifying the flow rate. It maintains a consistent suspension density (turbidity level). A turbidity sensor controls the nutrient supply via a control system. All nutrients are typically present in excess within the culture vessel, allowing the culture to grow near its maximum specific growth rate. The turbidostat is used at dilution rates close to the maximum growth rate (μ_max), although it is technically more complex than the chemostat.

3. *In the plug-flow reactor system*, the culture medium flows continuously through a tubular reactor without back-mixing. The cell mass is removed at the reactor outlet and reintroduced at the reactor's inlet.

During continuous fermentation, the removal of cells along with the solution quitting the system should be balanced with the growth of microorganisms.

The substrate concentration at the inlet is designated as S_0. With a slow flow, the substrate (S) is almost completely consumed by cells $(S > 0)$, and their concentration is $X = S_0 / Y$. If D increases, then X slowly decreases at the beginning, and then, when $D = \mu_{max}$, it drops sharply to 0. The value of S gradually increases and becomes S_0 for $D = \mu_{max}$.

Thus, biomass balance is described as:

$$\frac{dX}{dt} = \mu \cdot X - D \cdot X$$

where

μ—specific growth rate (h^{-1})
D—dilution rate (h^{-1})
X—biomass $(g \cdot L^{-1})$

Substrate balance is calculated:

$$\frac{dS}{dt} = D(S_0 - S) - \frac{\mu \cdot X}{Y_S}$$

where

Y_S—yield coefficient (kg/kg)
S—substrate concentration in bioreactor $(g \cdot L^{-1})$
S_0—substrate concentration in feed $(g \cdot L^{-1})$

For the stationary state it follows:

$$\frac{dX}{dt} = \frac{dS}{dt} = 0$$

and it further follows that $\mu = D$ using latter equation with previous two:

$$X = Y_S \left(S_0 - S\right)$$

$$D = \mu_{max} \frac{S}{K_S + S}$$

$$S = \frac{D \cdot K_S}{\mu_{max} - D}$$

where

$\mu_{\max}$—maximum specific growth rate (h^{-1})
K_S—Monod constant (g·L^{-1})

In conclusion, both types of fermentation processes—batch and continuous—have their advantages and disadvantages.

3.3.1   Advantages of the Batch System

Cost efficiency: It offers a cost-effective solution in terms of both equipment and control systems, making it budget-friendly.

Flexibility: The batch system allows the production of a variety of different products, making it highly adaptable.

Resistance to contamination: It is less prone to contamination, ensuring greater product purity.

Suitability for small-scale production: It is convenient for producing small quantities of products, making it ideal for research and pilot-scale applications.

Precise control over cultivation conditions: Cultivation conditions can be adjusted precisely as needed.

Secondary metabolite biosynthesis: Batch fermentation is particularly well-suited for the biosynthesis of secondary metabolites, offering enhanced efficiency in such processes.

3.3.2   Disadvantages of the Batch System

Inoculum preparation: Each batch requires fresh inoculum preparation, which can be time-consuming.

Extended fermentation time: Fermentation periods tend to be longer, reducing overall productivity.

Frequent sterilization: Sterilization must be performed between batches, increasing downtime and operational complexity.

Reduced yield: Compared to continuous processes, batch systems generally yield less biomass and product due to interruptions in growth and production phases during the cycle.

3.3.3   Advantages of a Continuous Fermentation System

Extended operational duration: Allows for longer production runs without interruption.

Reduced laboratory costs: Less manual intervention is required compared to batch processes.

Efficient reactor utilization: The reactor operates continuously, maximizing usage.

Higher productivity: Productivity per unit time is generally higher.

Consistent product quality: The system offers more stable and uniform production parameters.

3.3.4   Disadvantages of the Continuous System

Higher contamination risk: Continuous operation increases the risk of contamination over time.

Microbial mutation risk: There is a greater chance for spontaneous mutations to occur in microbial populations.

Limited applicability: Some bioprocesses are not suitable for continuous operation due to biological or technical constraints.

Fermentation Types

Since the biosynthesis of any product depends on energy metabolism, three types of fermentation processes are distinguished.

Type I   The product results from primary metabolism aimed at obtaining energy. Growth, carbohydrate catabolism, and product formation proceed concurrently. An example of this type of fermentation is cell mass formation, such as single-cell protein (SCP), baker's yeast, and algae, as well as alcohol fermentation (Fig. 3.17).

Type II   The product is also formed from the substrate used in primary metabolism but appears in the second phase, which is temporally separated from the first (Fig. 3.18). In the periodic fermentation characteristic of this type, two peaks are observed. Initially, there is

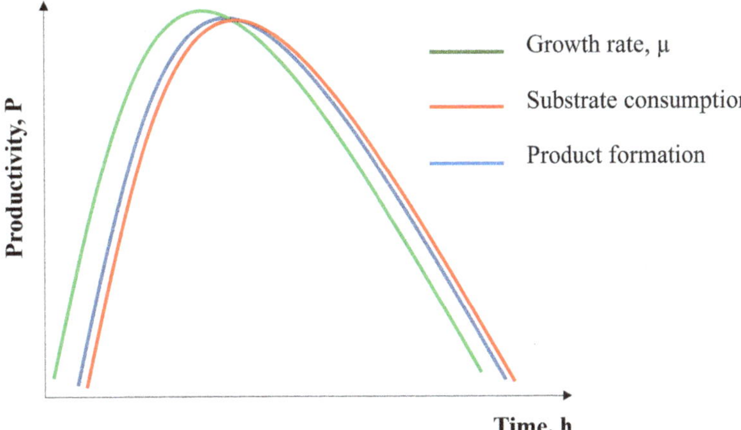

Fig. 3.17   Fermentation type I

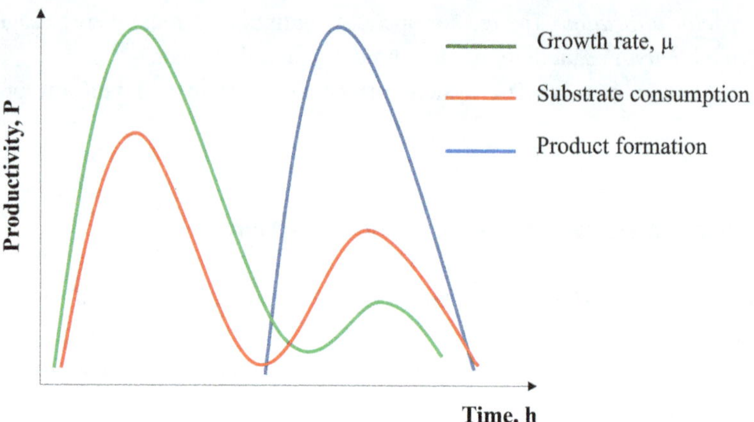

Fig. 3.18 Fermentation type II

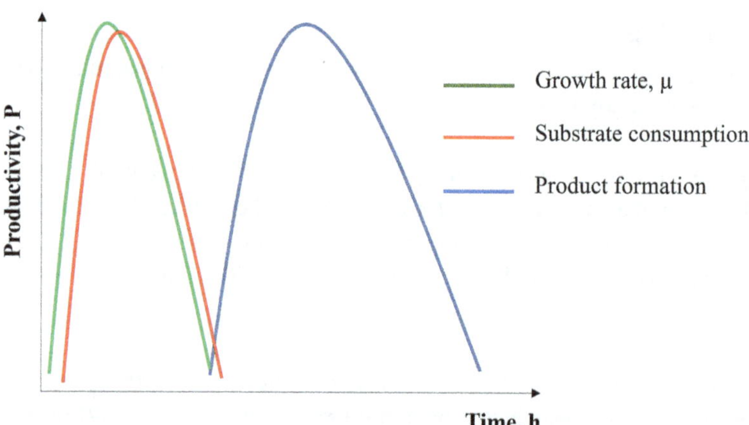

Fig. 3.19 Fermentation type III

strong microbial growth, which then slows down; this is followed by the start of product formation, accompanied by intensive substrate consumption. Examples include the production of citric acid and amino acids.

Type III Here, primary metabolism is completely separated from the formation of the final product (Fig. 3.19). In this case, growth and substrate consumption occur first, while the target product is synthesized later through intermediate metabolism reactions. Many antibiotics, vitamins, enzymes, and solvents are obtained via Type III fermentation.

It is not always possible to strictly classify all fermentations into these three types, as intermediate forms may exist. However, since the synthesis of most Type II products is also growth-coupled, modern practice usually restricts classification to Type I and Type III fermentations.

3.3.5   Quantitative Parameters of Fermentation Processes

Heat Generation

In order to obtain the optimum yield of product, fermentation must be carried out at a constant temperature regime, so the heat that is formed during mixing and the metabolic activity of microorganisms must be removed through unique cooling systems (Fig. 3.20).

The amount of heat released by biological agents is defined as:

$$Y_{\text{kcal}} = \frac{Y_S}{H_S - Y_S H_C}$$

where

Y_{kcal}—the amount of heat (*g* cells formed per kcal energy)
Y_S—biomass yield (*g* cells per *g* consumed substrate)
H_S and H_C—the heat released when the substrate and cells are combusted (kcal/g)

The rate of heat release per unit time is found from the growth rate, reactor volume (*V*), and Y_{kcal} coefficient. So, the growth rate is calculated as:

$$\frac{dx}{dt} = \mu x$$

and the rate of heat release per unit time:

$$Q_\omega = V \mu x \frac{1}{Y_{kcal}}$$

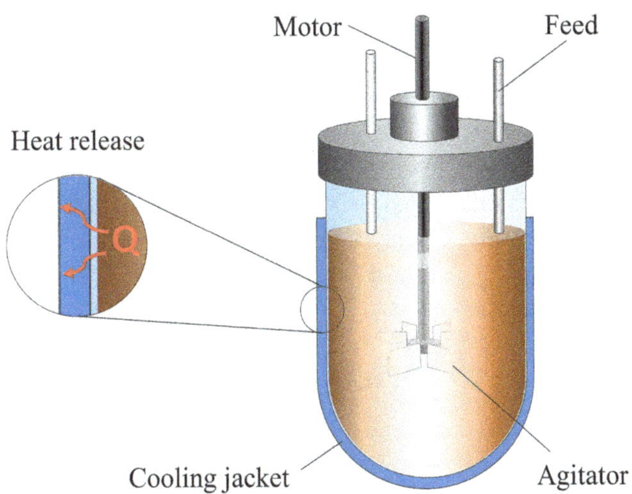

Fig. 3.20 Controlled combustion

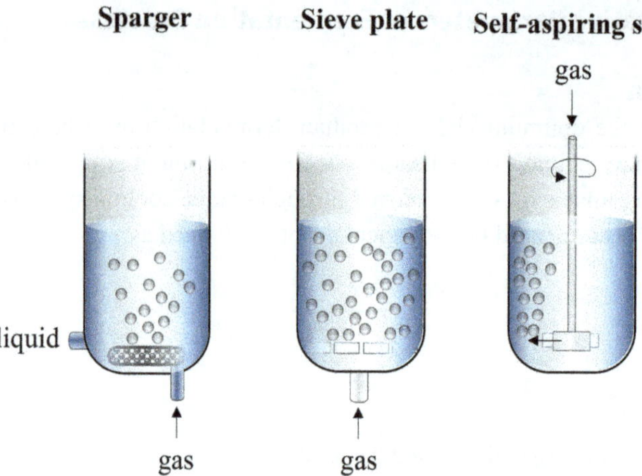

Fig. 3.21 Examples of aeration types

Aeration

The critical factor of fermentations is the mass transfer of gases: oxygen and carbon dioxide.

Oxygen often acts as a limiting factor because it dissolves insufficiently: only 0,3 mmol of O_2 is dissolved in 1 liter of water at 20 °C. At the same time, in the nutrient medium, its content is even lower. To optimize its content, several factors (biological and technical) should be addressed (Fig. 3.21).

The optimal oxygen transfer is correlated with the specific maximum oxygen uptake rate of microorganisms:

$$Q_{O2} = X_{qO2} = k_L a\left(\acute{C}_{O2} - C_{O2}\right)$$

where

q_{O2}^{max}—maximum O_2 resorption rate ($mol \cdot g^{-1} \cdot h^{-1}$)
q_{O2}—specific O_2 resorption rate ($mol \cdot g^{-1} \cdot h^{-1}$)
$k_L a$—volumetric transfer coefficient (h^{-1})
X—biomass concentration ($g \cdot L^{-1}$)
C_{O2}—dissolved O_2 concentration ($mol \cdot L^{-1}$)
$\acute{C}_{O2}$—saturation concentration O_2 ($mol \cdot L^{-1}$)

Thus, O_2 resorption rate follows as:

$$q_{O2} = q_{O2} \max \frac{C_{O2}}{K_{O2} + C_{O2}}$$

where

K_{O2}—Michaelis constant O_2 ($mol \cdot L^{-1}$)

For O_2 transfer, several phase limits must be overcome:
1. From the gas bubble to the phase boundary surface
2. Through the fluid to the boundary around the biological agent
3. Through the phase boundary surface into the fluid
4. Into the biological agent

Oxygen transfer depends on a broad range of technical and design conditions (reactor volume, stirrer performance, aeration, pressure), physicochemical parameters (medium composition, density and viscosity, antifoams, temperature) and biological factors (type of microorganisms, their growth form and rate, etc.).

The valuable parameter for the interpretation of O_2 transfer is the volumetric transfer coefficient $k_L a$:

$$k_L a = k \left(\frac{P}{V_R} \right) \alpha \left(u_{GO} \right) \beta$$

where

k, α, β—constants
P—impeller performance (W)
V_R—reactor volume (m^3)
u_{GO}—gas superficial velocity (m/s)

Carbon dioxide is one of the most decisive metabolic products, but its role in fermentation processes is far less known than that of oxygen. Carbon dioxide is considered a factor that directly affects the fermentation process. If ORR2RR usually limits fermentation kinetics for aerobic organisms, then CORR2RR affects the pH value and the acid-base equilibrium in the medium. It has a double effect on biological agents, as its low concentration stimulates cell growth and its high concentration inhibits it.

A profound lack of CORR2RR usually harms the growth of most bacteria species; typically, its limiting condition is created at the beginning of growth. In certain amounts exceeding the limits of the partial pressure of 25–100 kPa, CO2 slows the growth of most organisms, although resistance to it is very different in different species.

Agitation/Mixing
The process is achieved by impellers, stirrers, pumps, or agitators, resulting in a turbulent current in the direct surrounding of the stirrer, which is described by its *Reynolds number Re*. The factor for Re is viscosity, and it depends on the concentration of biological agents, their morphological parameters, and the produced product type:

$$Re = \frac{\text{inertial force}}{\text{viscous force}} = \frac{d_{R2} n \rho}{\eta}$$

where

d_R—impeller diameter (m)
n—impeller revolutions (s^{-1})
η—dynamic viscosity (Pa·s)
ρ—density (kg/m^3)

The *power number Ne* determines the energy demand of stirrer bioreactors and is correlated with the Re number:

$$Ne = \frac{propulsion}{inertia} = \frac{P_O}{d_{R5}n_3\rho}$$

where

P_O—impeller performance (W)

The turbulence in the vital zone is distributed identically in an ideally agitated bioreactor; however, various factors can affect the proper function of stirrers.

Different stirrers and agitators have been developed on an industrial scale, including disk and turbine types, which encourage oxygen transfer and good mixing (Fig. 3.22).

Brainstorming
1. What is the main stage in biotechnological production?
2. How is the fermentation process classified?
3. What is the difference between submerged and solid-state fermentation?
4. What are the characteristics of solid-phase culture?
5. Provide a general description of periodic cultivation.
6. Describe the stages of microbial growth in periodic cultivation.
7. What are the main types and characteristics of continuous cultivation?

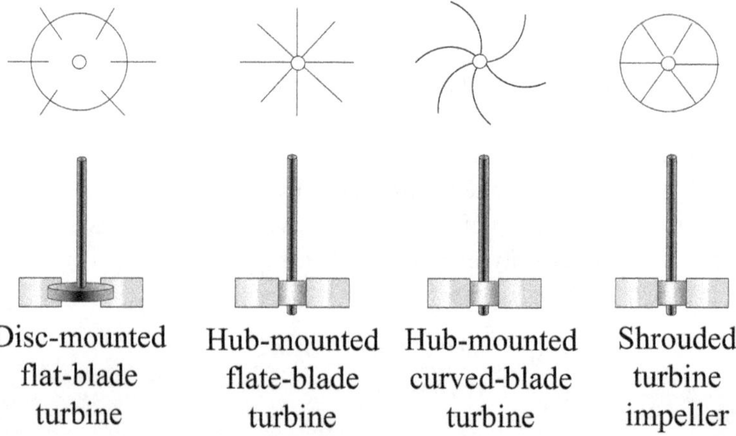

Fig. 3.22 Basic types of impellers used for mixing

8. What are the methods used in continuous fermentation?
9. What parameters must be set and monitored in a chemostat?
10. Describe the structure and working principle of a turbidostat.
11. What is a plug-flow reactor system?
12. What are the pros and cons of periodic vs. continuous cultivation?
13. What are the differences between fermentation types I, II, and III?
14. What factors influence the fermentation process?
15. Why is gas mass transfer (oxygen and CO_2) a critical factor in fermentation?
16. List the basic types of impellers used for mixing.
17. Describe the quantitative parameters relevant to fermentation processes.

Take-Home Messages
- Fermentation refers to any controlled microbial cultivation—aerobic or anaerobic—used to convert nutrients into desired products, including biomass, metabolites, and bioenergy.
- Fermentation modes include:
  1. Submerged fermentation (surface or throughout the liquid volume)
  2. Solid-phase fermentation (in low-moisture substrates)
  3. Batch (periodic) and continuous (chemostat, turbidostat, plug-flow) systems
- Batch fermentation is simple, flexible, and ideal for small-scale or secondary metabolite production, but has limitations like downtime and lower productivity.
- Continuous fermentation offers high productivity and stable conditions but is more vulnerable to contamination and microbial mutations.
- Microbial growth kinetics are described by parameters like:
  – Specific growth rate (μ)
  – Maximum growth rate (μ_{max})
  – Doubling time (tD)
  – Generation time (tG)
- Fermentation types are classified based on metabolic pathways:
Type I: Primary metabolism (e.g., alcohol, SCP)
Type II: Product formed in a second phase (e.g., citric acid)
Type III: Product from intermediate metabolism (e.g., antibiotics)
- Key process variables include:
  – Heat generation from metabolism (requires cooling)
  – Oxygen transfer (quantified by kLa and qO_2)
  – Carbon dioxide levels, which affect pH and growth
  – Agitation and mixing, crucial for mass and energy transfer
- Impellers and stirrers (e.g., turbine, disc-type) generate turbulence, described by Reynolds number (Re) and power number (Ne), influencing bioreactor efficiency.

3.4 Downstream Processing

The final stage of the biotechnological process essentially depends on whether the product accumulates in the cell, is released into the culture liquid, or is the cell mass itself. If a product accumulates in a cell (intracellular compound), it is complex and difficult to isolate, while a compound released to the environment (extracellular product) is easily recovered. Therefore, the isolation and purification technology also depends on various characteristics and the nature of the final product.

3.4.1 Biomass Separation

When the target product accumulates in the cells, it is necessary to separate the cells from the culture liquid, destroy them, and then remove the final product from the mass of destroyed cells (Fig. 3.23).

The division of culture liquid and biomass is called separation. Usually, before separation, the culture is pre-treated (shifting pH, heating, adding coagulants, proteins, or flocculants).

3.4.2 There Are Different Types of Separation

Flotation, used if cells accumulate on the bioreactor fluid's surface layers due to low wettability. The main driving force of flotation is the bubbles of gas that pop up, grab the cells, and carry them to the surface. Flotators of various designs are used; they are capable of decanting, pumping out, or scraping the foam in which the cells are located. The technique is economical, highly productive, and can be employed in continuous fermentation.

Filtration. Drums, disks, belts, trays, vacuum filters, filter presses, membrane filters, etc., are commonly used (Fig. 3.24).

All filters operate on the same principle: biomass retention on a porous filter layer. As the liquid passes through, the diameter of the capillary channels narrows due to particle

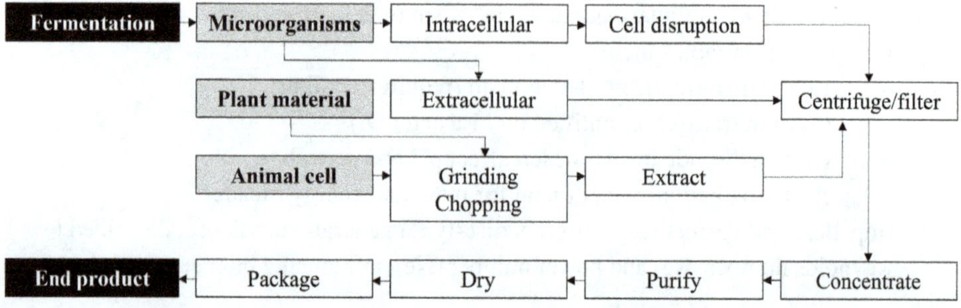

Fig. 3.23 Survey on the recovery of bio-products

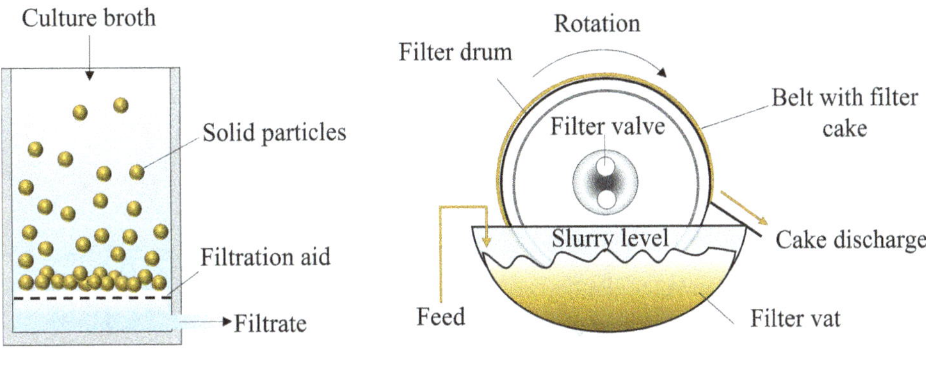

Fig. 3.24 Examples of filtration

adherence, preventing biomass from leaking through the filter. If filters operate continuously, automatic removal of the biomass layer is done using compressed air or a special knife. For this purpose, drum vacuum filters are often used. The drum's internal volume is separated from the liquid by fine-pored filtering material. A vacuum pump creates a liquid flow through the filter layer by drawing air inside the drum. The suspension adheres to the drum and is cut off with a knife.

Membrane filters are widely applied in periodic-mode processes, especially for highly diluted cell suspensions. However, such filters are often clogged by cells, proteins, and colloidal particles. To overcome this, membranes are coated with a hydrophilic layer, which prevents proteins from contacting the membrane surface.

Centrifugation is based on the precipitation of particles suspended in a liquid using centrifugal force (Fig. 3.25). In some production processes, centrifugation and filtration are used simultaneously.

Cell Disintegration Approaches

Cell disintegration can be achieved through *physical*, *chemical*, or *chemical-enzymatic* techniques.

- Physical methods such as ultrasound, rotating blades, and high-pressure homogenization are commonly used. Although these methods are relatively simple and cost-effective, they may sometimes compromise product quality.
- Chemical methods involve substances like toluene or butanol to produce yeast autolysate and release certain enzymes.
- Chemical-enzymatic methods, on the other hand, use agents like lysozyme with EDTA to treat Gram-negative bacterial cells. Antibiotics (e.g., polymyxins, nystatin) and surfactants can also induce cell lysis.

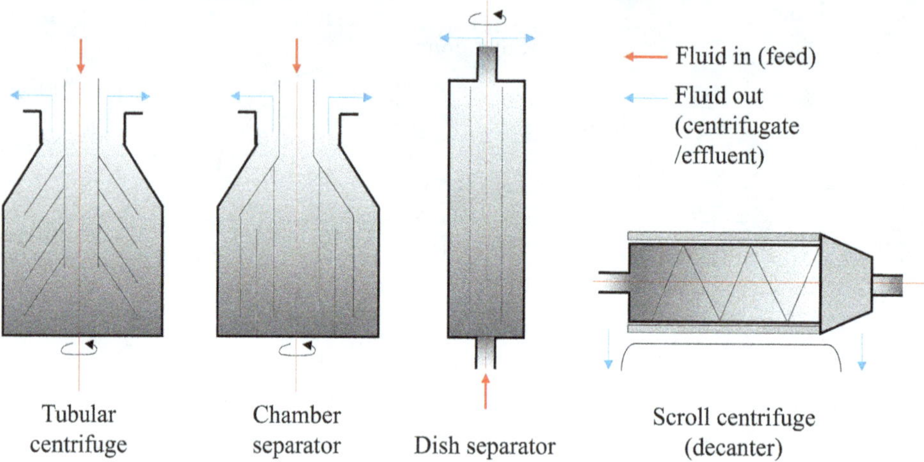

Fig. 3.25 Centrifuge types

After disintegration, fragments of cell walls are removed using separation techniques such as centrifugation or filtration. High-speed centrifuges and fine-pore filters are especially effective.

Separation and Purification Techniques

1. *Precipitation*: This method involves adding a reagent to a liquid, causing the dissolved product to precipitate into a less soluble form. Both physical and chemical approaches can be used to induce precipitation. For instance penicillin can be converted into its crystalline form by adding potassium compounds, while proteins are typically precipitated using ammonium sulfate or ethanol.

2. *Extraction*: This technique involves transferring a product from one phase (e.g. aqueous) to another (e.g.) organic. Typically, hydrocarbons or other organic liguds are used to extract target compounds. Extractions are generally classified into two categories.

   Solid-liquid phase: The product is extracted from a solid matrix to a liquid phase. For example: extraction of salts from bacterially treated ores.

   Liquid-liquid phase: The compounds is transferred between two liquid phases. This method is particularly useful for extracting vitamins, antibiotics and lipids.

3. *Adsorption*: In this method substances are captured on the surface of a material from either a liquid or gas phase. Used adsorbents include charcoal and silica gel. Adsorption is often employed to extract antibiotics and vitamins.

4. *Chromatography*: This technique separates substances based on their distribution between two phases. There exist types of chromatography methods that are typically based on specific ligands (Fig. 3.26).

• Ion exchange chromatography: It utilizes absorbent granules for trapping ions or aiding in the purification of neutral compounds. Sulfonated polymers and amino group-containing polymers are choices for this type of chromatography.

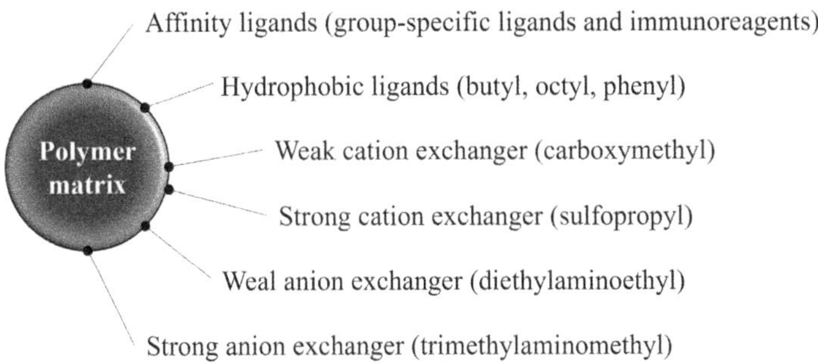

Fig. 3.26 Chromatography types

- Affinity chromatography: this method captures substances that strongly bind to ligands on carrier particles. Immunochromatography is a subtype. It is not commonly employed due to its high expenses.
5. *Electrophoresis*: This approach separates components of a mixture by taking advantage of their movement in a field. It involves using columns or beds filled with substances such as agarose or sepharose, which form gel matrices.

Concentration and Product Stabilization

1. Role of product concentration: Initially, during biotechnological stages, the product concentration is roughly 1%. As the process advances, this concentration escalates, eventually reaching near 100% after the concentration phase.
2. Methods of concentration:
- *Reverse osmosis*: Solvent is forced through a semipermeable membrane under pressure greater than the osmotic pressure, concentrating the solute.
- *Ultrafiltration*: Membrane filters separate molecules by size. Larger pore membranes capture viruses; finer pores retain smaller molecules like organic acids. This method is used for concentrating antibiotics, organic acids, and enzymes.
- *Evaporation*: Solvent is removed by heating. Vacuum evaporators are used to avoid heat damage. These systems can operate in batch or continuous mode, and may include disc or film evaporators that distribute liquid over a heated surface. Disk and film evaporators, which spread the liquid thinly over a heated surface, are also popular (Fig. 3.27).
3. Storing the dehydrated product: Once dehydrated, the product can undergo modifications or be shaped for commercial presentation. Through modification, it's possible to bestow specific optical asymmetry on the product, a feat hard to achieve chemically.

3.4.3   Product Stabilization in Biotechnology

Stabilizing products is a key step in biotechnology:
1. Drying bolsters resistance to external elements.
2. Fillers such as wheat bran, cornmeal, and glycerin are used for stabilization.

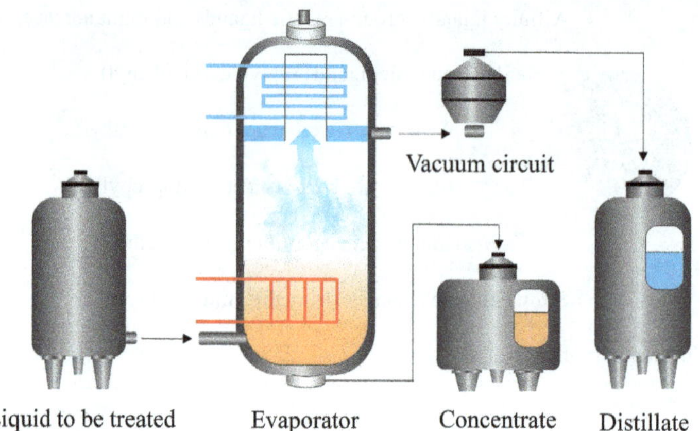

Fig. 3.27 Typical process of vacuum evaporation

3. Inorganic ions like cobalt, sodium, and magnesium also aid stabilization.
4. Thermal processing is another stabilization method.

Isolation and Separation of Extracellular Products

The isolation of extracellular products from the culture liquid is simpler and cheaper. First, the producer biomass is separated, then, depending on the product's properties, the methods for isolation, concentration, and purification are selected (Fig. 3.28).

- There are various types of isolation:*Distillation* is the transformation of the product into a vapor state followed by condensation. Examples are ethanol, acetone, and butanol.
- *Dehydration*, used to obtain lysine feed concentrate and some essential antibiotics.
- *Freeze-drying* (lyophilization), freezing of solution or suspension of cells and further sublimation in a vacuum; widely used for starter cultures, vaccines, hormones, etc. production.
- *Precipitation* in the form of insoluble salts by adding chemical precipitant; citric and lactic acid are presented.
- *Crystallization* is a process based on the different solubility of substances at different temperatures. Slow cooling allows crystals to be formed from solutions of the target products, and their purity is usually extremely high. It is possible to obtain an even purer product if the crystals are dissolved in water or a solvent and then crystallized again, i.e., through recrystallization. They are used to produce itaconic, glutamic, and other acids.
- *Sorption* is applicable for isolated amino acids, enzymes, antibiotics, etc.
- *Extraction* adds solvent to the solution, which absorbs the product, then the emulsion is separated from the target substance, e.g., antibiotics, vitamins, etc.

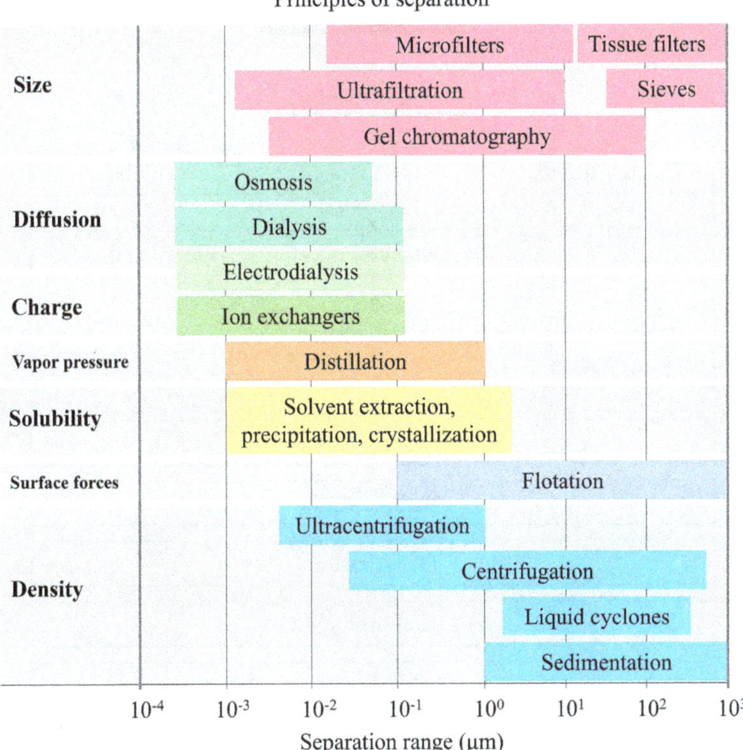

Fig. 3.28 Classification of recovery techniques based on separation range. Products obtained by biotechnological methods differ not only in nature or chemical composition but also in what place in the technical flow chart they occupy

- *Ultrafiltration* uses membrane filters with specific pore sizes to isolate desirable products, including enzymes and other proteins.
- Biotechnology production is completed by the formulation of the finished product. The product acquires its marketable form through different methods, including granulation, coatings, tableting, bottling, and packing (Fig. 3.29).

- Products of biotechnological production can be:Gases (Carbon dioxide, methane, hydrogen, etc.)
- Fermentation medium (culture liquid or solid substrates)
- Liquid, obtained after biomass separation, or its concentrate (supernatant, solution, filtrate, permeate, or supernatant)

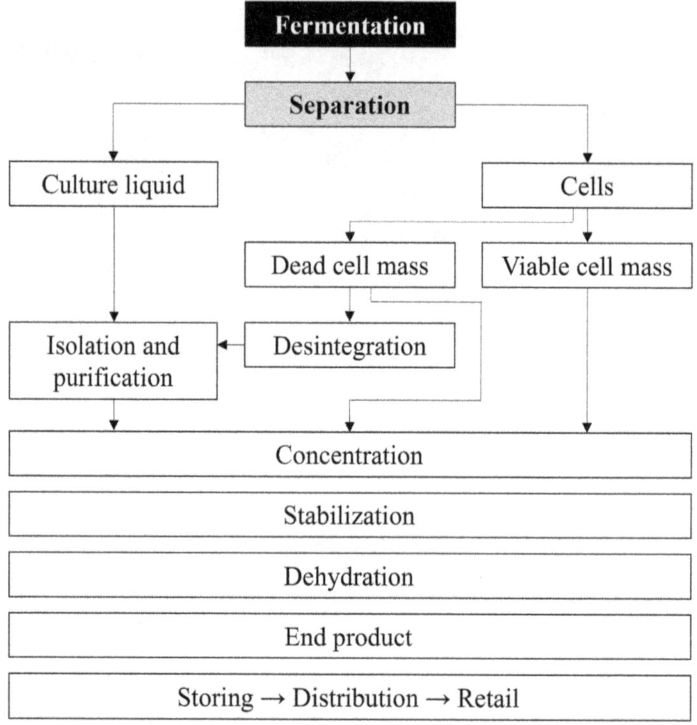

Fig. 3.29 Downstream processing flowchart for microbial production

- Inactivated biomass (Single cell proteins)
- Viable biomass of microorganisms (bio-destructors, bio-fertilizers, starter cultures, etc.)
- Extracellular and intracellular bio-products (liquid or solid substances)
- Processed microbial biomass (hydrolysates, ferment lysate, cell membranes
- Purified liquid (in wastewater treatment) or solid (in terrestrial remediation)

Brainstorming

1. What is the last stage of biotechnology production, and what is its purpose?
2. How many products can be obtained at the end of the fermentation step?
3. What methods should be used to isolate the products obtained at the fermentation stage?
4. What methods can be used to separate culture fluid from biomass?
5. What are the differences in the separation of extracellular and intracellular products?
6. Describe the ways of cell destruction techniques.
7. Why is the stage of product purification necessary in the biotechnological process?
8. What methods are used to purify the product?
9. What happens to the product of biotechnology production at the concentration stage?
10. List the products obtained in the process of biotechnology production.

Take-Home Messages

- The final stage of biotechnological production—downstream processing—is essential for isolating, purifying, concentrating, and stabilizing the desired product. It prepares the product for its final form and commercial use.
- Products at the end of fermentation may be gases, liquid culture media, concentrated supernatants, whole cells (biomass), viable microorganisms (e.g., probiotics, starter cultures), or extracellular/intracellular compounds (enzymes, antibiotics, acids, proteins).
- Extracellular products are easier and more cost-effective to isolate since they are already released into the culture liquid. Intracellular products require cell separation, disintegration (lysis), and subsequent purification steps, making the process more complex.
- Key biomass separation techniques include filtration, centrifugation, and flotation, often preceded by pretreatment to enhance efficiency (e.g., pH shift or coagulation).
- Cell lysis can be performed using physical (e.g., ultrasound, high-pressure), chemical (e.g., toluene, EDTA), or enzymatic agents (e.g., lysozyme). Post-lysis, fragments are separated by centrifugation or filtration.
- A wide range of purification techniques is used:
1. Precipitation (e.g., ammonium sulfate)
2. Extraction (solid–liquid or liquid–liquid)
3. Adsorption (e.g., charcoal, silica gel)
4. Chromatography (ion-exchange, affinity)
5. Electrophoresis (especially for sensitive protein mixtures)
- Techniques such as ultrafiltration, reverse osmosis, and vacuum evaporation are used to increase product concentration. Final stabilization may involve drying, thermal processing, or the use of fillers and salts to protect product integrity during storage.
- After purification and concentration, products are shaped into usable forms—granules, tablets, liquids, or powders—via techniques such as tableting, coating, bottling, and packaging.
- The choice of separation and purification techniques depends on the chemical nature, physical properties, and localization (intra- or extracellular) of the target product, ensuring maximal yield and purity.

Products Based on the Live or Inactivated Microbial Biomass

In biotechnological production, microbial cells themselves can serve as the primary product. Various types of microbial biomass are produced on an industrial scale.

Industrial Biomass Categories

- Unicellular organism protein: Biomass from microorganisms like bacteria, yeasts, fungi, and algae, utilized as protein-vitamin concentrates or food supplements
- Probiotics: Microorganism biomass that aids in balancing gut flora
- Bioinsecticides: Biomass derived from bacteria, fungi, and viruses that target pests
- Bacterial fertilizers: Biomass composed of nitrogen-fixing bacteria
- Vaccines: Microorganism biomass meant to stimulate active immunity in vaccinated individuals or animals

4.1 Industrial Technologies to Produce Protein from Unicellular Organisms

Proteins play a vital role in the human body and are key structural components of cells. While plants and many microorganisms can synthesize all amino acids, humans and animals cannot produce several essential amino acids, such as valine and leucine. These essential amino acids must therefore be obtained through the diet. A deficiency in these nutrients can result in health issues and reduced productivity in livestock.

Typically, humans obtain these amino acids from both animal-based and plant-based protein sources, while animals mainly rely on plant-derived proteins. It is crucial that the proteins consumed contain a proper balance of essential amino acids.

I. Digel et al., *Introduction to Industrial Biotechnology*, Learning Materials in Biosciences, https://doi.org/10.1007/978-3-032-07918-3_4

When plant-based feeds are deficient in protein quality, essential amino acids are supplemented to avoid overfeeding and rising feed costs in animal husbandry. Such supplements often include meat meal, fish meal, or legumes.

Protein is a key component of human and animal nutrition, especially high-quality protein from animal sources. However, the current global supply of protein meets only about 40% of the total demand. As the global population increases, the demand for protein is also expected to rise, creating a growing gap in protein availability, especially for animal feed. This challenge is driving the search for new protein sources, including synthetic and microbial options.

The cultivation and processing of microorganisms for use in food and feed production has become a significant area of focus in modern microbiology.

The viability and economic potential of microbial protein production depend on several factors:

- Microorganisms can utilize a broad range of chemical compounds and widely available raw materials.
- They have a high protein synthesis rate.
- Microbial cultivation is relatively simple and fast, with some production cycles as short as 24 hours, making year-round production feasible.
- Microbial biomass products are rich in proteins, vitamins, carbohydrates, and lipids.
- These products typically contain higher levels of essential amino acids compared to plant-derived proteins.

  Using genetic engineering, the protein and vitamin content of microbial biomass can be further enhanced. Microorganisms grow significantly faster than crops or livestock. For example, the daily production of 500 kg of microbial protein corresponds to the production of only 0.5 kg of animal protein, 40 kg of soybeans, or around 50 tons of yeast biomass (Table 4.1).

Microbiological synthesis is efficient in terms of material and energy usage. It does not require vast land areas, is unaffected by weather conditions, and is environmentally friendly. Microbial proteins match the quality of animal proteins. Adding microbial proteins to feed improves the nutritional profile and digestibility of conventional plant-based feeds. For example, using one ton of fodder yeast can save up to five tons of grain and increase livestock productivity by 15–30%.

A modern, medium-sized facility producing 50 tons of microbial protein annually on just 0.2 hectares can meet the protein requirements of nearly 10 million people. By comparison, producing this amount of protein via traditional agriculture would require approximately 16,000 hectares of wheat or a farm yielding 400 piglets per day.

Microbial protein, also referred to as *single-cell protein (SCP)*, is derived from unicellular organisms and is commonly used as food or feed. It includes dried biomass from yeasts, bacteria, fungi, and algae. While the term "SCP" suggests a product composed only of protein, it actually contains other components such as sugars, fats, and nucleic acids.

For SCP to be effective in food or feed, it must meet certain criteria related to nutritional value, digestibility, and economic feasibility. Chemically, microbial protein is

Table 4.1   Composition of some microbes, plants, and animals

Source	Crude protein (% on dry weight basis)	Time required to double the mass
Microbial sources		
Bacteria	72–78	20–120 min
Yeast	47–53	
Filamentous fungi	31–50	2–6 h
Algae	47–63	
Plant sources		
Corn meal	7.0–9.4	1–2 weeks
Wheat flour	9.8–13.5	
Rice	7.5–9.0	
Animal sources		
Milk	22–25	
Beef	81–90	1–2 months
Egg	35	

similar to conventional protein sources, but its nutritional value depends on whether the digestive enzymes in humans or animals can effectively break it down.

There are, however, a few challenges. Some microbial cells have tough cell walls that must be broken down to make the protein digestible. Moreover, microbial biomass may contain high levels of nucleic acids. While animals can metabolize and excrete these compounds without problems, excessive intake in humans may lead to health issues such as kidney stones or gout due to uric acid accumulation. Therefore, before SCP is approved for human consumption, treatments are applied to disrupt cell walls and reduce nucleic acid content. Techniques such as acid precipitation and hydrolysis are used to lower the nucleic acid concentration to about 2%.

The economic viability of microbial protein synthesis largely depends on raw material and energy costs, which can make up 15–30% of the total production cost. Thus, one of the key challenges in developing new methods for microbial protein production is ensuring the consistent availability of affordable raw materials. This requires access to multiple alternative feedstock sources that can be used interchangeably without compromising product quality.

Modern industrial processes rely on both standardized substrates with well-defined chemical compositions and more complex raw materials, including industrial waste products. Using waste as a substrate not only lowers production costs but also supports environmental sustainability.

Microorganisms can metabolize a wide range of carbon-containing substrates:

- Carbohydrates (e.g., molasses, starch, whey, and cellulose-containing agricultural and industrial residues)
- Liquid and gaseous hydrocarbons
- Alcohols
- Carbon dioxide, including mixtures with hydrogen (see Table 4.2 for details)

Table 4.2 Substrate materials that support the growth of microorganisms in the production process of SCP

Substrates	Microorganisms
Molasses	*Saccharomycopsis cerevisiae, Candida utilis, Kluyveromyces fragilis*
Milk whey	*Candida utilis, Candida tropicalis, Rhodotorula glutinis*
Brewing waste (residual yeasts)	*Saccharomycopsis carlsbergensis*
Starch	*Saccharomycopsis fibuligera*
Cellulose	*Cellulomonas* spp.
Wastes from agriculture, food and woodworking industries (bard, hydrolysates, sulphite liquor)	*Candida utilis, Candida tropicalis, Candida scottii*
Lignin-cellulose materials	*Penicillium roquefortii, Fusarium graminearum*
Liquid hydrocarbons (C11-C18), n-alkanes, diesel fractions, Liquid paraffins, oil distillates, kerosene	*Candida scottii, Candida guilliermondii, Candida maltose, Candida intermedia, Candida lipolytica*
Gaseous hydrocarbons—methane (natural gas)	*Methanomonas methanica, Methylomonas clara, Pseudomonas rosea*
Methanol Ethanol	*Methylomonas methanica, Methanomonas methanica Candida utilis Acinetobacter calcoaceticus*
HRR2RR and CORR2	*Hydrogenomonas eutropha*
CORR2 RR and sunlight	*Chlorella pyrenoidosa, Scenedesmus quadricauda, Spirulina maxima, Spirulina platensis*

Several types of yeasts, bacteria, microscopic fungi, and unicellular algae serve as sources for food and feed protein. In selecting a specific microorganism, factors such as growth rate on a particular substrate, biomass yield, stability in continuous culture, and cell size are taken into account.

Microorganisms can utilize a wide range of carbon-containing substances, including carbohydrates, liquid hydrocarbons, alcohols, gaseous hydrocarbons, and CO_2 mixed with hydrogen. When selecting protein-producing microorganisms, several critical characteristics must be considered (Fig. 4.1):

- Rapid growth rate
- Efficient assimilation of carbon and energy sources
- High conversion efficiency of raw materials into biomass
- Ability to synthesize large amounts of protein, essential amino acids, and certain vitamins, with minimal formation of undesirable by-products
- Competitive advantage in non-sterile conditions (e.g., dominance in the fermenter), and ease of filtration, separation, or flotation
- Safety: the strain must be non-pathogenic and safe for animals and humans

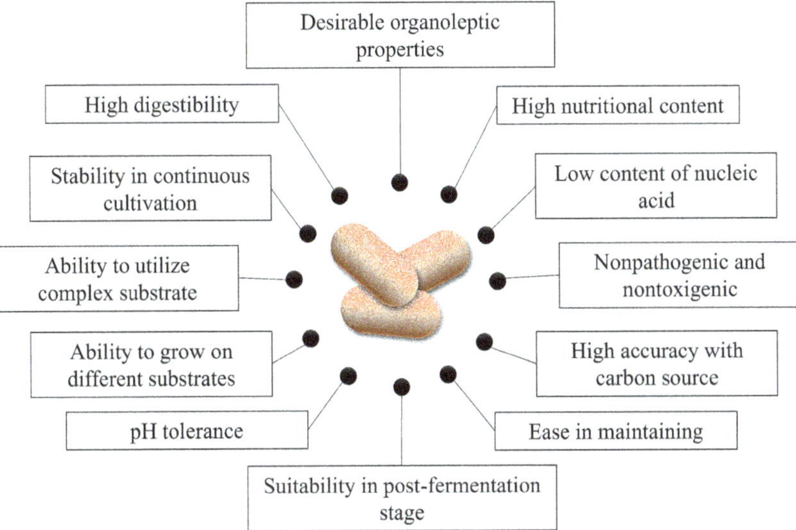

Fig. 4.1 Criteria in selection of microbes for protein production

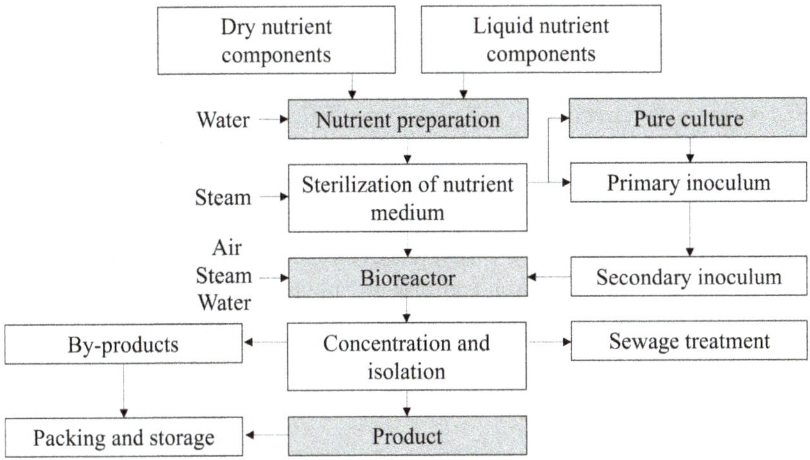

Fig. 4.2 Generalized scheme of biotechnological production of protein microbial product

Regardless of the raw material choice, the standard process for microbiological protein production involves several stages: obtaining and preparing the raw materials, creating the inoculum, fermenting, isolating, inactivating, concentrating microbial biomass, followed by drying and final product standardization (Fig. 4.2).

The quality of the inoculum is of primary importance. It is derived from a preserved culture and scaled up step-by-step. The prepared inoculum is combined with the primary growth substrate, necessary nutrients, and air (or gas mixture) in a fermenter.

Fermentation is the core stage of biotechnological protein production. This process follows specific technological protocols tailored to the chosen substrate and microorganism. It involves the controlled addition of nutrients and air, continuous maintenance of optimal process parameters, and timely removal of exhaust air, by-products, and excess heat.

Creating optimal environmental conditions is crucial to achieve maximum rates of protein synthesis in microbial cells. The most efficient single-cell protein (SCP) production occurs at peak cell growth rates. Therefore, precise control of fermentation parameters is maintained to promote targeted protein biosynthesis.

Non-sterile fermentation processes are frequently employed in the production of animal feed biomass, depending on factors such as substrate type, strain characteristics, and the intended application of the final product.

The fermentation broth typically contains 1–2.5% microbial biomass (equivalent to 10–25 kg/m^3). This is thickened in several stages to achieve a concentration of 12–16%. Then, it undergoes heat treatment at 75–90 °C for 10–40 minutes to inactivate nearly all microbial cells, including the producer strain.

After thermal treatment, the biomass concentration is further increased to 20–25% using vacuum evaporators. The resulting product is dried to reduce moisture content to around 10%, yielding a fine microbial cell powder. This powder is granulated, resulting in 25–30 kg of final product per batch, which is packaged in multi-layered paper bags.

A critical step in SCP production is the purification of gas-air emissions, which are generated during both fermentation and post-fermentation stages. These emissions contain large volumes of air contaminated with live cells, protein dust, and other microbial byproducts. Additionally, the residual culture liquid—after biomass removal—is also purified and returned to the internal water supply system of the production facility.

Currently, microbial protein production is a leading branch of industrial biotechnology. It provides important feed additives and protein supplements for livestock, fur-bearing animals, poultry, and aquaculture, as well as food-grade proteins, all based on a variety of available raw materials and substrates.

4.1.1 Preparation of Yeast Protein

From a technological perspective, yeasts excel as producers of both fodder and food protein. Their superiority can be attributed to several "process-friendly" features:
- Yeasts thrive effortlessly under industrial conditions.
- Due to their larger size compared to bacteria, yeast cells can be more efficiently separated from the liquid during centrifugation.
- They possess a rapid growth rate and exhibit resilience against external microbial contamination. Furthermore, they can utilize various nutrient sources, can be separated with ease, and do not release spores that could contaminate the air.
- Yeast cells comprise up to 25% solid matter.

Yeast biomass's primary treasure is its protein content, which surpasses the protein found in cereal grains in terms of amino acid composition. It is only marginally less nutritious than milk proteins and fishmeal. The richness in essential amino acids underlines the high biological value of yeast protein. In terms of vitamin content, yeast outshines all other protein foods, including even fishmeal. Beyond these, yeast cells incorporate trace elements and a notable quantity of fats dominated by unsaturated fatty acids. When livestock, such as cows, are fed with fodder yeast, there's a marked improvement in milk yield and its fat content. Likewise, the quality of fur in animals like minks improves. Yeast strains suitable for fodder include *Candida, Rhodotorula, Torulopsis,* and *Trichosporon.* Depending on yeast metabolism, protein production can occur via oxidative or reductive pathways (Fig. 4.3).

Different types of raw materials have been used in the production of microbiological yeast proteins. Initially, sugar-rich materials known as first-generation substrates (refer to Table 4.3) were utilized. These substrates come from waste produced by various industries such as food, dairy, alcohol, sugar, and cellulose, as well as the processing of plant materials like wood, straw, and peat and non-edible parts such as stems and husks. Molasses is a commonly used substrate for fermentation due to its affordability and its high content of easily degradable nitrogen compounds, vitamins and micronutrients. It is a thick syrup obtained by extracting residues from sugar beet or cane with hot water.

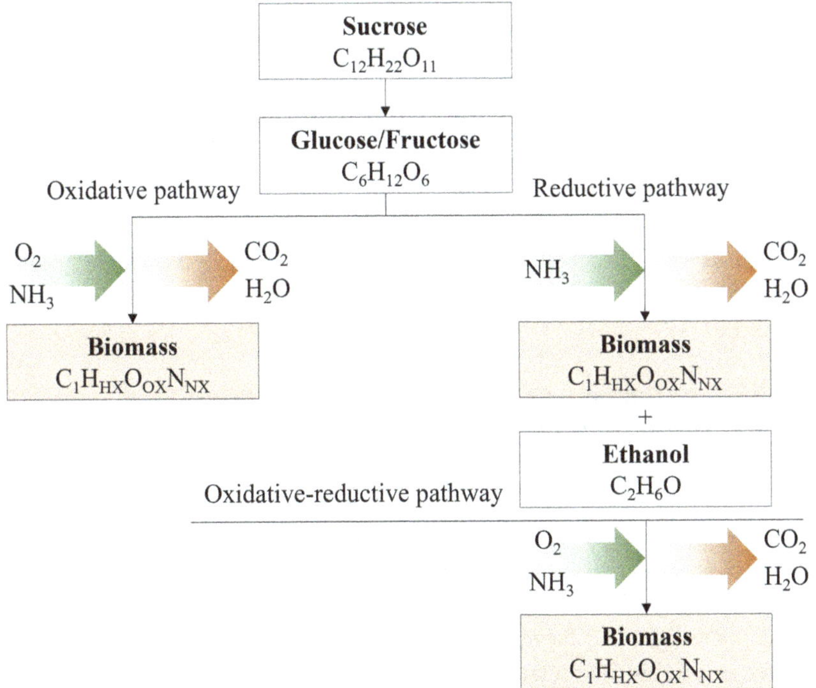

Fig. 4.3 Yeast metabolism pathways of biomass formation

Table 4.3 Carbohydrate raw material utilization in yeast protein production

Substrate	Producer	Application area
Molasses	*Saccharomyces cerevisiae*	Protein additive to food products
Milk whey, liquid waste permeate	*Candida utilis—Mucoprotein thoratein, Candida tropicalis, Rhodotorula glutinis*	Protein additive to food products and food colors
Brewing waste (residual yeast)	*Saccharomyces carlsbergensis* casein substitute in sausages, aromatization of meat, cottage cheese	Protein additive to food products
Agricultural waste, food and woodworking industry (hydrolysates, sulphite lye, draff)	*Candida utilis, Candida tropicalis, Candida scottii*—pentose and hexose, resistant to alcohol, furfural, and other hydrolysis products	Fodder additive
Lignin-cellulose materials	*Penicillium roqueforti, Fusarium raminearum*	Mycoprotein-soy proteins—50–60%

Molasses contains around 40–50% sucrose, which undergoes transformation into glucose and fructose by the Saccharomyces cerevisiae yeast's invertase enzyme. The efficiency of this conversion is measured by the H_{27} parameter, which indicates that approximately 27% of the dry yeast content in molasses can be produced from 1 kg of molasses.

The Swedish "Simba" method utilizes potato starch and involves two yeast species: *Endomycopsis fibuliger* and *Candida utilis*. In this process, *Endomycopsis fibuliger* breaks down the starch, while Candida utilis is responsible for cellular protein synthesis.

When using other carbohydrate-based nutrient media, these are typically complex mixtures containing mono- and disaccharides, organic acids, alcohols, various other organic compounds, and essential mineral elements. Two main criteria dictate the choice of producer strains for these media. First, the strains should have the capability to absorb both pentoses and hexoses. Secondly, they should exhibit resilience to potentially inhibitory substances like alcohols, furfural, and other products resulting from plant biomass hydrolysis (Table 4.3).

Optimizing Yeast Cultivation Strategies for Carbon-Containing Component Utilization

The most widespread species for microbial protein production belong to the genus *Candida: C.utilis, C.scottii, and C.tropicalis*. These yeasts can sequentially utilize the carbon-containing components of hydrolysates and sulfite liquor in the following order: glucose, acetic acid, mannose, xylose, galactose, and arabinose. The completeness of utilization of these carbon sources varies depending on the selected yeast cultivation scheme, with maximum utilization achieved when mixed cultures are used. The two most effective schemes for combining fermentation apparatuses in the joint cultivation of *C.scottii* and *C.tropicalis* are two-step sequential and parallel-sequential.

In the first variant, refined hydrolyzate (wort) with a concentration of reducing substances (RS) of 30–35 g/l (by weight) is used as the initial nutrient medium. In the first fermenter, about 70% of RS is utilized due to easily assimilated hexoses to a residual concentration of RS of about 10–15 g/l, mainly pentose. The yeast obtained in the first apparatus is isolated from the yeast suspension and processed until a final product is formed. The separated culture liquid transforms the second apparatus, in which the remaining pentoses are utilized by more yeast strains adapted to them. In the second variant, two connected fermenters are used: the first receives a diluted wort with an RS concentration of about 15–18 g/l; during yeast fermentation, mainly hexoses are utilized. Further, the yeast suspension enters the second apparatus, in which the remaining sugars are digested without adding the substrate. The total yield of yeast reaches 70–80% concerning RS.

Yeast cultivation on these substrates typically takes place in airlift-type fermenters. As the nutrient medium is saturated with air, a gas-liquid emulsion forms, circulating throughout the apparatus and ensuring efficient mixing. To control foam formation during aeration, mechanical defoaming techniques are applied. In some facilities, bubble-airlift fermenters are used, which distribute air across 4 to 5 zones.

Yeast growth is a continuous process, with cultivation temperatures depending on the strain, ranging from 30–35 °C to 38–40 °C. Any drop in pH during fermentation is automatically corrected using ammonia water. Heat exchangers integrated into the fermenters remove excess heat generated during the process, using circulating cooled water.

The resulting yeast suspension, with a concentration of 20–40 g/L and 75% moisture, is sent to the processing and concentration phase. This phase includes flotation, three-stage separation, heat treatment, and drying. To enrich the biomass with vitamin D_2, ultraviolet irradiation is applied. During this step, the ergosterol in the lipid fraction of the yeast cells is converted into vitamin D_2. The thickened yeast suspension is passed through quartz tubes for this purpose. From wood waste, a final yeast yield of 46–48% can be obtained.

Current limitations in chemical hydrolysis methods for plant raw materials constrain the expansion of fodder yeast production using wood hydrolysates. However, the pulp and paper industry offers certain advantages, such as the low cost of its by-products, including sulfite lye and hydrolysates. From one ton of cellulose, about 37 kg of yeast can be produced. The process achieves a productivity rate of 2.4 kg/m^3/h, with the yeast biomass containing about 48% crude protein (see Fig. 4.4).

Milk whey is also a suitable substrate for cultivating fodder yeasts, particularly from the Torula and Kluyveromyces genera. One ton of whey typically contains about 10 kg of protein and 50 kg of lactose. Proteins are first separated using ultrafiltration, and the resulting lactose-rich solution is used to cultivate yeast. Species from at least 14 different *Candida* strains are known to convert whey into vitamin- and protein-rich biomass. This enables the production of functional products such as acidophilus-yeast milk and cottage cheese.

In addition, the yeast *Rhodotorula glutinis* can produce carotenoids, which are used in natural food dyes.

The potential for producing fodder yeast using combined hydrolysis of plant raw materials and sewage sludge is receiving increasing attention. This approach involves enriching the nutrient media with amino acids from plant and animal sources, which improves both yeast yield and protein content. Furthermore, the raw material base for microbial protein

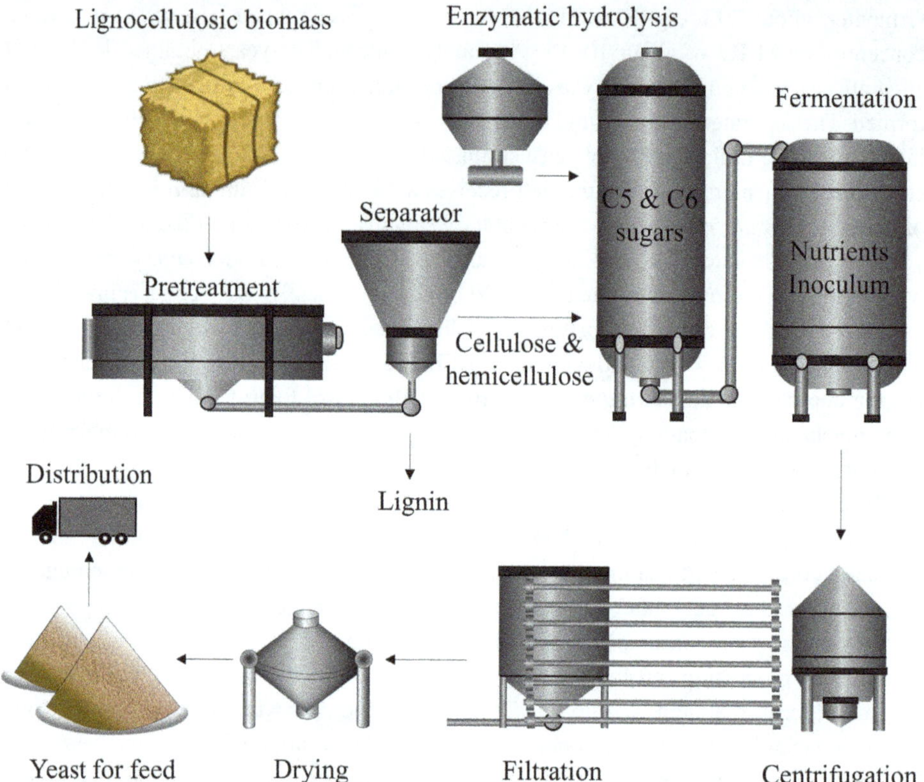

Fig. 4.4 Generalized overview of yeast protein production

production is expanding through the use of peat hydrolysates, which are rich in easily digestible monosaccharides and organic acids. This allows yeast yields to reach 65–68% of the RS in hydrolysates, and the quality of biomass surpasses that of yeast grown on vegetable waste-derived hydrolysates.

Yeast has the capability to utilize hydrocarbons as a carbon source for growth—these are categorized as 2nd-generation substrates. These substrates consist of straight-chain hydrocarbons containing between 10 and 30 carbon atoms per molecule. Among them, the liquid fraction derived from petroleum, with a carbon chain length of C_{11} to C_{18} and a boiling point range of 200–320 °C, is especially suitable for microbial cultivation. The initial development of this method was driven by the need to manage surplus paraffin, which remains in quantities of approximately 10–15% after gas oil purification.

Candida species such as *C. guilliermondii*, *C. maltosa*, and *C. scottii* are commonly employed for utilizing these substrates. Through selective breeding and genetic engineering, rapidly growing strains have been developed that can outcompete other microorganisms in non-sterile industrial cultures.

However, one challenge in protein biomass production from hydrocarbons is the presence of cyclic hydrocarbons in the raw paraffin. Therefore, highly purified paraffin, with less than 0.01% aromatic hydrocarbons, is required.

Because paraffin is insoluble in water, cultivation on this substrate requires the formation of an emulsion with hydrocarbon droplets no larger than 5 microns. This results in a four-phase system: gas, liquid, hydrocarbon, and microbial cells. In addition to mixing, surface tension plays an important role in dispersing hydrocarbons, requiring precise control over the medium's composition and rheological properties.

In this system, paraffin serves as both an energy and carbon source for the yeast. The nutrient medium must be supplemented with macro- and microelements, such as ammonium sulfate, superphosphate, potassium chloride, and surface-active agents. Ammonia water is used to adjust pH and supply nitrogen. Typically, the initial nutrient medium contains 3–5% paraffin.

As the carbon concentration increases, oxygen requirements also rise, as hydrocarbon metabolism requires intensive aeration.

Protein production from yeasts grown on liquid hydrocarbons is carried out in specialized 12-section reactors shaped like tori, with a total volume of 800 m^3 and a working volume of 320 m^3. Each section has a self-priming turbine mixer and an ejection device for mixing. The yeast suspension passes sequentially through all 12 sections. In the first nine, active growth occurs with continuous carbon substrate input. In the final three, no more substrate is added; instead, residual hydrocarbons are oxidized and digested by the yeast.

This design enables near-complete utilization of the substrate and achieves residual hydrocarbon levels below 0.01%. The oxidation process consumes significant oxygen and releases 2.5–3.5 kcal/kg of heat, which is removed by built-in heat exchangers with surfaces up to 3000 m^2 per section.

The total residence time in the reactor is about 8 hours, with a medium flow rate of up to 0.22 h^{-1}. The pH is maintained at 4.0–4.5, and the temperature at 32–34 °C. The system achieves daily productivity of up to 27 tons. After fermentation and drying, the protein mass is granulated and used as a protein-vitamin concentrate in feed.

Yeast cells can also utilize lower alcohols, specifically methanol and ethanol, which are categorized as third-generation substrates (see Table 4.4). These alcohols can be derived from natural gas or plant residues.

When cultivated on alcohols, yeasts demonstrate:
- Higher protein content (56–62% of dry mass)
- Fewer impurities, such as benzene derivatives, D-amino acids, abnormal lipids, toxins, or carcinogens, compared to those grown on oil-derived paraffin
  The ability to metabolize methanol is a natural feature of yeasts from the Hansenula and Candida genera.

Advantages of methanol over hydrocarbons include:
- Full solubility in water
- High purity
- Lack of carcinogenic impurities
- High volatility, making it easy to remove residues during heat treatment and drying

Table 4.4 Technological characteristics of fodder yeast of different groups

Type of fodder yeast	Medium for cultivation of yeast cells	Ready fodder product		Feed protein yield per 1 t of dry raw material, kg
		Structure	Color	
Hydrolytic	Woody and agricultural waste	Powder, granules	Yellow, dark yellow	240–450
Fodder classic	Dried distillers grains (DDGS)	Scaly powder, granules	Light brown, brown	260–400
Protein-vitamin concentrate (PVC)	Petroleum paraffins, lower alcohols, natural gas	Powder, granules	Light yellow, light brown	600–800

Moreover, methanol's biological activity suppresses unwanted microorganisms, promoting the growth of the target yeast strain.

However, it also presents some hazards:

- It is flammable and forms explosive mixtures with air at 6–35% concentrations by volume
- It is toxic, requiring strict safety precautions

The nutrient medium contains alcohol (8–10 g/L) and all essential elements for unrestricted cell growth. To complement standard macro- and microelements, yeast extract (50 mg/L) is added as a source of nitrogen and vitamins. The choice of fermentation regime and equipment depends on the physiological characteristics of the producer strain. For example, when cultivating *C. boidinii* and *H. polymorpha* under aseptic or partially non-aseptic conditions, fermenters equipped with ejection-based liquid-phase energy input are used. Cultivation is carried out at 34–37 °C, pH 4.2–4.6, and a medium flow rate of 0.12–0.16 h^{-1}. The productivity of such systems can reach 75 tons per day, with cell concentrations in the suspension of 30 g/L.

In addition to being used as feed additives, yeast proteins are also studied for human consumption. The idea of using microbial biomass as a dietary protein source dates back to 1890, when Delbrück and his colleagues proposed cultivating brewer's yeast (*Saccharomyces cerevisiae*) on molasses. They suggested the resulting biomass could be used to supplement food proteins. During World War II in Germany, food yeast—especially *Candida arborea* and *C. utilis*—became an important protein source. Even today, *S. cerevisiae* (Carlsbergensis) is cultivated to obtain edible protein and vitamins.

These yeasts are well absorbed by the human body, with more than 90% bioavailability of nutrients. Brewer's yeast contains 14 vitamins, primarily B vitamins. It is used to enhance the nutritional value and flavor of sausages as a substitute for casein, and it can also improve the taste and aroma of meat, cottage cheese, and other food products.

To obtain protein suitable for human consumption, a thorough purification process is required. First, the yeast cell walls are broken using mechanical, alkaline, acid, or enzymatic methods. The resulting mixture is extracted with organic solvents, and the proteins are dissolved in an alkaline solution. Through dialysis, proteins are separated from other cell components. Once purified, the proteins are precipitated and used in food products such as sausages, pâtés, and meat fillings after removing low-molecular-weight impurities.

Yeast proteins are also used in the production of artificial meat. The process begins by heating the proteins, followed by rapid cooling or extrusion through fine nozzles. Then, polysaccharides and other components are added to the protein paste.

Production methods differ depending on the desired artificial meat product. Isolated proteins are pushed under high pressure through spinnerets into acid-salt solutions, where they harden and are stretched into protein filaments. These filaments are enriched with amino acids, vitamins, flavorings, and colorants, and then shaped using heat and pressure. Once hydrated, these fibers resemble natural meat.

Candida arborea and *C. utilis* are used to grow yeast on hydrolysates of plant raw materials. These strains are excellent sources of protein for use in various food products. For instance, in the USA, *C. utilis* is used to produce "Torutaine," a high-protein dietary supplement. In the UK, glucose fermented from corn starch is used to produce a mucoprotein product used in sausage manufacture as a meat substitute.

Ethanol is also a refined base for microbial protein production. Strains such as *C. utilis* and *Hansenula anomala* can yield up to 60% protein content under optimized conditions. While historically unprofitable due to the high cost of alcohol, advances in alcohol production and rising protein demand have renewed interest in this method.

Yeast biomass is rich in amino acids and vitamins, making it useful for producing *yeast autolysate*, which is important for formulating both natural and semi-synthetic media in laboratories and industrial microbiology. However, since many beneficial compounds are locked in protein complexes, adding untreated yeast directly to media is inefficient.

Protein hydrolysis can be done using acids, alkalis, or enzymatic methods. Alkaline hydrolysis may convert or destroy certain amino acids and deactivate some vitamins. Acidic hydrolysis can degrade tryptophan and B vitamins. Enzymatic hydrolysis using proteolytic enzymes is a gentler method. In some cases, yeast autolysis (self-digestion by internal enzymes) is induced under specific conditions.

To produce yeast autolysate:
- A yeast paste with 65–76% moisture is prepared.
- This paste is mixed with water (50 °C) at a 1:1 ratio in a reactor to form a suspension.
- The suspension is held at 45 °C for 1–2 days to allow autolysis.
- Autolysis is enhanced by adding phosphates or 2.5% sodium chloride.
- Afterward, 100 L of suspension is acidified with 0.25 L of diluted sulfuric acid and boiled for 15–20 minutes.
- After cooling, the autolysate is ready for use.

After autolysis, the liquid fraction contains up to 5% solids, half of which are nitrogenous compounds like tyrosine and tryptophan. B vitamins are also present in the filtrate. This liquid can be concentrated and freeze- or spray-dried to form a storable dry mass (see Fig. 4.5). Specially treated autolysate can also serve as a source of amino acids and vitamins for *parenteral nutrition*.

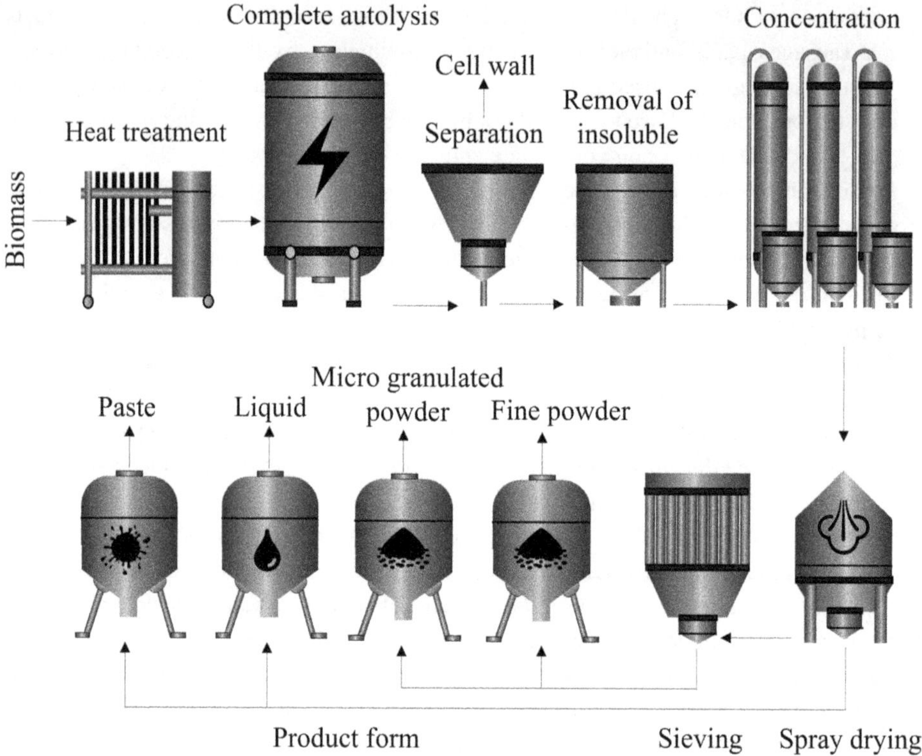

Fig. 4.5 Schematic representation of yeast autolysate production

4.1.2   Fungal Protein

Fungi represent another important group of protein producers, with the ability to utilize a wide range of organic materials such as molasses, whey, plant juices, root crops, peat, manure, and lignin- or cellulose-rich solid wastes from the food, woodworking, or hydrolysis industries. These substrates are saccharified through conventional physical, chemical, or biotechnological methods using cellulolytic enzymes or microbial cells. Microbial entities such as yeast, bacteria, and white rot fungi degrade cellulose during their growth, enriching the resulting protein product with amino acids (Fig. 4.6).

The pool of fungal protein producers continues to expand, notably including fast-growing species from the *Trichoderma*, *Cellulomonas*, and *Aspergillus* genera. These species offer faster growth rates and improved amino acid profiles compared to yeast. The Finnish Pekilo method, for example, utilizes the fungus *Paecilomyces variotii* to produce cell protein from carbohydrate-rich residues left after cellulose production. Similarly, the Canadian Waterloo method uses *Chaetomium cellulolyticum* grown on agricultural and timber industry waste such as straw and sawdust. These methods were particularly attractive when soybean flour was expensive, although they have faced challenges due to the availability of more affordable vegetable proteins.

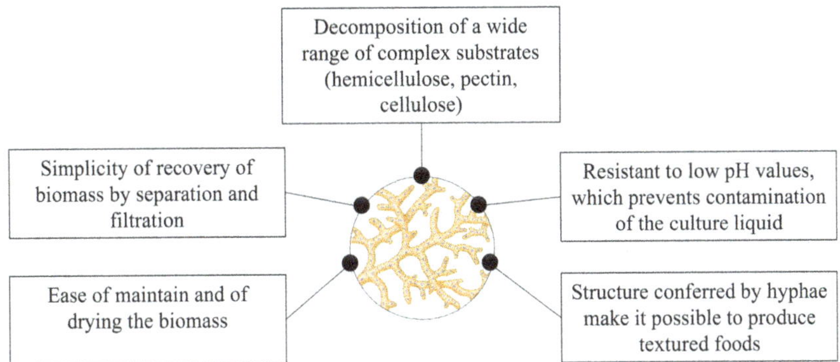

Fig. 4.6 Properties that determine the choice of fungi as protein producers

Dried fungal mycelium contains around 45% protein. This protein-rich fungal mass closely matches soy protein in essential amino acid composition. Fungal proteins are especially high in lysine, which is often deficient in cereal-based proteins. As a result, balanced food and feed mixtures can be developed using both grain and fungal biomass. Fungal proteins have relatively high biological value and are easily digestible. Additionally, the fibrous texture of the cultured fungi resembles natural meat-like products, and with the addition of flavor and coloring agents, a similar taste and appearance can be achieved. Mycelium is usually stored frozen.

Fungal protein biomass is highly digestible in animals and contains less nucleic acid than yeast biomass, making it suitable for use at higher concentrations in animal feed. In the diets of adult animals, up to 50% of the vegetable protein can be replaced with fungal protein. However, before newly developed Single Cell Protein (SCP) products can be approved for commercial use, they must undergo expensive safety evaluations mandated by quality control institutions. These high costs often limit SCP-based products from entering the market, particularly for direct human consumption. Consequently, there has been a shift toward using SCP for animal feed rather than food.

Mycoprotein remains the only officially approved microbial-origin protein product for human food. It is primarily derived from fungal mycelium. In its production, strains such as *Fusarium graminearum* (isolated from soil) and *Chaetomium cellulolyticum* (a cellulose-degrading fungus) are used. The fungus is cultivated using inexpensive glucose syrup obtained through the hydrolysis of wheat or corn starch. Both submerged and surface cultivation methods are employed. After fermentation, the culture is heat-treated to reduce its ribonucleic acid content. The mycelium is then separated by vacuum filtration, as illustrated in Fig. 4.7.

Mycoprotein serves as a high-quality meat substitute, offering advantages in terms of protein (44%), minerals, vitamins, and lipid content compared to animal protein. Sausages containing microscopic fungi have excellent protein digestibility due to the activity of pepsin and trypsin. Typically, microbial biomass is added at 5% concentration in minced meat products. Fungal biomass is also appreciated as a vegetarian ingredient, prized for its pleasant flavor and unique texture, making it popular in many countries.

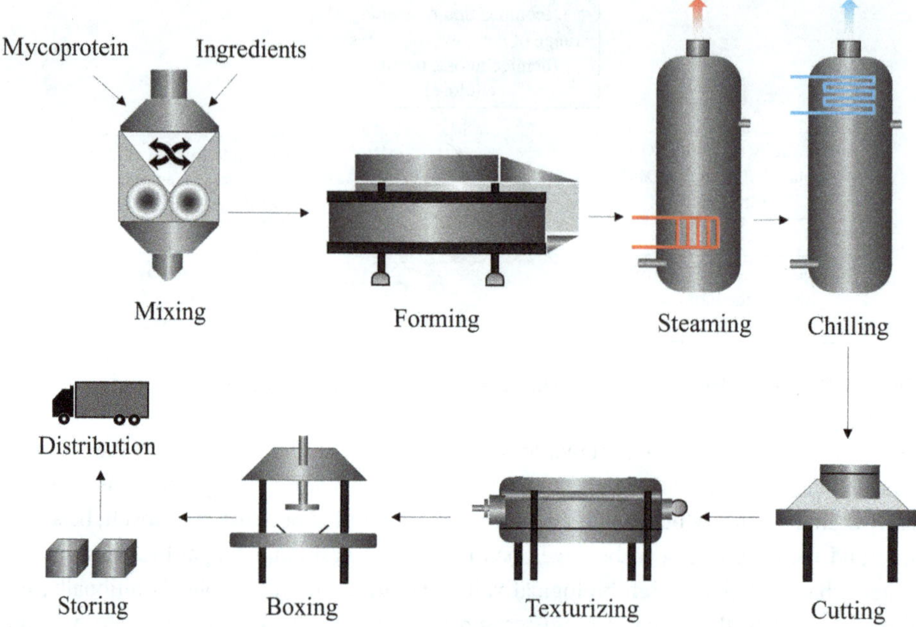

Fig. 4.7 Outline of mycoprotein manufacturing process

The cultivation technique for microscopic fungi varies depending on the processing of the plant material. Growing fungi on hydrolysates of plant waste and liquid by-products from woodworking and paper industries results in a higher substrate utilization rate compared to cultivation on solid media. After deep (submerged) cultivation, the fungal biomass typically contains 50–60% protein in dry matter. To enhance raw material utilization, *symbiotic cultivation*, using both fungi and bacteria, is also employed.

4.1.3 Production of Bacterial Biomass

Over 30 bacterial species can serve as protein sources. These organisms grow rapidly, and their biomass can contain up to 80% protein. Additionally, bacterial proteins are of high biological quality, enriched with amino acids such as cysteine and methionine. Bacteria are amenable to selective breeding, allowing for the development of high-yielding strains. However, there are some challenges: their small cell size makes sedimentation difficult, they are prone to bacteriophage infections, and their biomass has a high nucleic acid content. Bacteria can be cultivated using both natural and gas-associated raw materials, including hydrogen and alcohols like methanol, ethanol, and propanol. Methane, a component of natural gas, can also serve as a substrate for bacterial protein production. Biomass yields using this method can reach 66% of the substrate mass.

Methane has several advantages:
- It's widely available.

- It's cost-effective.
- Methane-oxidizing microorganisms efficiently convert it to biomass.
- The resulting biomass has a protein that's well-balanced in terms of amino acid composition.

  Moreover, methane contains few contaminants that inhibit microbial growth, supporting high biomass productivity and reducing the need for extensive purification of the substrate or final product.

  Bacteria from the *Methylococcus*, *Pseudomonas*, *Mycobacterium*, and *Methanomonas* genera are commonly used to produce microbial protein from methane. These bacteria oxidize methane as both a carbon and energy source in a sequence of steps, converting it to carbon dioxide:

$$CH_4 \rightarrow CH_3OH \rightarrow HCHO \rightarrow HCOOH \rightarrow CO_2$$

However, using methane as a growth substrate presents several technical challenges. Methane's low solubility in water (around 0.02 g/L under standard conditions) limits its bioavailability and restricts microbial growth rates. Additionally, methane oxidation releases by-products (0.2–0.6 g carbon/g biomass), which can inhibit the primary producer strain. To overcome this, mixed microbial cultures are used, including *Methylomonas* (methane consumers), *Hypomicrobium* and *Pseudomonas* (methanol consumers), and other supportive bacterial strains. These cultures show high productivity and resilience, tolerating acidic pH and elevated temperatures, and are less susceptible to microbial contamination.

Methane's high reduction potential requires large amounts of oxygen for oxidation—five times more than carbohydrates and 2–3 times more than hydrocarbons. This oxygen demand complicates fermenter design. Methanotrophic bacteria are generally cultivated in flow systems at 34–38 °C and neutral pH, using media supplemented with mineral salts and various nitrogen forms. In oligotrophic systems, nitrogen concentration is kept low (20–30 mg/L), and oxygen demand can exceed methane needs by a factor of 2–3. Because methane is explosive, maintaining the optimal gas mixture is difficult; thus, fermentation typically proceeds under oxygen limitation and methane excess.

Effective methane-based cultivation requires specialized fermenters with strong gas-dispersion and mass-transfer capabilities. Enhancing methane utilization involves recycling gas, increasing pressure, and using pure oxygen instead of air. These improvements can raise methane assimilation efficiency to 95%. Fermentation proceeds at a medium flow rate of 0.25–0.30 h^{-1}, with biomass concentrations up to 10 g/L at the fermenter outlet. Producing one ton of biomass requires 1.8–2.2 tons of methane and 4.5–5.0 tons of oxygen. The final biomass contains up to 75% crude protein, 10% nucleic acids, 5% lipids, up to 10% ash, and ≤10% moisture. The protein quality is comparable to fish or soybean meal.

The technological potential of this method is clear, but economic feasibility depends on advances in fermenter design and process intensification. While animal feed protein from methane is complex and costly, alcohols—especially methanol—are promising alternative substrates due to easier handling and scalability.

Methanol can be obtained by oxidizing methane and supports microbial growth through a three-step oxidation process:

$$CH_3OH \rightarrow HCHO \rightarrow HCOOH \rightarrow CO_2$$

- Methanol offers advantages over hydrocarbons:It is water-soluble and highly pure.
- It lacks carcinogenic impurities.
- Its high volatility allows easy removal from the final product.
- Methanol generates less heat during fermentation and has antimicrobial properties that support monoculture growth.

  However, methanol is flammable and forms explosive mixtures with air (6–35% by volume), necessitating careful handling.

In methanol-based fermentation, the growth medium contains 8–10 g/L alcohol and standard macro- and micronutrients. Yeast extract (50 mg/L) is added for extra nitrogen and vitamins. The fermentation regime and equipment are selected based on the physiological traits of the bacterial strain. For example, *Methylomonas clara* and *Pseudomonas rosea* are cultivated in jet-type fermenters at 32–34 °C, pH 6.0–6.4, and a flow rate of 0.5 h^{-1}. Methanol efficiency (yield) reaches 0.45. Bacterial biomass typically contains up to 74% crude protein and 10–13% nucleic acids—more nitrogen-rich than yeast biomass.

Historically, methanol-based protein production was initiated in the UK under the trade name Pruteen (see Fig. 4.8). In Russia, a similar methanol-derived protein product known as Meprin was developed.

Ethanol serves as a refined substrate for producing nutritionally valuable microbial protein. A method to produce feed protein from ethanol using Acinetobacter bacteria—known as the "Eprin" biologic—has been developed. Although the economic feasibility of microbial protein production from alcohols was once questioned due to the high cost of the substrate, recent advancements in alcohol production technologies and the growing demand for protein products are making this approach increasingly viable.

Cultivation of Bacteria on Carbon Dioxide

Cultivating bacteria with carbon dioxide presents significant potential for protein production, particularly through the use of hydrogen-oxidizing bacteria. These bacteria derive energy by oxidizing hydrogen with atmospheric oxygen. Hydrogen-oxidizing bacteria that utilize carbon dioxide as both a carbon and energy source are especially noteworthy. The energy released during the oxidation process is used to fix and assimilate carbon dioxide. *Hydrogenomonas* species are commonly used in this biomass generation approach.

Currently, microbial protein production using hydrogen-oxidizing bacteria remains at the experimental stage. The process is carried out in continuous fermentation systems using apparatuses equipped with ejectors or self-priming turbine mixers.

- Key process parameters include:A medium flow rate of 0.4 h^{-1}
- A cell concentration of 10–20 g/L in the culture
- Resource requirements per ton of biomass:
  - 0.7 tons of hydrogen
  - 2.0 tons of carbon dioxide

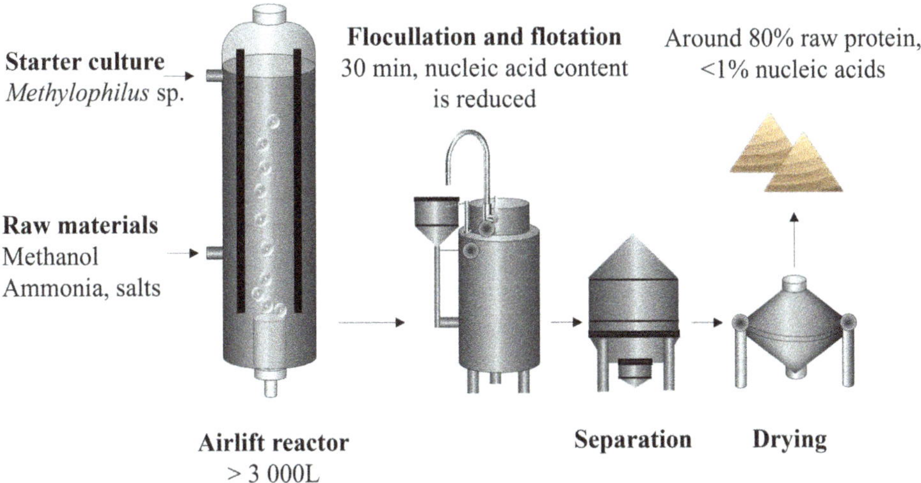

Fig. 4.8 Pruteen production outline

– 3.0 tons of oxygen

The method yields up to 20 g of dry biomass per liter of culture.

The appeal of hydrogen-oxidizing bacteria lies in their autotrophy, independence from scarce organic raw materials, rapid growth, and the high amino acid content of their protein. They produce no extracellular organic intermediates—water is the sole by-product of hydrogen oxidation. This contributes to both ecological safety and high product purity.

In addition to electrolysis, hydrogen can also be sourced from synthesis gas, waste gases from chemical and petrochemical industries, flue gases rich in CO_2, and CO_2 from biochemical sectors. Therefore, microbial protein production with hydrogen-oxidizing bacteria may also serve as a bioremediation or gas utilization strategy. However, the process is complicated by the use of poorly soluble and explosive gaseous substrates, posing challenges similar to those encountered in methane-based single-cell protein production.

Compared to the widely available and cost-effective natural gas, large-scale protein production using hydrogen is currently considered less economically viable. Nonetheless, with the rising interest in hydrogen energy and its environmental benefits, this biotechnology shows long-term promise. In the future, hydrogen-oxidizing bacteria could become a significant sustainable protein source for human consumption.

Bacterial-derived protein is often incorporated into animal feed, making up 2.5–7.5% of the total protein content for livestock and up to 15% for adult pigs. Notably, bacterial biomass contains high levels of nucleic acids.

In animals, there is no need to significantly reduce the nucleic acid content, since nitrogenous bases are broken down into uric acid and its salts, which are further converted into allantoin and easily excreted. However, in humans, excessive intake of uric acid can lead to health issues such as kidney stones or gout, making nucleic acid reduction a necessary step when bacterial biomass is intended for human consumption.

4.1.4 Unicellular Algae in Protein Production

A fundamentally new area in the production of proteins is photoautotrophic organisms, which use carbon dioxide as a carbon source and light as an energy source. Studies of algae as possible protein producers have been carried out for several decades. Attention to algae is determined by their nutrition, the chemical composition of biomass, and manufacturability. The process of growth of the biomass of algae occurs due to photosynthesis. Therefore, the main factor determining the efficiency is illumination. Initially, unicellular algae such as *Chlorella* and *Scenedesmus* were actively explored as promising sources for protein biosynthesis. These algae are typically cultivated in the surface layers of ponds. Algae, for their development, need specific lighting conditions, appropriate temperature, and large volumes of water. Therefore, they are usually grown in the natural conditions of southern regions in open-type cultivation pools. When growing algae in cultivators of the open type with 1 hectare of water surface, up to 70 tons of dry biomass can be produced per year, which exceeds the biomass yield in the cultivation of wheat, rice, soybean, and maize. The protein content of *Chlorella* and *Scenedesmus* cells is about 55% (based on dry weight). The protein mass from algae cells introduces production as a suspension, a dry powder, or a pasty preparation for food purposes. However, the process of separating algal cells from the mass of water is highly time-consuming. Moreover, these algae have indigestible cell walls; it is necessary to disintegrate cells to purify proteins from toxic chlorophyll.

A novel approach to protein production involves photoautotrophic organisms that utilize carbon dioxide for carbon and light for energy. Algae have been researched for several decades as potential protein sources. The interest in algae stems from their nutritional mode, biomass composition, and production feasibility. Their growth is powered by photosynthesis, making illumination vital for efficiency. Initially, unicellular algae like *Chlorella* and *Scenedesmus* were viewed as potential protein producers.

Typically, algae thrive in the topmost layer of ponds, requiring specific light conditions, temperatures, and vast water quantities. Hence, they are mainly cultivated in open pools in warmer regions. Using open-type cultivators, one can produce up to 70 tons of dry biomass per hectare annually, surpassing yields from crops like wheat, rice, soybean, and maize. The protein content in *Chlorella* and *Scenedesmus* is roughly 55% of their dry weight. This algal protein can be processed into suspensions, dry powders, or pasty preparations for consumption. However, separating the algal cells from water is challenging and time-consuming. Additionally, given their indigestible cell walls, there's a need to break the cells to extract proteins and remove the toxic chlorophyll.

Spirulina, a type of cyanobacteria, has gained attention as an excellent source of protein because it can grow naturally and absorb nitrogen from the atmosphere. The composition of *Spirulina*'s biomass consists of approximately 70% proteins with a wide range of amino acids, 19% carbohydrates, 4% nucleic acids, 4% lipids, 6% pigments, and 3% combined ash and fibers. Unlike other microalgae, *spirulina* has a cell wall that is easily digestible. Its protein profile is abundant in essential amino acids, except for methionine. Furthermore, Spirulina cells produce significant amounts of polyunsaturated fatty acids and β-carotene (Fig. 4.9).

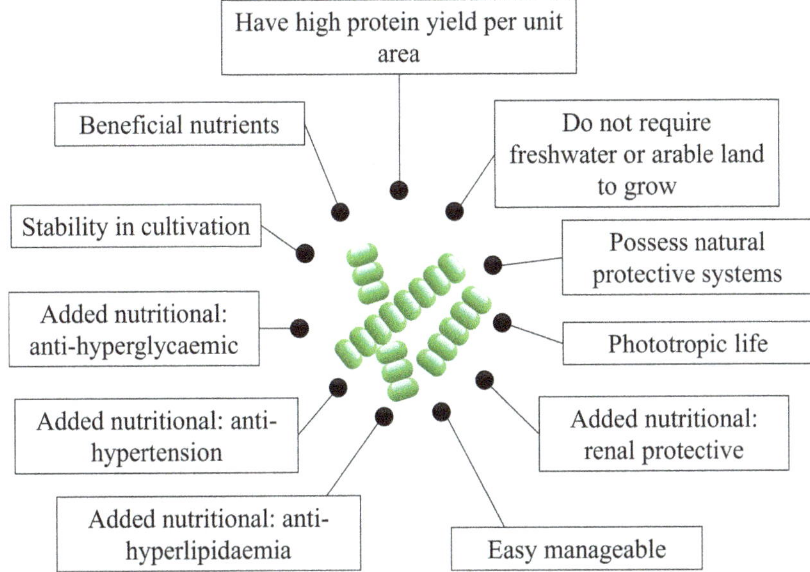

Fig. 4.9 Benefits of using microalgae as a source of protein

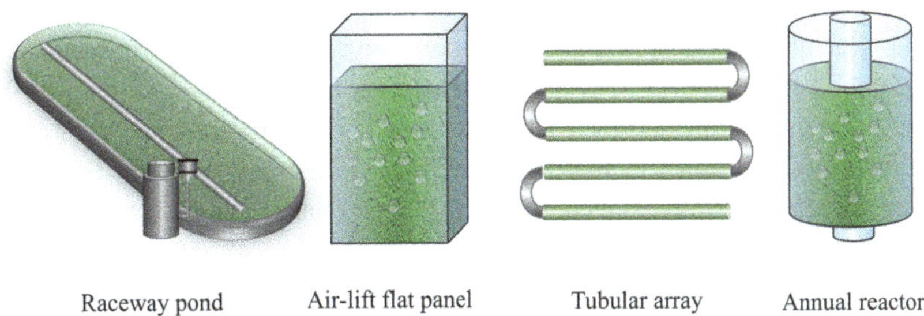

Fig. 4.10 Examples of photobioreactors

Spirulina's initial references date back to the early sixteenth century, where dried Spirulina maxima was sold in Mexico City markets. It thrived naturally in the alkaline waters of Lake Texcoco. By the mid-nineteenth century, a Belgian expedition in the Sahara found dried blue-green algae, identified as *Spirulina platensis*, being sold in village markets near Lake Chad. This species grew in ponds around the lake. Today, various open and closed photobioreactor designs are utilized for protein production (See Fig. 4.10).

Spirulina predominantly grows as a single species due to the high pH of its natural lake environment, reaching levels of 10.5–11.0. The algae's gas-filled vacuoles and spiral filaments allow it to rise to the surface. Consequently, the wind pushes these floating algae towards the shoreline.

The biomass of *Spirulina* doubles roughly every 3–4 days, and it can be harvested daily. In ideal conditions, the biomass yield reaches up to 20 g/m^2 daily. This yield significantly

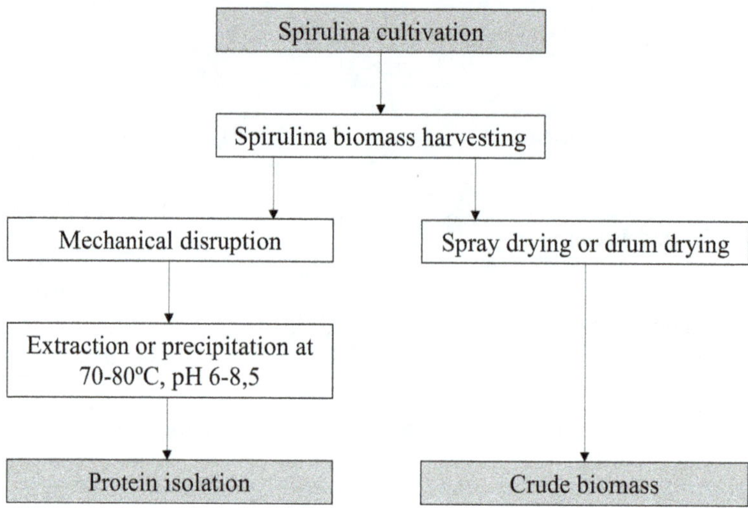

Fig. 4.11 Schematic representation of Spirulina SCP recovery process

surpasses that of wheat, and the protein quality from Spirulina exceeds typical vegetable proteins (Fig. 4.11).

Experiments on the study of the biological value of *Spirulina*, made by the French Petroleum Institute in collaboration with the company "Sosa Texcoco," culminated in the creation of the first experimental factory. In Israel, near Haifa, in the swamps of 12,000 m², *Spirulina platensis* is grown for feed and food. Genetic improvement of the existing strains of *Spirulina* can significantly increase their yield. In addition, mutants have been obtained, and if the growth rate is maintained, the pool of amino acids can be significantly higher than that of the initial one.

The French Petroleum Institute, in partnership with "Sosa Texcoco," conducted experiments to study the biological value of *Spirulina*. This research led to the establishment of the first experimental production facility. Near Haifa, Israel, *Spirulina platensis* is cultivated in 12,000 m² marshlands for both feed and human consumption. Genetic enhancement of current *Spirulina* strains can boost their production. Moreover, certain mutants have been developed that offer a more diverse amino acid pool, while retaining the original growth rate.

Similar production of *Spirulina* in artificial conditions exists in France and Italy. It has been shown that *Spirulina* can grow in artificial alkaline ponds, as well as in the warm wastewater from the cooling system of thermal power plants, which also release spent vapors rich with carbon dioxide. In southern Italy, a factory was built where *Spirulina* is cultivated in a closed system made of polyethylene pipes. These pipes act as solar collectors, thereby extending the productive season. In addition, the closed system prevents exogenous contamination.

Spirulina production in artificial conditions is also seen in France and Italy. Research has demonstrated the potential for cultivating *Spirulina* in man-made alkaline ponds and in the heated wastewater from power plant cooling systems, which release carbon dioxide-rich vapors. In southern Italy, a facility has been established where *Spirulina* is cultivated within a closed system of polyethylene pipes. These pipes act as solar collectors, extending the growing season, and the enclosed environment reduces the risk of external contamination.

After filtration, *Spirulina* is dried using hot air and processed into flour. This algal protein is valuable both as animal feed and as human food. The minimal nucleic acid content, the harmlessness of the phycocyanin pigments, and the digestibility of its protein qualify Spirulina as a premium food-grade protein source. Unlike some other proteins, Spirulina does not lead to cholesterol buildup, positioning it as a dietary supplement. It's offered in forms like powders and capsules as natural food additives.

In conclusion, addressing the current protein shortage can be achieved through innovative biotechnologies utilizing various substrates and microbes. The journey of microbial protein has just started. Although today's microbial proteins may not completely resolve the existing protein gap, their significance in human nutrition will undoubtedly grow in the coming years.

4.1.5   Brainstorming

1. What is the protein of unicellular organisms, what is its composition, and where does it find its application?
2. Discuss the prospects and economic feasibility of using microorganisms in food and feed protein production technologies.
3. What do standard requirements for proteins of unicellular organisms mean in terms of nutrients, digestibility, and economic efficiency?
4. Enumerate the substrate materials that support the growth of microorganisms in the production process of SCP.
5. List the criteria for the selection of microbes for protein production.
6. Describe the generalized scheme of biotechnological manufacture of microbial protein products.
7. Why are yeasts the best producers of fodder and food protein from a technological point of view?
8. Discuss the directions and possibilities of producing yeast protein on carbohydrates.
9. Discuss the directions and possibilities of producing yeast protein on hydrocarbons.
10. Discuss the directions and possibilities of using alcohols and hydrogen to produce protein.
11. How are yeast autolysates obtained and used?
12. Describe the properties and determine the choice of fungi as protein producers.
13. Please provide detailed information about mycoprotein, its receipt, and its use.
14. What are the advantages and disadvantages of using bacteria to produce protein?
15. What types of raw materials are used for the cultivation of protein-producing bacteria?
16. What bacteria are used to produce protein?
17. Analyze the advantages and disadvantages of using single-celled algae to produce protein.

Take-Home Messages

- Microbial proteins (*single-cell proteins, SCPs*) are produced from microorganisms like yeasts, bacteria, fungi, and algae. They offer a rich source of essential amino acids, vitamins, and lipids, and are used as food supplements, animal feed, biofertilizers, probiotics, and even vaccines.
- SCPs are economically viable and sustainable: Microorganisms grow rapidly, require minimal land and water, and utilize diverse, often low-cost substrates—including industrial waste, molasses, hydrocarbons, alcohols, and CO_2.
- Nutritional quality, digestibility, and safety are key requirements for microbial protein products. SCPs should have balanced amino acid profiles, be free from toxic substances, and be readily digestible by animals or humans.
- Substrate selection is critical: Carbohydrates (molasses, whey), hydrocarbons (paraffin, methanol), alcohols (ethanol, methanol), and even gases (methane, $CO_2 + H_2$) can serve as raw materials depending on the microorganism used.
- Yeasts are optimal producers of food and fodder protein because of their large cell size, ease of separation, fast growth, tolerance to contamination, and high nutritional value. Candida and Saccharomyces are widely used.
- Hydrocarbon- and alcohol-based protein production using yeasts and bacteria is technologically advanced, although safety, cost, and regulatory concerns (e.g., toxicity of residual compounds) need consideration.
- Autolysates of yeast are hydrolyzed protein products used in microbiological media and medical nutrition. They are obtained via autolysis or enzymatic treatment of yeast cells, followed by purification.
- Fungi (e.g., Trichoderma, Fusarium, Chaetomium) offer high protein yield and valuable amino acid content. Mycoprotein from *Fusarium graminearum* is approved for use in human food and is widely used in meat alternatives.
- Bacteria such as Methylomonas and Pseudomonas can convert methane, methanol, or CO_2 into protein. Their proteins are rich in sulfur-containing amino acids, but high nucleic acid content and culture safety require attention.
- Algae and cyanobacteria (e.g., *Chlorella, Spirulina*) use light and CO_2 for growth, providing an efficient and ecologically friendly protein source. *Spirulina* is particularly digestible and rich in vitamins and carotenoids. Spirulina and mycoprotein are among the few microbial proteins already widely accepted in human diets, demonstrating the growing potential of microbial biomass as a food source.
- A generalized biotechnological process for SCP production includes:
  - Substrate selection and preparation
  - Inoculum development
  - Fermentation
  - Biomass concentration
  - Drying and formulation

- – Product stabilization and packaging
- Challenges include:
- Removing nucleic acids for human use
- Ensuring safe handling of explosive or toxic substrates (e.g., methanol, methane, hydrogen)
- High initial cost of production and regulatory approval

4.2 Probiotics

4.2.1 The Microbiota of the Human Organism and Its Valuable Functions

The human body and the environment around it form a connected ecological system, where symbiotic microbes play a vital role. These microbes are known as the normal microbiota of the human body and are essential for maintaining the body's biochemical, metabolic and immune balance that is crucial for good health. They consist of approximately 600 species belonging to 17 families and 45 genera. Recent research has identified 395 distinct phylogenetic groups within these microorganisms. Anaerobic microbes outnumber aerobic ones by about 100 to 1. Surprisingly, the number of microbial cells in our body exceeds that of human cells. In fact, there are approximately ten times more microbial cells than human cells, making up around 90% of all cells in our body while only about 10% are human cells. These microorganisms collectively contribute to a weight of about 2.5 to 3.0 kg in our bodies. It is interesting to note that the genetic material of our microbiota, known as the "Microbiome," is three times larger than the volume of our own genes. When we combine both human and microbiota genes together, we get what is called the "Metagenome." This extensive collection of microorganisms is sometimes referred to as a "Superorganism" (Fig. 4.12).

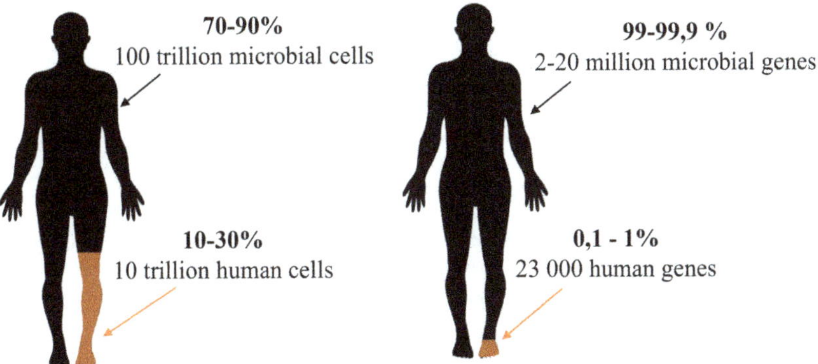

Fig. 4.12 Human microbiome in numbers

Microorganisms in our bodies form communities that work together with us to establish ecosystems, creating both permanent and temporary groups within specific environments. The makeup and ideal population of microorganisms in a healthy individual are influenced by various factors like where they live, their age, lifestyle choices, eating habits and the surrounding environmental and societal conditions.

The entire collection of microorganisms in the human body can be divided into three categories.

1. Obligate microflora, also known as autochthonous, indigenous, or resident microflora. They make up around 90% of the microbial community.
2. Facultative microflora, sometimes referred to as supplementary or concomitant. They comprise less than 9.5%.
3. Transient microflora, known as random, allochthonous or residual microflora. They account for up to 0.5%.

The primary location for these microbial communities is the large intestine (Fig. 4.13). The microorganisms in the intestines are divided into two main categories:

1. Mucous microflora (M): These are microbes that are closely associated with the lining of the intestine, residing within the mucus layer and between the villi. This creates a dense layer of bacteria. The mucous microbiota forms a biofilm within this layer, which is like a solid gel called mucin. Mucin acts as both a protective barrier and a source of nutrition for these microorganisms. Within this biofilm, the microbes attach firmly to the cells of the intestine, forming a complex of microbes and tissues. This system mainly consists of stable strains of microbes such as *Bacteroides, Bifidobacteria, Lactobacilli, Enterococci,* and *Escherichia.* Foreign microbes face difficulty in infiltrating this system due to its stability.

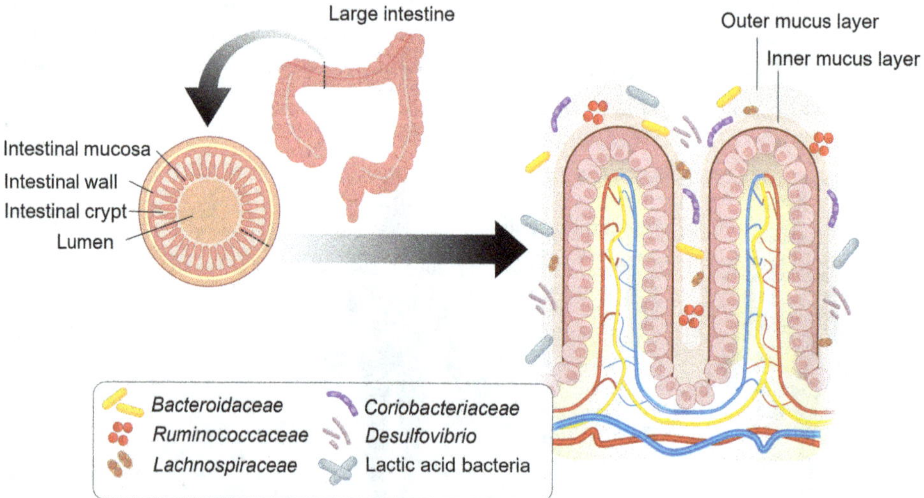

Fig. 4.13 Intestinal mucus and its typical microorganisms

Table 4.5 Beneficial effects of microbiota

Protective function	Structural functions	Metabolic functions
Pathogen displacement	Barrier fortification	Control of IEC differentiation and proliferation
Nutrient competition	Induction of IgA	Metabolize dietary carcinogens
Receptor competition	Apical tightening of tight junctions	Synthesize vitamins, e.g., biotin, folate
Production of anti-microbial factors, e.g., bacteriocins, lactic acids	Immune system development	Ferment non-digestible dietary residue and endogenous epithelial-derived mucus
		Ion absorption
		Salvage of energy

2. Luminal microflora (L): These microbes are found within the lumen, or central cavity, of the intestine.

Normal flora has three essential functions: *protective*, *structural*, and *metabolic* (Table 4.5).

Intestinal bacteria protect the host in several ways. They displace pathogens, compete for nutrients, and produce antimicrobial substances. These microbes prevent harmful organisms from settling in the gut by competing for attachment sites and essential nutrients—a phenomenon known as *competitive exclusion*. Some beneficial bacteria physically block or mask receptors, thereby preventing pathogen adhesion. This protective effect is not limited to the intestines but is also observed in other areas such as the mouth, skin, and vaginal mucosa. By consuming nutrients vital for pathogen survival, commensal bacteria limit their growth when these nutrients become scarce.

Normal flora may also antagonize other microorganisms by producing substances that inhibit or kill nonindigenous species. For example, intestinal bacteria produce a wide range of antimicrobial compounds, from relatively nonspecific fatty acids and peroxides to highly specific protein-based substances called *bacteriocins*. These bacteriocins, typically 18–90 kDa in size, are synthesized by many bacterial species and provide an adaptive advantage to the producers, which are naturally resistant to their own bacteriocins. Both Gram-negative and Gram-positive bacteria are known to produce bacteriocins.

A well-known example is *Escherichia coli*, which synthesizes bacteriocins known as *colicins*. These are encoded by genes located on various plasmids (e.g., ColB, ColE1, ColE2, ColI, ColV). Colicins may bind to specific receptors on the cell envelope of sensitive bacteria and cause cell lysis, target intracellular components such as ribosomes, or interfere with energy production. Other notable examples include *lantibiotics*, produced by genera such as *Streptococcus*, *Bacillus*, *Lactococcus*, and *Staphylococcus*. Many lactic acid bacteria produce similar antimicrobial peptides, which act as defensive effector molecules protecting both the microbiota and the host organism.

Some bacterial species are capable of altering the local environment to make it inhospitable for competitors. This is achieved by producing volatile fatty acids that reduce pH, releasing hydrogen sulfide, or modifying the oxidation-reduction potential. In addition, the indigenous gut microbiota contributes to *colonization resistance*—the ability to inhibit the establishment of harmful enteric pathogens. This is accomplished through the production of antimicrobial compounds, volatile fatty acids, and modified bile acids.

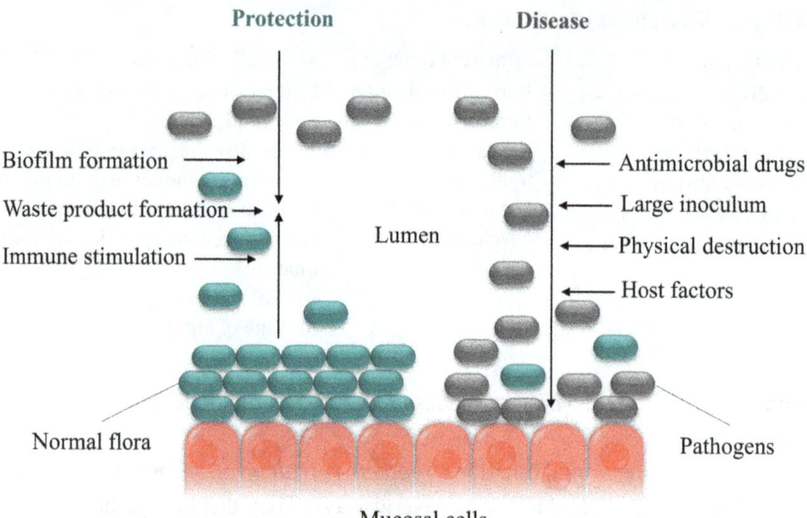

Fig. 4.14 Gut protection, performed by commensal normal flora

One important mechanism of colonization resistance is the formation of microbial *biofilms*. These biofilms act as a physical barrier, limiting access to epithelial surfaces and creating a frontline defense against invading organisms. In this way, the gut microbiota forms a protective shield. This shield can function in a *drastic* mode—completely eliminating invading strains—or in a *permissive* mode—allowing limited colonization of foreign microbes at a minimal level (Fig. 4.14).

Moreover, the intestinal flora plays a key role in host metabolism and is often compared to an additional metabolic organ akin to the liver. The human body provides a nutrient-rich environment, while the microbiota contributes unique biochemical functions that the host lacks. These functions include the degradation of resistant starches and polysaccharides, the synthesis of certain vitamins, and the biotransformation of bile acids.

In the final stages of food digestion, the colon serves as a reservoir for bacteria involved in the breakdown of complex carbohydrates. Bacterial enzymes facilitate the degradation of polysaccharides through colonic fermentation. This process primarily targets plant-derived polysaccharides—such as cellulose, xylan, and pectin—as well as those associated with intestinal cells. Colonic bacteria efficiently hydrolyze these compounds.

Metabolic functions and nutrient processing The intestinal microbiota plays a vital metabolic role, often likened to that of the liver. It contributes to the breakdown of complex starches, generates short-chain fatty acids (SCFAs), and enhances the absorption of essential minerals, including calcium, magnesium, and phosphorus.

- *Bile acid transformation*
  Gut microbes also participate in the modification of bile acids synthesized by the liver. Additionally, they convert dietary choline into trimethylamine (TMA), which is further metabolized in the liver.

- *Anti-cancer and immune support*
  The gut microbiota counters toxins, suppresses tumorigenic processes, and supports the immune system. Lactic acid bacteria, for instance, inhibit tumor cell proliferation and stimulate immune responses.
- *Vitamin synthesis*
  Commensal microbes synthesize essential vitamins such as vitamin K, important for blood clotting, and vitamin B_{12}, which the human body cannot produce independently.
- *Mucosal immunity development*
  Colonization of the intestinal mucosa by beneficial microbes promotes the maturation of the immune system, particularly through the stimulation of antibody production and the development of mucosal immunity. This interaction is mediated by pattern recognition receptors, such as toll-like receptors (TLRs), which help the immune system differentiate between harmless and harmful microbes.
- *Dysbiosis and its consequences*
  Dysbiosis refers to an imbalance in the gut microbial community, often caused by factors such as poor diet, medication use, illness, or stress. This disruption can adversely affect digestion, immune function, and overall health. Essentially, the gut microbiota acts as a dynamic extracorporeal organ that significantly influences host physiology.

Examples of Dysbiosis and Related Conditions:

1. Excessive bacterial growth in the small intestine
   An imbalance of bacteria in the small intestine can lead to *Small Intestinal Bacterial Overgrowth Syndrome* (SIBOS), observed in 70–95% of individuals with chronic intestinal disorders. SIBOS arises from impaired intestinal motility, anatomical abnormalities, or direct connections between the small and large intestines, facilitating colonization by microbes that are not typically present in the small bowel.
2. Altered microbial composition in the colon
   Dysbiosis in the colon involves a reduction in beneficial bacteria such as *Bifidobacterium* and *Lactobacillus*, along with an overgrowth of potentially harmful microorganisms like *Enterobacteriaceae*, *Enterococcus*, and pathogenic strains of *Escherichia coli*, *Staphylococcus*, *Streptococcus*, *Proteus*, and *Candida* species.
3. Health implications of dysbiosis
   Microbial imbalance can lead to abnormal colonization and is associated with increased intestinal permeability, endotoxemia, septicemia, systemic inflammation, insulin resistance, obesity, and diabetes.
4. Restoring microbial balance
   Addressing dysbiosis involves replenishing beneficial microorganisms and improving the condition of the intestinal epithelium that supports their growth. This is commonly achieved through the use of medications or supplements containing live cultures of symbiotic organisms (probiotics).

4.2.2   Medications for the Normalization of Microbiota. Characteristics of Probiotic Products

There are three ways to correct dysbiosis: intervention method (receiving probiotics), supporting method (use of prebiotics), and combination therapy (administration of synbiotics, which include both probiotics and prebiotics) (Fig. 4.15).

Preparations for the correction of microbiocenosis can be divided into six groups:

1. Monoculture preparations: These formulations consist of strains of live microorganisms found in the normal gut flora.
2. Complex microorganism preparations: These products contain a blend of live microorganisms.
3. Stimulating substances: When consumed orally, these formulations promote the growth of gut bacteria, particularly bifidobacteria and lactobacilli.
4. Microorganisms with growth stimulators: These preparations combine multiple microorganisms with substances that enhance their growth and reproduction.
5. Genetically engineered strains: These products feature microorganisms with genetic characteristics.
6. Multi-component preparations: Apart from microorganisms and growth boosters, these formulations also include compounds that impact the functions of human organs and tissues.

The term "probiotics" combines "pro" (for) and "biota" (life), originating from the Greek phrase "for life." Initially, it referred to substances produced by one microorganism that promote the growth of others. Over time, various definitions have emerged. The most recent definition defines probiotics as microbial cell preparations or components with positive effects on the host's health and well-being, regardless of their viability, and beyond just impacting the gut microflora.

Probiotics are designed to restore a balance of microorganisms in our bodies. They include types like *Bifidobacteria*, *Lactobacilli*, *Streptococci*, and others that naturally occur in our gut.

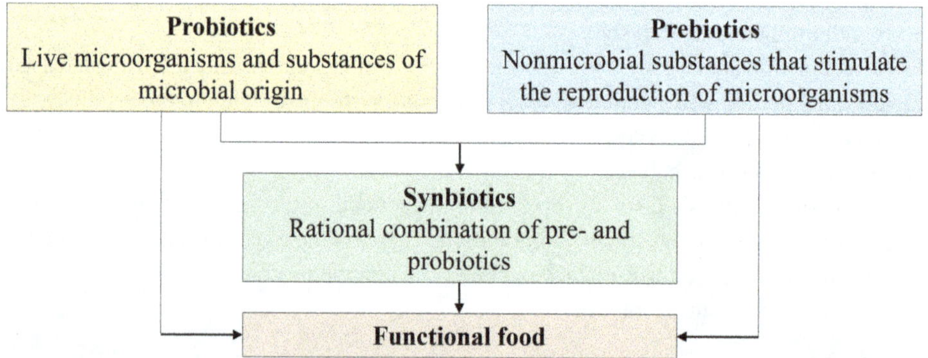

Fig. 4.15 Means for the treatment of dysbacteriosis

Probiotics can be divided into three generations:

1. Single-strain preparations: These contain one type of bacteria. They are used for mild imbalances in the gut as a preventive measure for up to two weeks. Examples include Bifidumbacterin, Bifidine *(Bifidobacterium bifidum)*, Lactobacterin (*Lactobacillus plantarum* or *Lactobacillus fermentum)*, Biobakton (*L. acidophilus*), Gastropharm (*L. bulgaricum*), Colibacterin (*Escherichia coli*), and Enterol (*Saccharomyces boulardii*).

2. Antimicrobial agents: Mainly represented by *Bacillus* species such as *B. subtilis* and *B. licheniformis*, these probiotics have antimicrobial properties. They help eliminate microorganisms from the gut, similar to antibiotics. Once they start growing in the intestines, they produce enzymes that aid absorption and tissue healing. An engineered version called Subalin produces human interferon type α-2. It is used in veterinary medicine.

3. Multi-strain preparations: These consist of bacterial strains or additives that enhance their effects. Polycomponent probiotics aim to replicate the microbiota of the intestine, offering a comprehensive approach to maintaining microecological balance.

The unique characteristics of poly-component probiotics are as follows:

- They combine the beneficial properties of individual strains, enhancing their overall biological effectiveness.
- These probiotics can selectively colonize the intestine with the most suitable beneficial microorganisms for an individual's microbiota, eliminating the need for mandatory microbiological tests.

Examples of poly-component and combination probiotics are Bifilong (*B.bifidum* and *B.longum* biomass), Bifacid (*B.bifidum, L. acidophilus*), Bifikol (*B.bifidum* and *E. coli* M-17 biomass), Acylact (3 strains of *L.acidophilus*), Bifiform (*B.longum* and *Enterococcus faecium*), Bifilis (*B. bifidum* and lysozyme), Linex (*Lactobacillus, Streptococcus faecium, B.infantis*), Acipol (*L.acidophilus* and kefir polysaccharide (restorative), Acidophilus (biomass of *L.acidophilus*, *L.bulgaricum*, and *Streptococcus thermophilus*), Kipatsid (biomass of 3 strains of *L.acidophilus* and lysozyme). Here's a simplified and condensed version of the text:

Polyprobiotics are employed to treat first and second-degree intestinal dysbiosis over a period of 7 to 14 days. Furthermore, their poly-component composition includes biologically active supplements, also known as para-pharmaceuticals, which exhibit significant clinical and microbiological effects in various illnesses.

4. Immobilized bacteria: Immobilized bacteria are utilized in a detoxification method known as enterosorption, which is highly effective for metabolic and immune correction. This method can extend and enhance the effects of various drugs, including probiotics. The therapeutic impact of sorbents on organs and tissues is amplified when they are enriched with biologically active substances like antibiotics, antiseptics, enzymes, and biological preparations. These enriched sorbents are often referred to as such.

For instance, a "Polysorbovit" is an innovative enterosorbent created from specially modified high-quality citrus pectin. It incorporates a complex of vitamins E, A, sodium selenite, and lactic acid bacteria. "Rekitsen-RD" is a highly biologically active biosorbent derived from heat-treated wine yeast culture *Saccharomyces vini* T-8, cultivated on wheat bran. "Phytosorbovit" includes lactobacilli, herbal extracts, and activated charcoal in its composition.

Natural complexes with sorbent properties serve a dual purpose, functioning as both a substrate for intestinal microflora and a sorbent. These preparations consist of a mixture of various components, including vitamin and mineral complexes, combined in specific proportions. This results in a poly-component powder.

Advanced probiotics employ a technique in which live strains of probiotic bacteria with antagonistic properties are immobilized on small particles (ranging from 10 to 160 μm) of activated carbon obtained from various sources. Known examples include Bifidumbacterin Forte and Probifor, both utilizing "Carbolongum" activated carbon derived from peach pits and approved for medical purposes. Each particle of the sorbent contains between 20 and 180 Bifidobacteria cells. The only difference between these medications lies in their *bifidobacteria* content. Recommended dosage per unit.

Ecoflor is a probiotic created through the initial method of immobilizing *Bifidobacteria* cells on a sorbent with the inclusion of a protective medium. Another established drug is Bifidum-Biosorb, consisting of *Bifidobacterium* immobilized on the SUMS-1 sorbent surface.

Fundamentally, all sorbents help protect probiotic cells, ensuring they reach the colon, where their effectiveness is highest. The therapeutic use of immobilized probiotics in correcting various pathologies typically involves several stages: bacterial delivery, intestinal colonization, restoration of the intestinal environment, and detoxification (see Fig. 4.16).

In the first stage, bacteria are delivered to the large intestine. They adhere to the sorbent via interactions between the surface proteins of *Bifidobacterium* and *Lactobacillus* cells and the carrier material. This leads to the formation of probiotic microcolonies on the sorbent surface. These microcolonies display unique physicochemical properties. Notably, *Bifidobacteria* in such colonies demonstrate enhanced resistance to gastric acid, in contrast to free probiotic cells, which often lose more than 90% of their activity upon stomach passage.

The second stage is intestinal colonization. As the sorbent reaches the intestine, the bacteria gradually detach. Unlike conventional dry probiotic formulations, these immobilized bacteria are metabolically active and capable of more effectively adhering to the intestinal mucosa. The activated charcoal-bound microcolonies bind to the mucosal surface via chemical and electrostatic interactions. This rapid colonization helps rebalance gut microbiota, strengthens the gut's defensive functions, and supports mucosal healing. As a result, the proliferation of *Bifidobacteria* on the mucosal surface increases.

Once detached from the sorbent, probiotics initiate the processes of intestinal repair and detoxification. Sorbents play a key role in this phase by adsorbing and removing toxins, metabolic byproducts, pathogenic microbes, and allergens. This dual function—deliv-

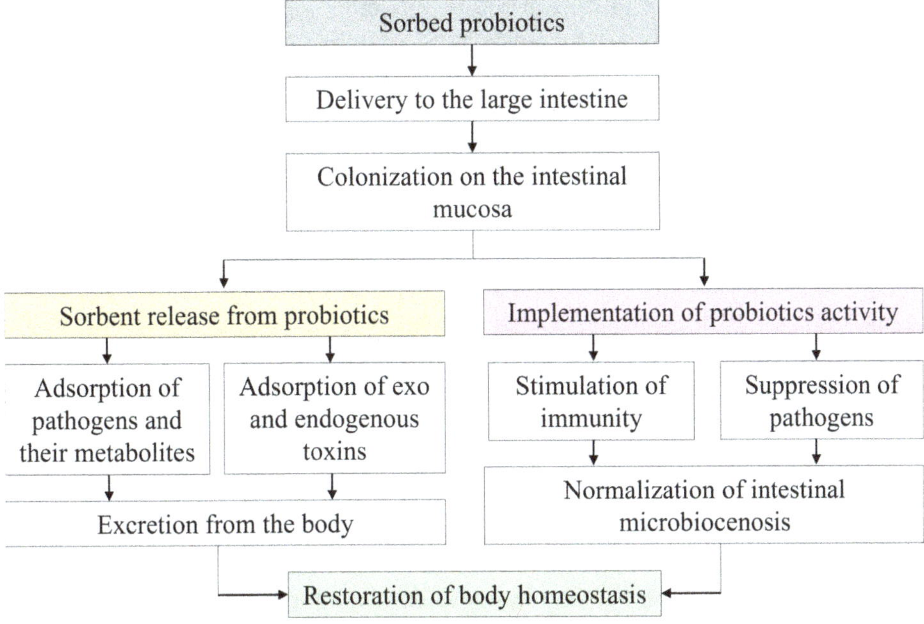

Fig. 4.16 Mechanisms of sorbed probiotics activity

ery and detoxification—not only facilitates intestinal healing but also reduces systemic toxic burden, alleviating stress on the liver.

Thus, sorbents in immobilized probiotics not only serve as carriers that protect and deliver viable cells into the intestine but also act as effective detoxifying agents. Their dual action—promoting microbial colonization and eliminating toxins—results in a potent therapeutic effect.

Probiotics are also included in active dietary supplements, known as biologically active substances (BAS). In Kazakhstan, the production, distribution, safety, and claimed benefits of BAS are regulated under national legislation concerning food safety and hygiene standards related to nutritional value. Although BAS are classified as food products, not medicinal drugs, their registration still requires scientific evidence of physiological benefits.

Dietary supplements containing probiotics or prebiotics may support gut microbiota balance but are not designed for disease treatment. Therefore, their packaging must clearly state that they are not pharmaceutical agents.

Probiotics that are part of immunobiological medicinal drugs (MIBDs) and possess official pharmaceutical approval are used in medical treatments. Unlike BAS, MIBDs are prescribed by healthcare professionals based on clinical needs. MIBDs may include bacterial strains not typically present in the healthy human intestinal microbiota, or even non-bacterial microorganisms like yeasts.

Microbial concentration in these products may differ significantly between dietary supplements and pharmaceutical formulations. Standardized thresholds include:

- $\geq 10^9$ microbial cells per unit in pure dry supplements
- $\geq 10^8$ cells per unit in mixed dry supplements
- $\geq 10^{10}$ cells per unit in concentrated liquid forms
- $\geq 10^7$ cells per unit in standard liquid forms

While specific intake durations for dietary supplements are often unspecified, general guidelines based on medical probiotics can be followed:

- For acute disturbances (e.g., intestinal infections, food poisoning, antibiotic-associated diarrhea): a 5–10 day course with high-dose or strongly antagonistic probiotics (especially *Lactobacilli* and *Bifidobacteria*) is recommended.
- In children with chronic dysbiosis, shorter courses (less than two weeks) are often insufficient for lasting benefits, regardless of the probiotic's potency.
- For moderate-strength probiotics or supplements, courses of 3–6 weeks are typical and generally effective.
- A 14-day break is advisable between repeated probiotic courses.

The beneficial effects of probiotics, when used without addressing the underlying cause of dysbiosis, typically last about three months. After this period, disturbances may return, possibly varying in type and severity. Regarding storage, lyophilized (dry) probiotics can be kept in a home refrigerator for up to one year. Imported varieties—particularly those from the United States—may have a shelf life of up to two years. In contrast, liquid forms are usually viable for only 1 to 3 months.

Probiotics are categorized as both supplements and health products. They are often referred to as either "drugs" or "products" containing live microorganisms. When consumed in adequate amounts, these microorganisms can confer health benefits by influencing the host's microbiota. Consuming probiotic-rich foods and supplements helps maintain a balanced microbial population. However, the survival rate of probiotics during gastrointestinal passage varies. While present in the gut, they temporarily influence the microbial composition during the course of consumption. Research shows that these bacteria can transiently colonize the lower intestine, but their numbers decline once intake stops. This phenomenon applies to all probiotics currently available on the market. Therefore, to sustain their health benefits, it is important to consume probiotic-containing foods regularly.

Probiotics derived from the normal human microflora are integral to the concept of healthy nutrition. They are widely used for the prevention and treatment of gastrointestinal disorders and associated disturbances in the digestive, immune, and endocrine systems. A growing area of research is the development of fermented dairy products with probiotic properties. It is projected that by the early twenty-first century, probiotic-enriched dairy products could capture up to half of the pharmaceutical market aimed at preventing and treating various human diseases.

Milk and dairy products containing probiotic cultures, such as *Lactobacilli* and *Bifidobacteria*, are among the most popular functional foods worldwide. These dairy-based probiotics account for approximately 65% of the global functional food market.

Milk and its derivatives—including yogurt and cheese—serve as ideal carriers for delivering beneficial microbes to the human digestive tract, offering a supportive environment for probiotic viability. The appeal of dairy-based probiotic products lies in their accessibility, affordability, nutrient-rich composition, and ease of processing.

Fermented dairy products incorporating "bio" cultures are widely used in the food industry. Common probiotic species include, but are not limited to: *Lactobacillus acidophilus, L. casei, L. bulgaricus, L. plantarum, L. salivarius, L. rhamnosus, L. reuteri, Bifidobacterium bifidum, B. longum, B. infantis,* and *Streptococcus thermophilus.* These lactic acid bacteria are well known for their role in the production of yogurt and other fermented dairy products.

For individuals who cannot tolerate dairy-derived probiotics, a variety of non-dairy options are available. These alternatives are free of casein and lactose. Examples of dairy-free fermented foods include kombucha, sauerkraut, kimchi, sour pickles, non-dairy kefir, and plant-based yogurts. While dairy products are commonly used to deliver probiotics, non-dairy options offer important benefits for people with dairy allergies or intolerances. However, it's important to note that probiotic strains in such foods may be sensitive to heat and prone to degradation. Therefore, to maximize their health benefits, these products should be consumed shortly after purchase.

4.2.3 Prebiotics and Synbiotics

A balanced microbiome is essential for maintaining overall health. When developing both dairy and non-dairy products that contain beneficial bacteria, it is important to consider the role of *prebiotics* in creating an optimal environment for *Bifidobacteria* and *Lactobacilli.* Prebiotics are not living organisms but are dietary substances that selectively promote the growth or metabolic activity of beneficial gut microbes. Unlike probiotics, which introduce new microbial strains into the body, prebiotics serve as nourishment for the existing beneficial microbiota. By enhancing the growth and activity of these bacteria, prebiotics support overall health and well-being.

Essentially, prebiotics are non-digestible components in food that reach the colon intact and serve as substrates for specific beneficial bacteria, ultimately leading to improved health outcomes.

Prebiotics are a type of fiber that is not digested in the small intestine but instead reaches the colon, where it is selectively fermented by certain beneficial microbes. While many fibers can be fermented by various bacteria, prebiotic fibers are specifically utilized by health-promoting bacteria such as *Bifidobacteria* and *Lactobacilli.* This unique fermentation process is referred to as the *prebiotic effect.*

During fermentation, these bacteria produce gases and short-chain fatty acids (SCFAs), which serve not only as an energy source for the microbes themselves but also benefit the epithelial cells lining the colon. This process, known as *selective stimulation,* gives these

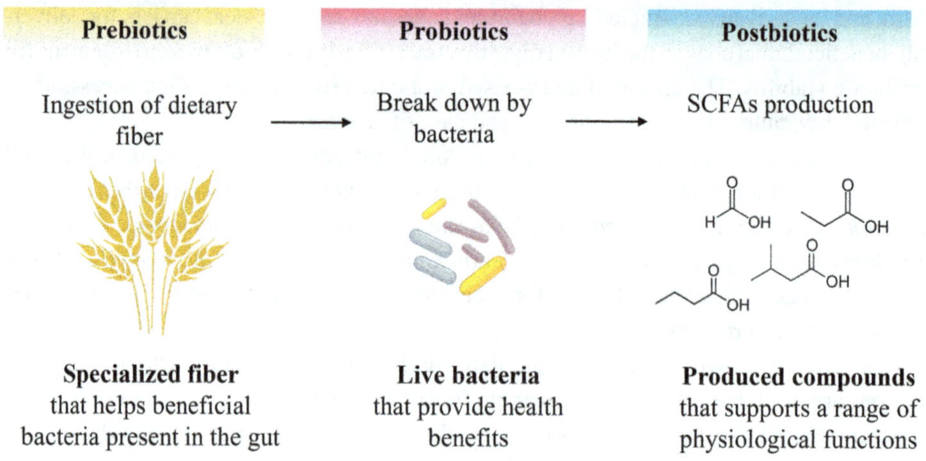

Fig. 4.17 Prebiotics, probiotics and postbiotics

microbes a competitive advantage and supports a balanced intestinal microflora, contributing to numerous health benefits (Fig. 4.17).

Prebiotics primarily include oligosaccharides consisting of 2–10 carbohydrate units. These oligosugars can occur naturally or be synthesized. Examples of commonly recognized prebiotics include:

- *Disaccharides*, such as lactulose and lactitol
- *Oligosaccharides*, including fructooligosaccharides (FOS), transgalactosylated oligo-saccharides (TOS), and soybean oligosaccharides
- *Polysaccharides*, including inulin and resistant starch

Among these, lactulose, FOS, galactooligosaccharides (GOS), and cyclodextrins are the most widely studied and accepted as effective prebiotic agents.

Prebiotics can be sourced from both natural and artificial origins:

- *Natural prebiotics*: These occur naturally in certain foods. Examples include inulin (from chicory root), acacia gum, and GOS (found in soybeans and legumes).
- *Artificial prebiotics*: These are synthesized through chemical or biotechnological processes. Examples include lactulose and polydextrose.

Prebiotic substances are typically derived from food-based raw materials. They can be either extracted directly from their natural sources—for instance, GOS from soybeans—or produced biotechnologically using specific carbohydrate-converting enzymes.

Lactulose, a well-known artificial prebiotic, is synthesized by modifying lactose (milk sugar), where the glucose residue is converted to fructose via isomerization. This structural change alters its metabolic behavior: unlike lactose, which is absorbed in the small intestine, lactulose passes undigested into the colon. There, it becomes a substrate for beneficial gut bacteria, promoting their growth and activity and improving the microbiota composition and colonic microecology.

Lactulose offers multiple health-promoting effects and holds promise as a therapeutic agent in gastrointestinal and metabolic health. Its key benefits include:
- Stimulation of the growth and activity of *Bifidobacteria* and *Lactobacilli*
- Restoration of disrupted intestinal microflora
- Mucolytic and tissue-reparative effects
- Maintenance of an optimal colonic pH
- Increased fecal mass and improved intestinal motility
- Reduction of gas production
- Enhanced synthesis of B-group vitamins and vitamin K
- Antibacterial effects through nonspecific immune mechanisms
- Antimutagenic and potential anticancer effects

Natural indigestible oligosaccharides (NIOs) are carbohydrates found in plants, algae, bacteria, fungi, yeast, breast milk, and honey. They are produced via enzymatic transferase systems that convert disaccharides such as sucrose, lactose, and raffinose into NIOs. When metabolized, NIOs generate acids that lower intestinal pH, creating an environment that inhibits pathogen growth. Notably, these pathogens are unable to break down NIOs, which supports their elimination from the body.

Common gut bacteria such as *Lactobacillus* utilize carbohydrate compounds and dietary fiber (DF), which may originate from various food industry byproducts like bran, cereal husks, fruit pulp, sugar beet pulp, oilcake, and potato residue. While DF supports the growth of gut microbes beyond just *Bifidobacteria*, its regular consumption improves gut motility and waste elimination.

Some pharmacotherapeutic agents also promote the development of beneficial intestinal microbiota. Examples include certain oligosaccharides, calcium pantothenate, and p-aminomethylbenzoic acid (PAMBA).

Prebiotics and probiotics are not mutually exclusive in use. Prebiotics enhance the gut environment but may have a limited effect in the absence of sufficient beneficial microbes. In such cases, *synbiotics*—a combination of prebiotics and probiotics—offer a synergistic benefit by supporting both the growth and survival of introduced and resident bacteria. This approach is known as *supplementation* and is associated with enhanced gut health.

Synbiotics may include not only probiotics and prebiotics, but also dietary fibers, enzymes, trace elements, and herbal extracts. Prebiotics improve the stability and survival of probiotics both in the product and in the gastrointestinal tract. As more live bacteria reach the colon, internal and external microbial growth is enhanced, further activating metabolic functions and strengthening the microbiota's defense against harmful microorganisms.

Examples of Synbiotic Products:
These are formulations combining viable probiotic bacteria with substrates that stimulate the growth and development of the normal microflora:
1. Normophlorines: Contain *Lactobacilli* and *Bifidobacteria*, along with vitamins, antioxidants, and trace elements.

2. Euflorins (B and L): Bio-complexes made up of:
   - Enzymatic protein hydrolyzates (amino acids and low molecular weight peptides of natural origin).
   - *Bifidobacteria* and *Lactobacilli* (titers of 10^8–10^{10} CFU/mL).
   - Metabolites of these bacteria, including vitamins, enzymes, organic acids, and antimicrobial compounds.
3. Biovestin-Lakto: Contains *B. bifidum*, *B. adolescentis*, *L. plantarum*, and bifidogenic factors.
4. Maltidifilus: A combination of *B. bifidum*, *L. acidophilus*, *L. bulgaricum*, and maltodextrin.
5. Bifido-Bak: A mixture of *Lactobacilli*, *Bifidobacteria*, and Jerusalem artichoke extract.
6. Bifidumbacterins—Multi 1, 2, and 3: Contain *B. bifidum*, *B. longum*, and *B. adolescentis*, tailored to different age groups, and supplemented with vitamins and antioxidants.
7. Bifiform:
   - Bifiform Adult: Live cultures of B. longum ATCC 15707 and Enterococcus faecium SF68.
   - *Bifiform Baby*: *B. lactis* Bb-12 and *L. rhamnosus* GG, each at 10^9 CFU per tablet. These supplements often include vitamins B1 and B6, xylitol, isomaltose, FOS, microcrystalline cellulose, and stearic acid.
8. Soya-Laktum/Soya-Bifidum: Contain probiotic bacteria and hydrolyzed soy milk.
9. Laminolact: Contains *Enterococcus faecium* 3, amino acids, pectins, and seaweed extracts.
10. Lactofeiber: Combines *Lactobacillus coagulans* with soluble and insoluble dietary fibers from cereals, beets, and citrus.
11. Lactofiltrum: Includes lactulose (as a prebiotic) and hydrolytic lignin, which acts as a sorbent.
12. LactoSpore: A formulation with *L. sporogenes*, natural lactase, FOS, and inulin.

The development of functional foods and supplements incorporating probiotics and prebiotics is a promising and active area of research. Their combined application—as synbiotics—is increasingly recognized as an optimal strategy for maintaining and restoring intestinal microbiota balance. Prebiotics enhance the stability and survivability of probiotics, resulting in better microbial colonization, metabolic activation, and suppression of harmful organisms, ultimately leading to improved intestinal health.

4.2.4   Probiotic Preparation Technology

The development of probiotic products begins with in-depth research on the strains to be used. These strains are typically isolated from healthy human bodies and are identified at the species level. Selection criteria for lactic acid bacteria intended for probiotic production are based on their documented beneficial properties.

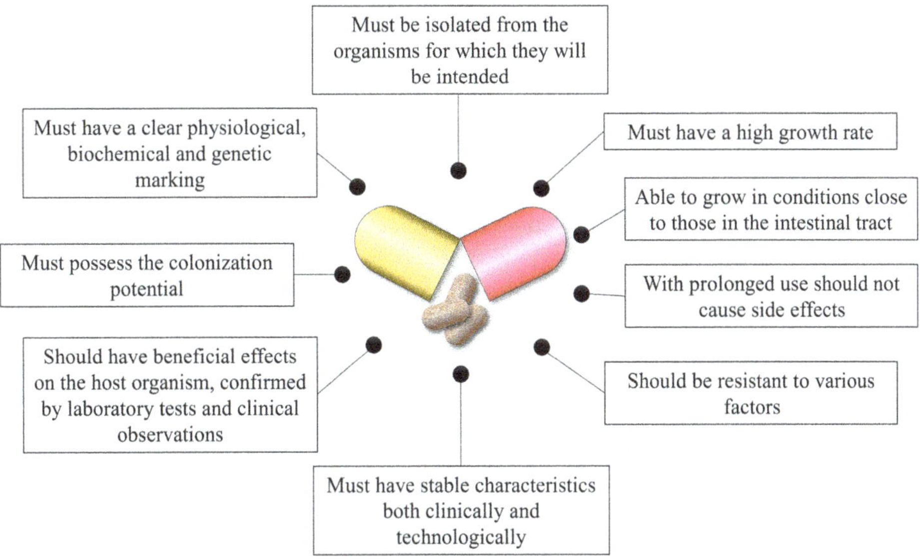

Fig. 4.18 The basic requirements for microorganisms-probiotics

While specific criteria may vary depending on the purpose of the study, several essential parameters are always considered. These include the strain's ability to produce acids, antagonistic activity against harmful microorganisms, resistance to digestive enzymes and antibiotics, and its colonization ability, also known as cytoadhesion (Fig. 4.18).

Antagonistic activity is particularly important in the selection of strains, as it reflects their ability to support and restore healthy microflora. As knowledge about the biological properties of lactic acid bacteria expands, including the widespread presence of bacteriocin-producing strains, new evaluation methods continue to emerge. Therefore, it is increasingly important to consider both traditional properties and bacteriocinogenic potential when selecting promising *Lactobacillus* strains—assessed both in vitro and in vivo.

Recent studies have broadened the scope of evaluation criteria for potential probiotics. These now include the strain's ability to modulate the immune system. Additionally, some experts suggest assessing lysozyme production, a nonspecific defense factor, as a selection criterion.

Since antibiotics can disrupt microbial balance, probiotic strains should ideally be resistant to them. However, this resistance must be carefully managed. While antibiotic resistance may be innate or acquired during antibiotic therapy—either through mutation or plasmid acquisition—the latter poses a risk. Resistance plasmids can be transferred between bacteria, potentially turning the gut into a reservoir for antibiotic-resistant organisms. Although no direct cases of plasmid transfer from probiotic strains have been documented, it is safer to select strains with chromosomally encoded resistance.

Only a few external microorganisms can survive the hostile environment of the upper digestive tract and reach the colon. To be effective, probiotics must endure multiple challenges: digestive enzymes in saliva and stomach acid, high concentrations of bile salts, and immune barriers in the small intestine. Thus, selected probiotic strains must show resilience to all of these conditions.

Furthermore, probiotic strains should exhibit *symbiotic properties*—working synergistically to amplify their overall effect during treatment. Strains should also be selected for their *technical and biological suitability*: their ability to grow under production conditions, maintain consistent properties across processing cycles, and remain viable during manufacturing.

To be considered for industrial-scale production, probiotic strains must meet strict requirements. They must be *safe, processable*, and able to withstand *technological stress* during manufacturing. Their *viability and stability* throughout production, storage, and application—whether frozen, dried, or incorporated into food products—are critical.

Standard preparation methods such as *freezing and spray drying* have proven effective for many strains, yet probiotics remain sensitive to environmental factors like oxygen and acidity, commonly encountered during production, storage, or in specific food matrices. Moreover, many probiotic strains have demanding nutritional needs, requiring expensive growth media and added factors to support robust development.

Probiotics are most commonly delivered through *fermented dairy products*, either added after fermentation or used as part of the fermentation process. Their ability to grow in milk and coexist with standard starter cultures is essential. Selecting strains that can be included in mixed starter cultures is a key goal in probiotic dairy production. Factors like fermentation time, product consistency, taste, sugar tolerance, and post-acidification behavior are all important considerations. The addition of probiotics should not negatively affect the taste or texture of the product. Even if a strain has significant health benefits, technological limitations may hinder its commercial application. Thus, the *technological suitability* of probiotics must be evaluated for scalability and practicality.

Like all microorganisms used in food or medicine, probiotics must undergo rigorous *safety assessments*. A recent WHO working group outlined new safety evaluation criteria, even for species traditionally considered safe. These include:
- The potential to transfer antibiotic resistance genes
- Production of harmful or undesirable metabolic byproducts
- Toxin or hemolysin production
- Risks to immunocompromised individuals

For *new probiotic strains*, epidemiological monitoring is recommended as an additional safety measure. Given the increasing prevalence of food allergies—particularly to milk, a common base for probiotics—it is also advised to *reduce or clearly label allergenic components* in probiotic formulations.

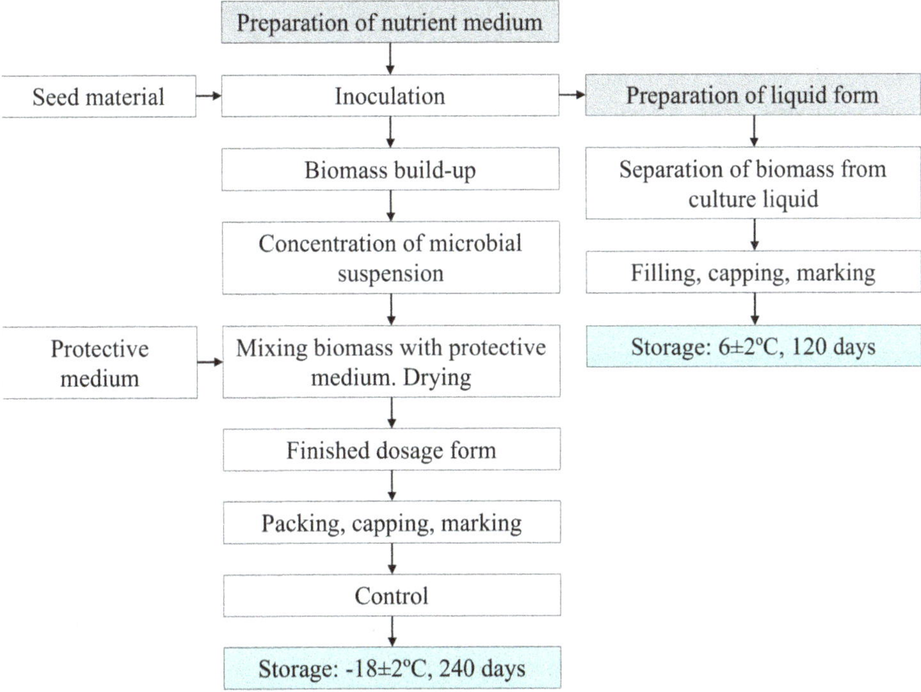

Fig. 4.19 General technological scheme of probiotics production

The development of probiotics generally follows a typical *biotechnological production process*, starting from nutrient selection and cultivation, through processing and formulation, and ending with packaging of the final product (see Fig. 4.19).

Effective probiotics require a robust raw material base and carefully optimized nutrient formulations to cultivate high-density, viable cell biomass. Microbial cultures used in probiotic production have complex nutritional needs essential for constructing cellular components such as nucleic acids, polysaccharides, and amino sugars. For instance, lactic acid bacteria depend on externally supplied organic nitrogen sources, as well as a range of vitamins. The presence of yeast or corn extracts and other components in the nutrient medium significantly influences their growth. In addition, the medium must contain essential macro- and microelements such as sodium, magnesium, and manganese. Lactose or glucose typically serves as the primary energy source.

The complexity of nutrient media used for cultivating lactic acid bacteria is justified by these requirements. Efficient growth has been achieved using complex media containing plant decoctions, meat and yeast extracts, and protein hydrolysates. Given the central role of nutrient media in determining product quality and production efficiency, research continues into the development of new media and the refinement of established ones—such as milk-yeast medium, hydrolyzed milk medium, and casein-yeast medium. Biogenic stimulants are also being explored as medium additives. For example, the growth of *Lactobacillus acidophilus* is stimulated by a combination of lactic and succinic acids. Studies have also shown that yeast cells can protect *Bifidobacteria*, which has led to the inclusion of yeast autolysates and extracts in some probiotic formulations, further promoting lactic acid bacterial growth.

The cultivation process begins with the initial growth of bacterial cultures in test tubes on nutrient media at 38 °C for 24 hours. These cultures are then transferred to 100 ml flasks, and subsequently scaled up to 1000 ml flasks, where they are incubated for another 24 hours under the same temperature conditions. This mature culture is then introduced into a fermenter at an inoculation rate of 5–10%.

To accumulate microbial biomass and ensure cell viability during subsequent processing and storage, controlled cultivation techniques are employed. These involve either periodic or continuous biomass accumulation under regulated conditions—maintaining stable pH, oxygen levels, redox potential, and substrate concentration. In such systems, microbial growth is usually limited not by nutrient depletion but by the accumulation of inhibitory metabolic byproducts. The effectiveness of cultivation depends on the quality of the nutrient medium, the seed culture, and the technological capabilities of the equipment used.

In bioreactors, fermentation typically lasts 10–15 hours and is halted at the end of the logarithmic growth phase, just as the stationary phase begins—when the microorganisms display maximum biological activity. After cultivation, the biomass is evaluated, with the goal of achieving a bacterial concentration of at least 108 cells/mL and ensuring the absence of contaminating microflora.

The next critical step in probiotic production is *culture stabilization*, which ensures that cells remain viable throughout the product's shelf life. Freeze drying (lyophilization) is the most widely used method for this purpose. It allows cultures to retain their properties and high cell viability for several years when stored at appropriate temperatures and is now the standard method for probiotic stabilization (Fig. 4.20).

The lyophilization process begins with freezing, which may damage cells due to ice crystal formation. Prior to freezing, a drying step called pre-freezing is carried out, during which some cell death can occur. Additional drying of the frozen biomass can also lead to further viability loss, depending on drying temperature and the degree of water removal.

To mitigate such effects, *protective agents*, also known as *cryoprotectants* or *geroprotectors*, are added to preserve cell membrane integrity. These protectants should be compatible with the microbial cells and ideally resemble bacterial cell surface components. Typical protective media include:

1. *Natural and biological derivatives*, such as whey, milk, broth, peptone, and gelatin
2. *Carbohydrates*, like glucose, sucrose, lactose, trehalose, and dextrins, which protect cells during drying and at elevated temperatures
3. *Synthetic compounds*, such as polyvinylpyrrolidone and hydroxyethyl starch

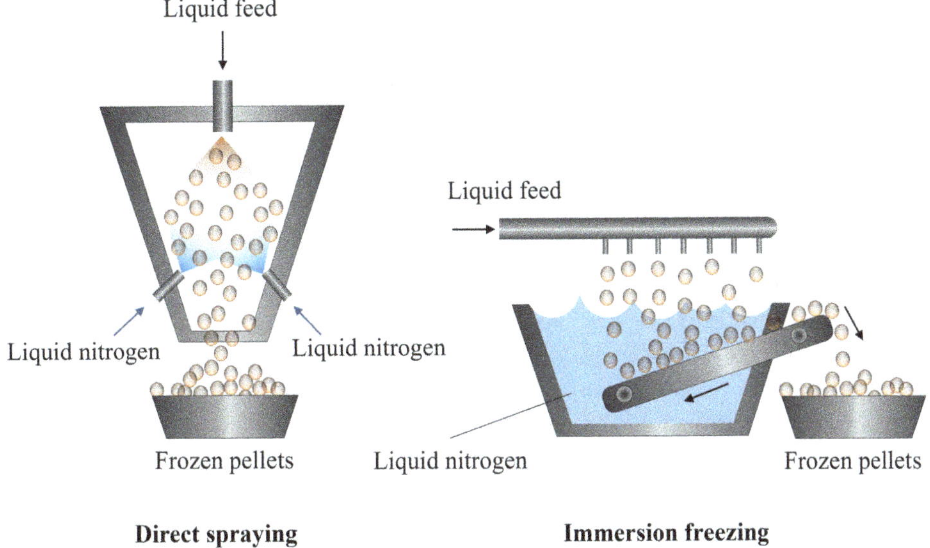

Fig. 4.20 Common examples of freeze-drying

Such substances are vital for preserving biological activity during drying and freezing.

In the production of probiotics like *Lactobacterin* and *Acylact*, protective media such as sucrose-gelatin and sucrose-gelatin-lactic acid blends are commonly used. Following fermentation, biomass is concentrated via centrifugal separation or membrane filtration. This concentrated biomass is mixed with the drying medium using mechanical stirrers, then transferred into sterile containers and moved to the filling station where it is dispensed into ampoules. These ampoules, enclosed in sterile gauze-wrapped cassettes and sealed with metal lids, are then frozen at $-40 \pm 2\ °C$ for 48 ± 2 hours.

The drying process is conducted at a pressure of 13.3 ± 0.13 Pa and a temperature of $30 \pm 4\ °C$ for 24 ± 2 hours. After this, the temperature is gradually increased to $+28 \pm 2\ °C$ over 16 ± 2 hours and maintained for 20 ± 2 hours. Once drying is complete, ampoules are sealed under a nitrogen atmosphere within 72 hours using a specialized sealing machine.

Despite its effectiveness, *lyophilization* is expensive, time-consuming, and energy-intensive. This has prompted efforts to explore alternative drying methods. One such method is *spray drying*, which is also used in probiotic production. However, due to significant losses in cell viability, spray drying is not yet widely used for preparing probiotic formulations. Nevertheless, research aimed at improving this method continues.

There are two categories of traditional probiotics, which are classified based on how they are produced.

1. *Therapeutic probiotics (Group 1):*
   - These probiotics are made by freeze-drying bacteria.
   - They come in forms like powder, tablets, capsules, or suppositories.
   - The good thing about them is that they have a long shelf life of up to 2 years and can withstand short-term temperature changes without being affected.
   - However, there is a downside. The lyophilization process used in their production can reduce their qualities and cause a delayed activation period of 8–10 hours for the bacteria to become active.
   - Additionally, lyophilization may lead to the loss of receptors (adhesins) in bacterial cells, which affects their ability to colonize the intestinal biotope for an extended period.
2. *Liquid probiotics (Group 2. biologically active additives—BAA):*
   - These probiotics contain living cells and exist in a suspended state.
   - When consumed, they can start colonizing the tract within 2 hours.
   - However, compared to therapeutic probiotics, their storage duration is shorter and typically ranges from 1 to 1.5 months.
   - There is also a risk of contamination by other microflora when it comes to liquid probiotics.

Each group has its advantages and limitations, making them suitable for different preferences and purposes when it comes to consuming probiotics.

Probiotic formulations that are considered innovative often make use of microencapsulated probiotics. The materials used for microencapsulation are carefully selected to ensure they are non-cytotoxic, produce biodegradation products, and do not require the use of organic solvents during the preparation of microcapsules. These microencapsulation matrices typically consist of polymer networks that are cross-linked either covalently or ionically. In some cases, non-cross-linked polymer granules, such as those produced through spray drying, are also utilized as materials for microencapsulation.

Microcapsules come in single or multiple layers, with both outer and inner shells accommodating one or more active substances. Each microcapsule can contain over 10^3 bacteria. These formulations offer significant benefits compared to traditional forms. Specifically, the acid-resistant microcapsule shell enables probiotic bacteria to survive the stomach's acidic environment, and it also controls the gradual release of the active components, prolonging the preparation's effects, optimizing resource usage, and enhancing efficiency.

The final product undergoes assessment based on various parameters, including physical characteristics, visual appearance, absence of foreign microorganisms, the quantity of live bacteria in a single dose, acidity levels, residual moisture content, pH, and seal integrity.

Traditionally, live cell counts in probiotics involve inoculation into a nutrient medium and counting colony-forming units (CFU) after 24–72 hours. Recent advancements offer quicker ways to assess the specific activity of these preparations. Laser microscopy enables visual distinction between live and dead microorganisms in probiotics within 30 minutes. Flow cytometry is a fast, sensitive method for microorganism quantification and physiological analysis. Potentiometry is used to determine active and titrated acidity, while antagonistic activity against test strains is evaluated. Furthermore, probiotic preparations must be safe with a moisture content not exceeding 4%.

Currently, we are gaining an understanding of the advantages of having a mutually beneficial relationship with helpful microorganisms. It's becoming increasingly clear how crucial it is to preserve these relationships. Utilizing knowledge in human microecology and innovative bacteriotherapy methods plays a significant role in improving public health. We can expect to see probiotic and prebiotic treatments being used frequently for various diseases, with more precise prescriptions and extensive research into how probiotics interact with the body's microorganisms. Additionally, there will be advancements in developing strains, forms, and combinations tailored to different age groups of patients. Thanks to production techniques, we can ensure that probiotic products are preserved effectively while maintaining their maximum efficacy.

4.2.5 Brainstorming

1. Clarify the concept of normal microflora. Describe the characteristics of the microbial cenosis of the macroorganism.
2. Discuss the functions of the gastrointestinal tract's normal microflora in terms of symbiotic microorganisms' positive role.
3. Explain how the molecules produced by the microflora participate in the activity regulation of the macroorganism systems.
4. Describe the mechanisms for ensuring colonization resistance.
5. What are the causes of dysbiosis?
6. Enumerate the basic requirements for microorganisms-probiotics.
7. Describe the modern generations of probiotics.
8. Identify the categories of medicinal forms of probiotic drugs.
9. Give examples of functional probiotic foods.
10. What are the most common ways of delivering probiotics to the intestines?
11. Explain the high therapeutic efficacy of sorbed probiotics.
12. Discuss the advantages and mechanisms of the action of sorbed probiotics.
13. What are microencapsulated probiotics? How are they created?
14. What are probiotics, prebiotics, and synbiotics?
15. Describe the prebiotics' mechanisms of action.
16. Describe the technological aspects of obtaining probiotics and the main production stages.

Take-Home Messages:

- The human microbiota is a complex ecosystem essential for health. The gut microbiota contains trillions of microbes—mostly anaerobic—that support digestion, metabolism, immunity, and protection against pathogens.
- Normal intestinal flora performs protective, structural, and metabolic functions. It blocks pathogen adhesion (competitive exclusion), produces antimicrobial substances (e.g., bacteriocins, organic acids), and forms biofilms that resist colonization by harmful microbes (colonization resistance). Intestinal bacteria contribute to metabolism and immunity. They ferment dietary fibers to produce SCFAs, synthesize essential vitamins (e.g., K and B12), transform bile acids, support immune regulation, and inhibit carcinogenic processes. Dysbiosis disrupts microbiota balance and promotes disease. Factors such as antibiotics, poor diet, and illness can lead to microbial imbalances (e.g., SIBOS, dysbacteriosis), contributing to inflammation, infections, metabolic disorders, and chronic disease. Restoring balance often requires targeted microbial supplementation.
- Probiotics, prebiotics, and synbiotics offer tools for restoring gut health. Probiotics are live beneficial microorganisms that improve or restore gut flora. Prebiotics are non-digestible food components that selectively feed beneficial microbes. Synbiotics combine both, enhancing the survival and activity of probiotics in the gut.
- Probiotics come in various forms, from single strains to complex combinations. They include monoculture preparations, genetically engineered strains, multi-component systems, and immobilized formulations on sorbents for improved delivery, colonization, and detoxification.
- Probiotics are widely incorporated into fermented dairy products, functional beverages, and capsules. Regulation distinguishes food supplements from medicinal probiotic drugs, with the latter requiring clinical validation. Prebiotics such as FOS, GOS, lactulose, and inulin selectively nourish beneficial gut bacteria.
- Synbiotic formulations are optimized combinations of live probiotics and targeted prebiotics. These combinations improve probiotic survival and colonization, leading to more robust and lasting effects. Products include multi-strain blends with vitamins, enzymes, and bifidogenic nutrients.
- Developing effective probiotics requires careful strain selection and production technology. Ideal strains exhibit strong acid and bile tolerance, antimicrobial activity, and colonization ability. They must be safe, genetically stable, and technologically suitable for industrial-scale production.
- Freeze-drying (lyophilization) is the gold standard for probiotic stabilization.

4.3   Biopesticides

Protecting plants from pests and diseases is extremely important for improving productivity and ensuring high-quality products. However, the use of chemical pesticides, even though they are highly effective, has some downsides:

- Emerging resistance: Pests and diseases are becoming resistant to pesticides over time.
- Negative impact on beneficial species: Pesticides can harm insects and microorganisms that're beneficial for plant growth.
- Consequences: Pesticides persist in the environment, accumulate in plants, and contaminate soil, water, and air.

Therefore, there is a growing emphasis on utilizing methods of control known as biopesticides. Biopesticides are derived from sources such as microorganisms, plants, animals, and minerals. They can be categorized into three groups:

- Microbiological products: These products are made from microorganisms, like bacteria, fungi, viruses, or their byproducts.
- Natural extracts: These products are based on biologically active substances found in plants and other natural materials.
- Phenols: natural compounds that influence the behavior of organisms, often used in traps and baits to control insect populations.

Biopesticides have benefits. They specifically target organisms while having lower toxicity towards non-target species. Additionally, biopesticides leave residue in the environment due to their biodegradability as microbial pesticides.

Overall, biopesticides are recognized as friendly and safe methods for protecting plants. The use of microorganisms as biopesticides is a new but promising area in biotechnology, with notable advancements such as employing spore-forming bacteria, fungi, and viruses for developing microbiological insecticides.

Entomopathogenic bacteria and fungi are typically cultivated on artificial nutrient media. A specialized technique has been devised to mass-produce viral drugs by accumulating viruses within insects for infection purposes. Microbiological preparations generally consist of microorganisms in various forms, such as spores, conidia, polyhedra, and other components. These microbial pathogens are typically available as moist powders, pastes, and, less commonly, granules, spore suspensions, and crystals.

4.3.1   Obtaining Entomopathogenic Bacterial Products

The commercial production of pathogens mainly consists of bacterial products. These bacteria rely on toxins to cause diseases, which makes it unlikely for insects to develop resistance against these biopreparations. Researchers have documented over 90 species of insect-infecting bacteria, with the majority belonging to families such as

Pseudomonadaceae, *Enterobacteriaceae*, *Lactobacillaceae*, *Micrococcaceae*, and *Bacillaceae*. Among these families, the genus Bacillus is the most commonly used in industrial strains. In particular, Bacillus thuringiensis is responsible for more than 90% of microbial preparations and effectively controls a wide range of pests, like caterpillars, mosquitoes, and midges.

Different strains of *Bacillus thuringiensis* have ways of harming insects. Apart from forming spores that cause septicemia when consumed by insects, these bacteria also produce a variety of exotoxins with unique effects.

- α Exotoxin (Phospholipase C): This toxin, generated by growing cells, is believed to be lethal to insects because it disrupts essential phospholipids in their tissues.
- β Exotoxin (nucleotide structure); Composed of adenine, ribose, and phosphorus, this toxin likely functions as a nucleotide complex with ribose and glucose. It binds to allosteric acid and potentially inhibits RNA synthesis in insects.
- γ Exotoxin: The structure and activity of this toxin are not well understood. It seems to be associated with phospholipids.
- Crystalline σ Endotoxin: non-toxic. These crystals are produced along with spores. However, when ingested in an insect's tract, they become activated by alkaline proteases, which lead to the destruction of the insect. *Bacillus thuringiensis* preparations primarily act as toxins, causing paralysis in the intestines, cessation of feeding, overall paralysis, and eventual death of the insect (refer to Fig. 4.21).

The crystals produced by *Bacillus thuringiensis* exhibit a range of effects on different insect species. An insect's susceptibility to the toxin depends on the presence of specific protease enzymes in its digestive system. Insects lacking the appropriate proteases are less affected, which explains the selectivity of this toxin.

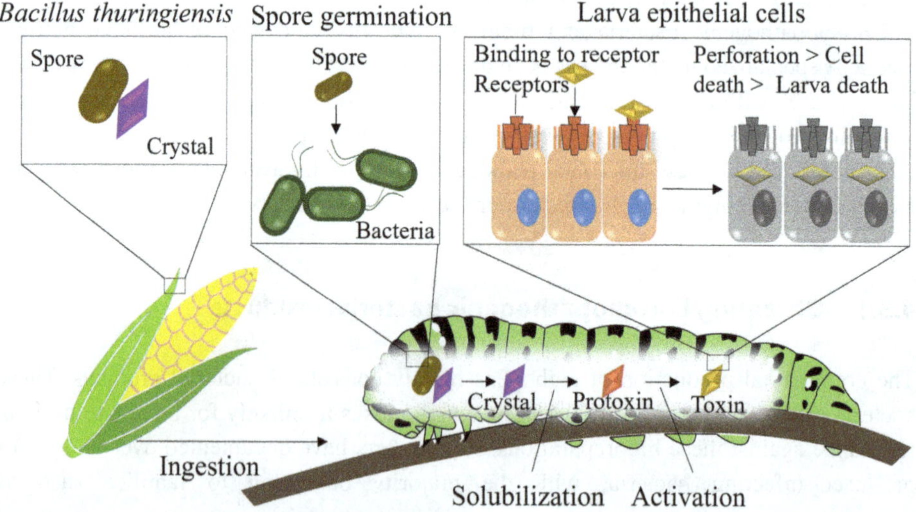

Fig. 4.21 Bacillus thuringiensis mode of action

The *Bacillus thuringiensis* group has proven highly effective against approximately 400 insect species, including pests that threaten crops in fields, forests, orchards, and vineyards. These microbial agents have demonstrated particular success in controlling pests that feed on leafy vegetation.

4.3.2   Production Technology of Entomopathogenic Bacteria

The industrial production of entomopathogenic bacteria (EB) is typically carried out via submerged cultivation. The main goals are to achieve a high cell concentration in the culture fluid and to accumulate the insecticidal toxins effectively. Industrial EB strains must meet specific requirements, such as belonging to a defined serotype, exhibiting high virulence, thriving in industrial nutrient media, and resisting phage infections.

The production process includes all standard steps of biotechnological manufacturing (see Fig. 4.22).

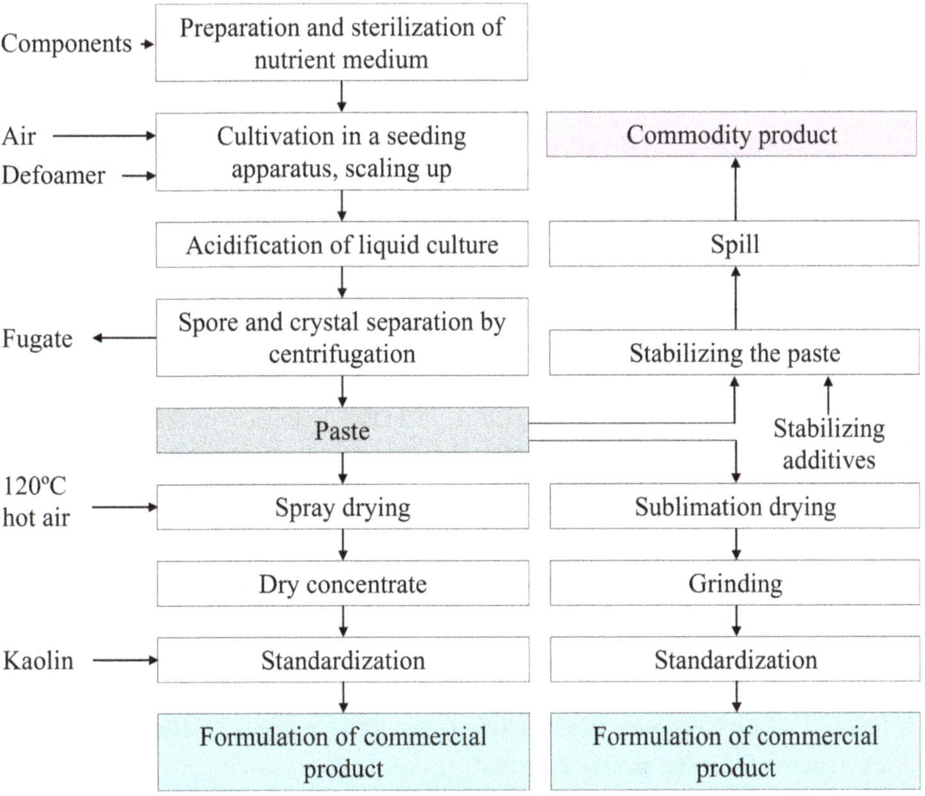

Fig. 4.22 General scheme for obtaining bacterial entomopathogenic products

The production process for bacteria (EB) involves several important stages:

- *Inoculum preparation*: The inoculum, in the form of spores, is grown in a nutrient solution that is also used for industrial cultivation. This solution typically contains fodder yeast (2.3%), corn flour (1.5%), and sperm whale fat (1%) as a defoaming agent. For the inoculation machine, a more concentrated medium is used to prevent foam formation.

- *Industrial fermentation*: The pH of the culture medium is not adjusted before fermentation and is initially around 6.3. During the fermentation process, the pH naturally rises to 8.0–8.5, which may lead to fragmentation of the crystals. To prevent this, the culture fluid is acidified to a pH of 6.0–6.2 before further processing. Fermentation is considered complete when the sporulation level reaches 95% and the spore concentration reaches 10^9 spores per milliliter. The temperature is maintained at 28 °C, and the process typically lasts 35–40 hours.

- *Concentration of culture fluid*: After fermentation, the culture fluid is separated to obtain a paste with approximately 85% moisture content, yielding about 100 kg of biomass per cubic meter of culture liquid. The paste is stirred for 30 minutes to evenly distribute the spores and crystals. Samples are taken to determine the titer, humidity, virulence, and phage contamination. For long-term stabilization, *carboxymethyl cellulose (CMC)* is added due to its strong sorption properties. This creates a three-dimensional matrix that evenly distributes the preservative throughout the paste.

- *Drying*: If a powdered product is desired, the paste is dried in a *spray dryer* until it reaches a residual moisture content of 10% and a titer of 100–150 billion spores per gram. The dried product is then stabilized with *kaolin*. Alternatively, a stabilized liquid paste can be produced by mixing with CMC. This final product is a light-gray, odorless suspension with uniform consistency that resists spoilage. The dry form can contain up to 30 billion spores per gram.

- *Standardization and packaging*: The finished product undergoes standardization and is then packaged for distribution. Various commercial products are developed using different strains of *Bacillus thuringiensis*, including Entobacterin, Insectin, Alestin, Exotoxin, Toxobacterin, Dendrobacillin, and Bitoxybacillin (see Table 4.6 for details). Recent developments for enhancing strains and products:

- Genetically modified *E. coli* and *B. subtilis* have been engineered to express *B. thuringiensis* toxins. Mycogen uses this approach to produce biopesticides for potatoes, eggplants, and tomatoes, significantly reducing production costs.

- Hybrid strains combining protein genes from multiple *B. thuringiensis* varieties have been created to expand insect host range. Sandoz's "Javelin" is effective against gypsy moths. Another new strain targets the Colorado potato beetle.

- Tobacco plants have been genetically engineered to produce insecticidal proteins, resulting in fully insect-resistant plants.

- Monsanto and Agrocetus are testing cotton varieties that inherit *B. thuringiensis* genes, ensuring insect resistance in subsequent generations.

Table 4.6 Entomopathogenic products based on *Bacillus thuringiensis*

Name	Bacteria strain	Commercial form	Spectrum of activity
Baciturin	*B. thuringiensis* var. *darmstadiensis* 24	Paste	Pests of protected soil: spider mites, cabbage whiting, larvae of the Colorado potato beetle
Entobacterin	*B. thuringiensis* var. *galleria*	Wettable powder	Leaf lepidoptera: apple moth, leafworm, sawfly
Insectin	*B. thuringiensis* var. *insectus*	Wettable powder	Pests of the forest: gypsy moth, oak leaf roller
Bithoxybacillin	*B. thuringiensis* var. *thuringiensis*	Wettable powder	Aphids, cabbage moth, Colorado beetle
Dendrobacillin	*B. thuringiensis* var. *dendrolymus*	Wettable powder	Owlet moth, black vein, moth, geometer moth
BIP (Bacterial insecticide product)	*B. thuringiensis* var. *darmstadiensis*	Dry powder, paste	Pests of fruit: apple and fruit moth, geometer moth, leaf-roller, silkworm
Bactulocid	*B. thuringiensis* H14	Liquid spray	Larvae of mosquitoes and midges
DiPel	*B. thuringiensis* var. *kurstaki*	Powder, liquid concentrate, granulated bait	Owlet moth, black vein, moth, geometer moth, meadow moth

- Transgenic potatoes and tomatoes expressing *B. thuringiensis* genes have demonstrated resistance to Lepidoptera pests.
- A genetically modified poplar tree has been developed with an *antitrypsinase gene*, granting insect resistance.
- In the U.S., a genetically modified bacterium is used to treat maize seeds, where it produces a toxin targeting corn moth larvae.

Summary:
Bacillus thuringiensis is the cornerstone of microbial pest control. Modern biotechnology techniques are enhancing its effectiveness, host range, and delivery methods, making biopesticides safer, more specific, and more sustainable alternatives to chemical pesticides.

4.3.3 Manufacturing of Fungal and Viral Entomopathogenic Products

Entomopathogenic products based on microscopic fungi are effective against a wide range of insect pests by inducing a disease known as *mycosis*. Fungi differ from bacteria and viruses in several key ways:
- They infect insects through their outer surface rather than via the digestive system.
- They are effective during both the *pupal* and *adult (imago)* stages of insect development.

- Fungi grow rapidly and reproduce extensively; their spores can persist in the environment for long periods without losing virulence.
- Their efficacy and pathogenicity depend significantly on the specific fungal strain used.

When an insect comes into contact with fungal spores, the spores penetrate the insect's cuticle and germinate, forming thread-like structures called hyphae. These hyphae develop into a mycelial network inside the insect's body. The mycelium then produces reproductive structures known as conidia, which serve as infective units.

Once inside the host, the conidia disseminate through the insect's circulatory fluid (hemolymph). In most cases, toxins released by the fungus lead to the insect's death. In instances where toxins are absent or the insect exhibits resistance, death may result from the mechanical damage caused by extensive fungal growth, especially within the musculature. Ultimately, the fungus consumes the insect's tissues. As the infection progresses, conidiophores may emerge from the insect's cuticle and cover the larval corpse with fungal structures. Depending on the size and susceptibility of the insect, death typically occurs within 2 to 8 days (see Fig. 4.23).

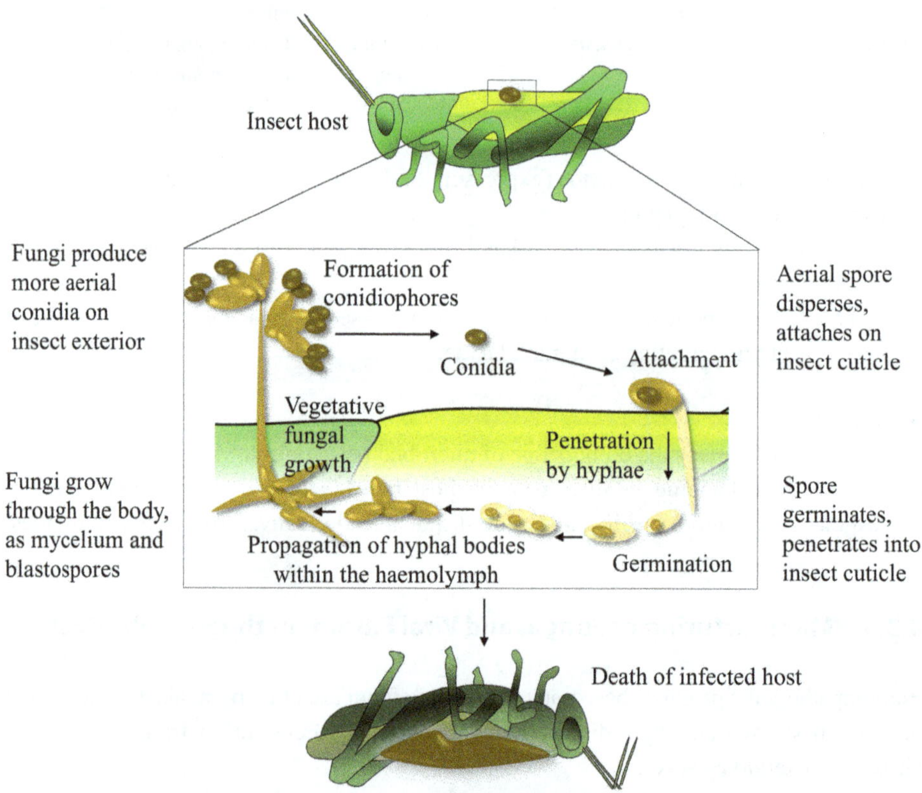

Fig. 4.23 Mode of action of entomopathogenic fungi against insects

Table 4.7 Entomopathogenic39T39T products based on fungi

Name	Causative agent	Commercial form	Spectrum of activity
Boverin	*Beauveria bassiana*	Dry granular powder, liquid based on blastospore	Greenhouse whitefly and some species of aphids, Colorado beetle
Nematophagin	*Arthrobotris oligospora*	Liquid	Gall nematode
Verticillin	*Cefalosporum (Verticillium) lecanii*	Liquid	Greenhouse whitefly, tobacco thrips, aphids
Entomofluorine	Entomophthoraceae	Liquid	Copper, aphids, thrips, beetle, butterfly

Currently, there's a growing interest in the industrial production of fungal-based products. This focus is on specific strains from the following genera: *Beauveria* (responsible for white muscardine), *Metarhizium* (causing green muscardine), and *Enthomophthora* (impacting probosci's insects) (Table 4.7).

Commercial products based on *Beauveria* fungi are primarily derived from *Beauveria bassiana* Vuill., which is effective against more than 60 insect species, and *Beauveria tenella*, which targets up to 10 species, particularly beetles. The most widely used fungal entomopathogenic product is Boverin, which contains conidiospores of *B. bassiana* (Bals.) Vuill.

The production process uses submerged cultivation and consists of these steps:

1. Nutrient medium preparation. The medium includes (%): yeast fodder (2); starch (1); sodium chloride (0.2); manganese chloride (0.01); and calcium chloride (0.05). Calcium chloride, which can vary up to 5%, enhances conidia resistance to adverse conditions.
2. Starter culture preparation and inoculation.
3. The culture grows at a pH of 4.5–5.6 and a temperature of 25–28 °C for 3–4 days with continuous mixing and aeration. The medium requires amino nitrogen; its absence slows down growth. As the fungus fully matures, it releases peak enzyme levels, causing mycelium lysis and boosting conidia accumulation.
4. The culture liquid undergoes separation and filtration, followed by directing 70–80% of the moisture content to a spray dryer. The result is dried spores, turned into fine powders with a 10% moisture content.
5. This powder is then standardized with a specified amount of kaolin. Sometimes, a wetting agent and an adhesive are incorporated into the final product. The end product is a white or cream-colored powder, with a conidiospore concentration of 1.5 to 6 billion per gram.

Manufacturing of Ciral Entomopathogenic Products

Entomopathogenic viruses, or insect viruses, are a class of biological pesticides that specifically target insects. Their most notable characteristic is their high host specificity, meaning they affect only particular insect species. This makes them generally safe for humans, plants, and other animals. Furthermore, these viruses remain infectious over extended periods and can effectively reduce pest populations to economically insignificant levels. In many cases, a single application is sufficient for long-term control and may even spread beyond the initially treated area.

The most commonly used viruses include *cytoplasmic and nuclear polyhedrosis viruses* as well as *granulosis viruses*. These typically affect foliage- and pine-feeding insects. Polyhedrosis viruses form crystalline structures within infected cells—either in the nucleus or cytoplasm—that are highly resilient to environmental stress, allowing the virus to persist for years.

- *Nuclear polyhedrosis viruses (NPVs)* replicate in the nuclei of intestinal cells, causing widespread infection.
- *Cytoplasmic polyhedrosis viruses (CPVs)* are generally less virulent and have a narrower host range.
- *Granulosis viruses (GVs)* form rod-shaped particles each enclosed in a protein matrix (granules), located in the nucleus or cytoplasm of infected cells.

Insects become infected by ingesting the virus. In the alkaline environment of the insect midgut, viral particles dissolve, releasing virions that invade host cells and begin replication. As infection spreads, the insect eventually dies.

Industrial Production of Viral Insecticides

Viruses replicate only in living tissues, making industrial production more complex than for bacterial or fungal biopesticides. Live insect hosts are required. The main production steps are illustrated in Fig. 4.24.

To produce a viral preparation, the insect host is first grown on specialized nutrient media. At a specific development phase, a viral suspension is added to infect the insects. The initial inoculum is sourced from a few infected larvae. After about 7–9 days, allowing for maximum viral growth in the insect tissues, the dead larvae are collected, dried at 30–35 °C, and then mechanically broken down to remove their body contents. Physiological saline or distilled water is added to the remains, usually at a rate of 1 ml per caterpillar, which is then mixed and either filtered or centrifuged. The final products are crafted as water or oil solutions, dry powders, or pastes.

Viral Preparations

Viral preparations can be applied in two ways: directly into dense insect populations in confined areas or by spraying or dusting infected areas during the early larval stages. Commercially available viral products include virin-GYAP (for apple moth caterpillars), virin-KSh (for ringed silkworms), virin-ENSh (for gypsy moths), virin-EKS (for cabbage moths), ABB (for fall webworms), and virin-diprion (for pine sawflies).

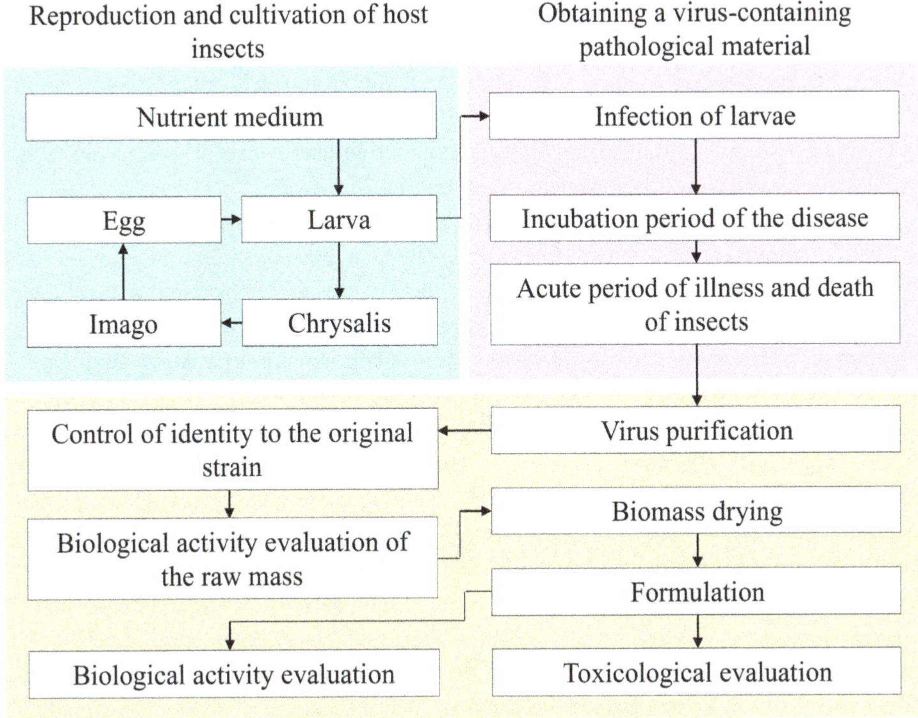

Fig. 4.24 Production technology of virus based biological products

The widespread adoption of viral pesticides is hindered by the challenges and costs of virus production. Insect pests are used to produce these viral insecticides, but they are not sterile and contain various microorganisms, including other viruses. Thus, producing viral insecticides demands rigorous quality control. Foreign microorganisms in the insects can degrade the product's quality. Moreover, the need to infect and then extract from numerous dead larvae further impacts the cost and quality of the pesticide.

Leveraging insect cell cultures may be the solution to these production challenges. Primary cultures, sourced from embryos, hemocytes, ovaries, and other insect parts, are used to generate these cell lines. Currently, transplantable cell lines from major agricultural and forestry pests like the gypsy moth and cotton bollworm have been developed. Using insect cell cultures for virus propagation is a promising technique. To optimize this, it's essential to produce high-yield cell lines, improve the nutrient media, and identify efficient "virus-cell" systems.

Genetic and cellular engineering can enhance biopesticides. By employing recombinant DNA methods, new sequences can be inserted into the envelope protein gene to produce novel proteins. For instance, integrating the toxins from *Bacillus thuringiensis* might amplify the virus's toxicity. Notably, baculoviruses with genes affecting insect water balance have shown great promise. Using such modified viruses causes insects to die within five days due to dehydration or water oversaturation.

A groundbreaking viral protein has been identified, increasing the efficacy of viral pesticides significantly. Sourced from the granulosa *Trichoplusiani*-baculovirus that impacts gypsy and owlet moths, this product is termed the viral enhancing factor (VEF).

Continued microbiological research on forest-dwelling insects will deepen our understanding of entomopathogenic virus hosts. Discovering new viral pathogens that can be used in controlling plant-eating insect populations is a possibility.

Bioinsecticides made from bacteria, fungi, and viruses are becoming increasingly popular in production. There are innovative methods being developed for cultivating entomopathogenic bacteria and fungi, as well as novel techniques for collecting viruses from insects. The *Bacillus thuringiensis Berliner* group is the most widely used type of bacteria for producing bioinsecticides. Notable fungal diseases include entomophthorosis. Cytoplasmic and nuclear polyhedra, along with granulosa viruses, are commonly found. Biopesticides are different from pesticides as they specifically target certain pests, have minimal impact on non-target species, and do not leave any residues in the environment. This makes them eco-friendly alternatives to the widespread use of chemical pesticides.

4.3.4 Brainstorming

1. Justify the advisability of developing and using biological agents to control pests and plant diseases.
2. Using microorganisms as biopesticides is a relatively new direction of biotechnology, but it has already made significant achievements. What kind?
3. Bacteria, fungi, and viruses are increasingly used as industrial biopesticides. Why?
4. Name and characterize the main types of entomopathogenic bacteria.
5. Describe the bacterial preparations for controlling insect pests and the mechanisms of their action.
6. Describe the technology for obtaining biopesticides based on entomopathogenic bacteria.
7. Name and characterize the main types of entomopathogenic fungi.
8. Describe the fungal preparations—biological insecticides and the mechanisms of their action.
9. Explain the technology of biopesticides based on entomopathogenic fungi.
10. Name entomopathogenic viruses as the basis of biopesticides. Clarify the mechanisms of their action.
11. Describe the technology of biopesticides based on entomopathogenic viruses.

Take-Home Messages

- *Biopesticides* are eco-friendly alternatives to chemical pesticides, derived from natural sources such as microorganisms, plants, animals, and minerals. They offer high target specificity, low toxicity to non-target organisms, and biodegradability, resulting in minimal environmental impact.
- Biopesticides can be grouped into:
  - Microbiological products (bacteria, fungi, viruses)
  - Natural extracts (plant-derived bioactive compounds)
  - Pheromones (behavior-modifying compounds used in traps)
- *Bacillus thuringiensis* (Bt) is the most widely used entomopathogenic bacterium, effective against over 400 insect species. Bt produces multiple exotoxins and endotoxins (e.g., Cry proteins) that paralyze and destroy insect gut cells. Different Bt strains vary in toxin composition and are tailored to control specific pests (e.g., caterpillars, mosquitoes). Insects must ingest the Bt toxin for it to be effective, which enhances species selectivity and minimizes collateral damage. Bt is cultivated in submerged fermentation systems, aiming for high spore and toxin yields.
- *Fungal entomopathogens* (e.g., *Beauveria, Metarhizium, Entomophthora*) infect insects through their cuticle, not the digestive tract. They are particularly effective at pupal and adult stages, producing mycelia and spores that invade and kill insects. The infection results in mycosis, with toxin production and systemic colonization of insect tissues. Commercial products like Boverin use spores of *Beauveria bassiana* and are effective against a broad range of insects.
- *Viral insecticides* (e.g., nuclear polyhedrosis viruses, granulosis viruses) are highly specific to target insects and safe for other organisms. They infect insects via ingestion, replicate inside living cells, and ultimately kill the host through systemic infection. Industrial production requires live insect hosts, making scalability and contamination control challenging. Viral products include Virin-GYAP, Virin-KSh, Virin-ENSh, among others.
- Biopesticides offer a sustainable solution to pest control with reduced environmental and health risks. They are ideal for integrated pest management (IPM) strategies and organic farming.

4.4 Production of Bacterial Fertilizers

Soil microorganisms are essential for maintaining soil fertility and promoting plant growth. They improve soil structure, mobilize nutrients, and transform compounds into plant-accessible forms. To enhance these beneficial processes, bacterial fertilizers are applied to the rhizosphere—the soil region surrounding plant roots (Fig. 4.25).

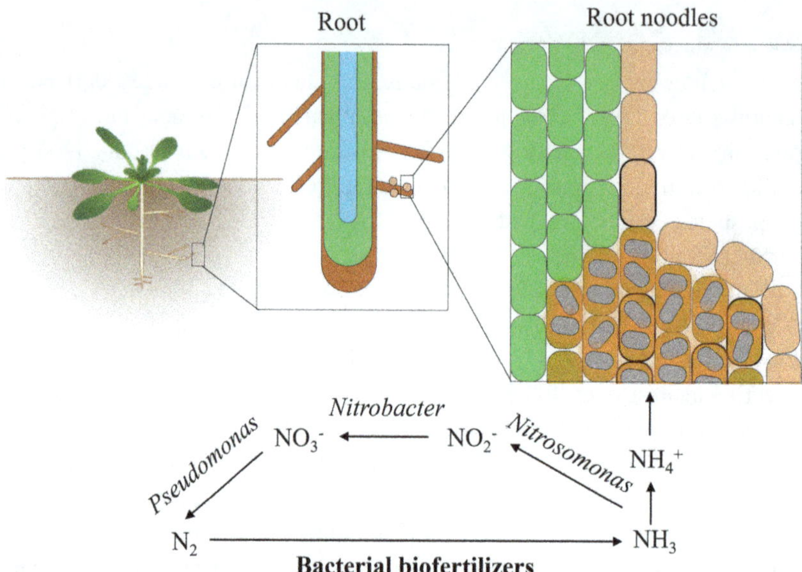

Fig. 4.25 Potential application of soil microbes

Microorganisms in these preparations provide essential nutrients and bioactive substances such as phytohormones and vitamins. Among the most widely used bacterial fertilizers are those based on nodule-forming bacteria and free-living nitrogen-fixing bacteria.

4.4.1   Products of Nodule Bacteria

Rhizobium bacteria form symbiotic relationships with legumes, enabling the fixation of atmospheric nitrogen for plant use. This process takes place in specialized structures on plant roots known as nodules, which develop following Rhizobium infection. The reddish or pink coloration of these nodules, caused by the pigment leghemoglobin (similar to hemoglobin in animals), indicates effective nitrogen fixation. Within the nodules, bacteria synthesize a suite of enzymes—most notably nitrogenase—that convert molecular nitrogen (N_2) into ammonia. In this mutualistic relationship, the plant supplies nutrients to the bacteria, which in return provide nitrogen, thereby enhancing soil fertility and crop productivity (Fig. 4.26).

Although the technology behind "Nitragine" production is complex, the manufacture of its dry form (see Fig. 4.27) is considered more promising. The first commercial Rhizobium seed inoculant, "Nitragine," was patented in the UK in 1896. In the United States, peat-based Rhizobium formulations were introduced. Today, two major products dominate the market: "Nitragine" and "Rhizotorphine."

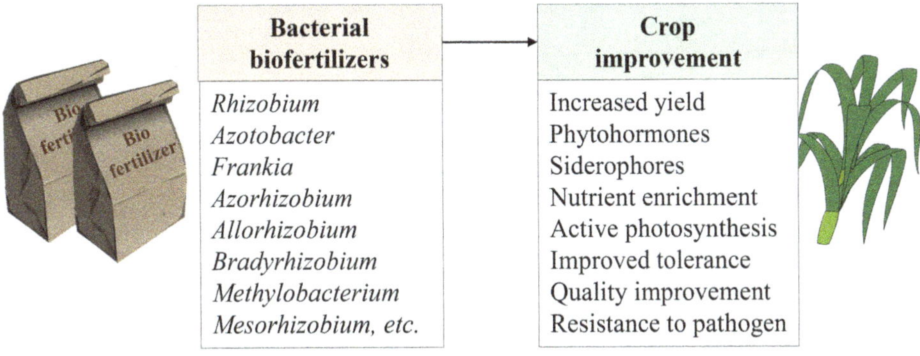

Fig. 4.26   Effect of biofertilizers on crop productivity

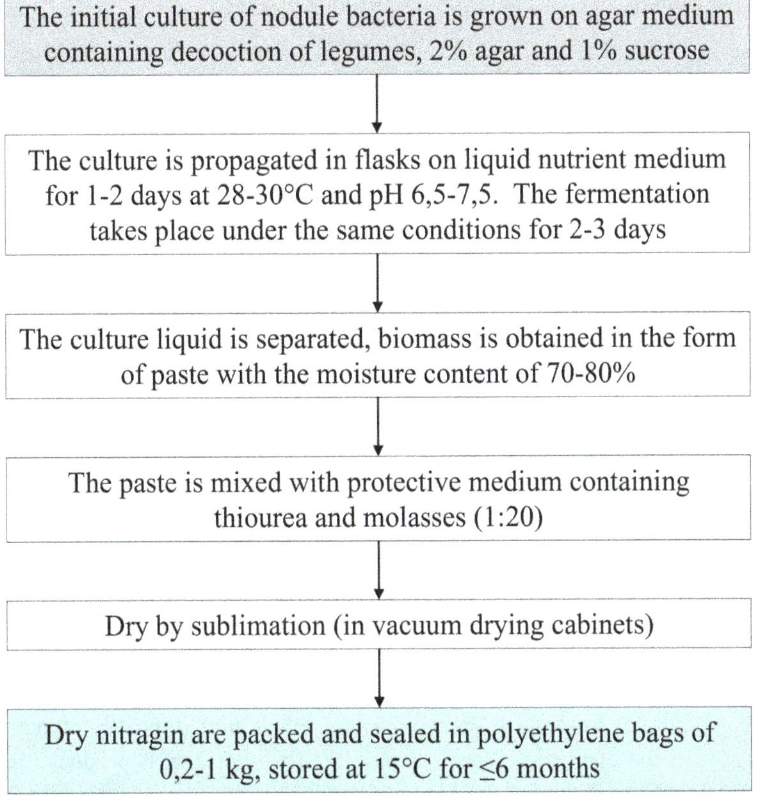

Fig. 4.27   Schematic representation of the technology for obtaining dry Nitragine

Dry Nitragine is a light gray powder in which each gram contains more than 9 billion viable bacteria, blended with a filler. Its moisture content ranges from 5 to 7%. This powder is applied to legume seeds prior to sowing, typically resulting in yield increases of approximately 15–25%.

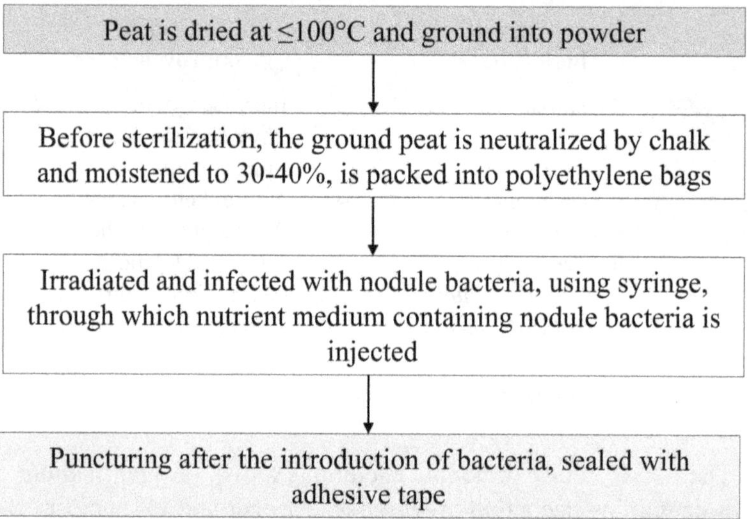

Fig. 4.28 Schematic representation of the technology for obtaining Rhizotorphine

Another widely used nodule bacteria formulation is Rhizotorphine. The standard industrial process for its production is shown in Fig. 4.28. Each gram of Rhizotorphine contains at least 2.5 billion viable cells with high nitrogen-fixing efficiency and strong competitiveness in the rhizosphere. The recommended application rate is 200 grams per hectare. Seeds are treated by mixing Rhizotorphine with water, straining the solution through a double layer of gauze, and then applying it to the seeds. This treatment can increase crop yields by up to 20%. Additionally, mulching seedlings with the dry form of Rhizotorphine may lead to a 15% yield improvement.

4.4.2 Products of Free-Living Soil Microorganisms

Azotobacterin is a bacterial fertilizer derived from the free-living soil bacterium *Azotobacter chroococcum*. It enhances soil quality by releasing biologically active substances that stimulate plant growth. These include organic acids, pantothenic acid, pyridoxine, biotin, heteroauxin, and gibberellins. Additionally, *Azotobacter* produces fungicidal compounds from the anisomycin group, which inhibit the development of fungal pathogens in the plant root zone. Industrially, *Azotobacterin* is produced in various forms, including dry soil-based and peat-based preparations.

The production of dry *Azotobacterin* follows a process similar to that used for dry *Nitragine*. The microorganism is cultivated via a deep culture method on a medium identical to that used for *Rhizobium*, but with additional components such as iron and manganese sulfates and a complex molybdenum salt. The medium typically has a pH of 5.7–6.5.

Fermentation continues until the culture reaches its stationary phase, at which point biologically active substances (BAS) are released into the medium. Although some BAS activity may be lost during drying, the viable bacterial cells quickly regain their metabolic functions. The final dried product is standardized to contain no fewer than 0.5 billion viable cells per gram. It is packaged in polyethylene bags weighing 0.4 to 2 kg and stored at 15 °C, with a shelf life of up to three months.

Azotobacter performs best in soils rich in phosphorus and micronutrients. *Azotobacterin* can be applied to seeds, seedlings, or composts and typically increases productivity by 10–15%. For seed treatment, 100 billion cells are used per hectare. In the case of potatoes or seedling root systems, a bacterial suspension is prepared by diluting 300 billion cells (enough for 1 ha) in 15 liters of water.

4.4.3   Soil and Peat Azotobacterin

To prepare these formulations, fertile soil or decomposed peat with a neutral pH is used as the base. The substrate is sieved and enriched with 2% lime and 0.1% superphosphate. Approximately 500 g of this mixture is placed in 0.5-liter bottles, moistened to 40% water content, plugged with cotton, and sterilized.

For inoculation, bacteria are first cultured on agar medium containing 2% sucrose and mineral salts. Once a mucous layer forms on the agar, it is rinsed with sterile water and added to the sterilized substrate.

The mixture is incubated at 25–27 °C until sufficient bacterial growth is achieved. The final product remains biologically active for 2–3 months. In practice, seeds are mixed with the damp substrate and then dried. The roots of seedlings are moistened with a suspension prepared from this product.

Recent advances in molecular biology and genetics have provided new insights into biological nitrogen fixation. The key enzyme responsible, nitrogenase, is encoded by 17 *nif* genes, which show structural similarity across all nitrogen-fixing organisms, indicating a high degree of gene homology.

Many genes involved in the symbiotic relationship between *Rhizobium* and legumes are located on easily transferable plasmids known as *nif* plasmids. These plasmids carry not only nitrogenase genes but also those required for nodule formation. The construction of self-transmissible plasmids carrying nitrogen-fixation genes has enabled their transfer into non-nitrogen-fixing organisms such as *Escherichia coli* and *Salmonella typhimurium*. *Nif* genes have also been successfully cloned in yeast and transferred using viral vectors.

A notable example is the nitrogen-fixing bacterium *Klebsiella pneumoniae* M5, which has been propagated using *E. coli* phage P1, enabling gene transduction of *nif* sequences.

These achievements lay the foundation for transferring nitrogen-fixation genes into plant cells. The next major challenge is selecting the most suitable host plants. With continued research into symbiotic microorganisms and gene transfer techniques, biotechnology may soon offer transformative solutions for improving plant nitrogen nutrition.

4.4.4 Brainstorming

1. What microorganisms and how do they play a role in increasing soil fertility?
2. What is the significance of the nitrogen fixation process?
3. Describe groups of free-living, associative, and symbiotic nitrogen fixers.
4. What is the role of nodule bacteria in increasing soil fertility?
5. What methods of obtaining biological products with nodule bacteria exist?
6. Describe the technology of obtaining biofertilizers based on bacteria of the genus *Rhizobium*.
7. What is the role of free-living nitrogen-fixing agents in increasing soil fertility?
8. How are biofertilizers obtained from free-living nitrogen-fixing bacteria?
9. Describe the technology of obtaining and using biofertilizers based on bacteria of the genus *Azotobacter*.
10. Give examples of new developments for obtaining nitrogen biofertilizers.

Take-Home Messages

- Bacterial fertilizers such as Nitragine, Rhizotorphine, and Azotobacterin enhance soil fertility and crop productivity by supplying plants with biologically fixed nitrogen and growth-promoting substances.
- Rhizobium species form symbiotic relationships with legumes, fixing atmospheric nitrogen in root nodules. Azotobacter species, as free-living soil bacteria, release phytohormones and antifungal compounds that support plant growth.
- Modern production of bacterial fertilizers involves scalable biotechnological methods and continues to benefit from genetic advances, including plasmid-based nitrogen fixation gene transfer. Genetic engineering of nif genes opens the possibility of developing crops with built-in nitrogen-fixing capabilities, potentially transforming agricultural sustainability.

4.5   The Basics of Modern Immunobiotechnology

Immunotherapy is a branch of immunology that uses targeted treatments to influence the immune system. These treatments include vaccines, antibody-based products, monoclonal antibodies, and cytokine therapies. In addition, immunotherapy encompasses cell therapies, immunomodulators, immunostimulants, as well as allergens and antigens.

Different types of immunomodulatory products (IMPs):

1. IMPs that are obtained from living or deceased microorganisms such as bacteria, viruses, fungi, or their byproducts. These are used for prevention or treatment purposes and include live vaccines, inactivated vaccines, subcellular vaccines, anatoxins, bacteriophages, and probiotics.
2. IMPs are derived from antibodies. This category includes immunoglobulins, sera, immunotoxins, enzyme-linked antibodies, and monoclonal antibodies.

3. Immunomodulators are utilized to adjust the system and treat various diseases and immune deficiencies. They can be either exogenous (obtained from microbes) or endogenous (such as interleukins, interferons, and peptides).
4. Adaptogens are compounds found in plants, animals, or other sources that have diverse biological effects, including impacts on the immune system. Examples of these include ginseng extracts, eleutherococcus extracts, and various bioactive food supplements.

4.5.1   Vaccines

Industrial biotechnology focuses on creating products for preventing and treating diseases through immunoprophylaxis and immunotherapy. This includes vaccines that stimulate a targeted response to establish active immunity as well as serums and immunoglobulins that offer immediate protection by introducing pre-formed antibodies capable of defending against toxins or infections.

The origins of vaccination can be traced back through history (see Fig. 4.29). Initially, substances that countered toxins were referred to as "antitoxins." Over time, the term "antibody" was introduced, and the substance responsible for their development was labeled as an "antigen."

Antigens carry foreign genetic signs. Antibodies are protective proteins known as immunoglobulins. They can bind to and neutralize the antigen that triggered them.

Antigen-antibody combinations are typically cleared by phagocytes or targeted by the complement system, which has about twenty specific interacting proteins.

B cells are responsible for generating antibodies when they encounter antigens. When an antigen is detected, it prompts the growth of B cells, which produce and release these antibodies. This process is known as immunity. On the other hand, when there are virus-infected cells that need to be eliminated, specific T cells are activated to take care of them through cellular immunity. Following an infection, certain B and T cells remain in the body. If exposed to the virus again, they rapidly multiply and combat the pathogen. This phenomenon is referred to as memory (Figs. 4.30 and 4.31).

Immunogenicity refers to the capacity to stimulate the production of antibodies. This occurs through regions on a protein's surface, typically consisting of six to eight amino acids. These regions establish a connection with immunoglobulins. It is possible for a single protein to possess epitopes, each recognized by lymphocytes that generate antibodies targeting those particular epitopes. Consequently, the immune system frequently generates immunoglobulins. (Table 4.8).

Vaccination (immunization, from the Latin *vaccus*—cow) is the creation of artificial immunity to certain diseases. Here, relatively harmless antigens are used, which are part of the microorganisms that cause disease. Microorganisms can be viruses or bacteria.

Vaccination, derived from the Latin word *vaccus* for "cow," creates artificial immunity to diseases. It uses harmless antigens from disease-causing microorganisms, which can be viruses or bacteria.

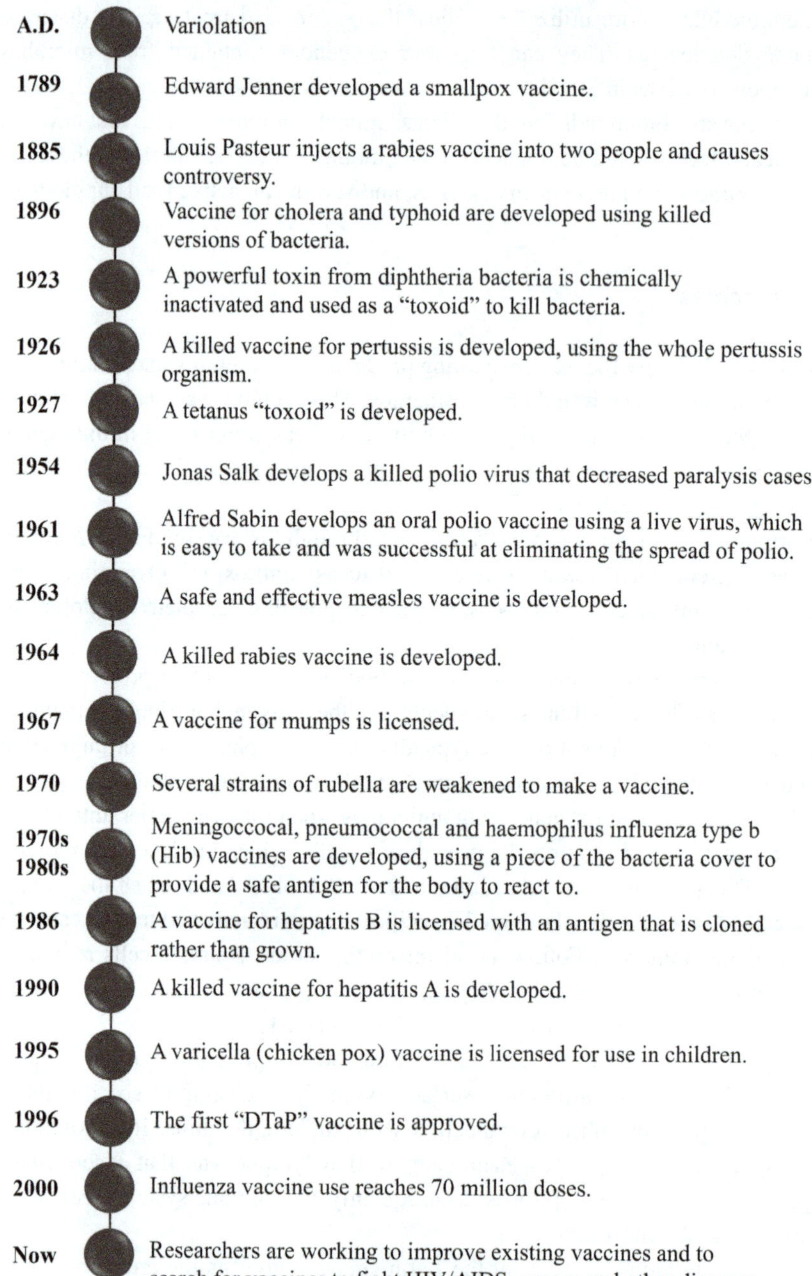

Fig. 4.29 Timeline of vaccine development

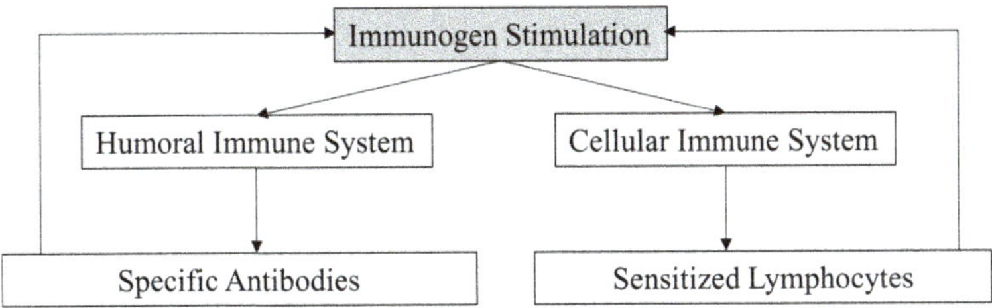

Fig. 4.30   Immune mechanism

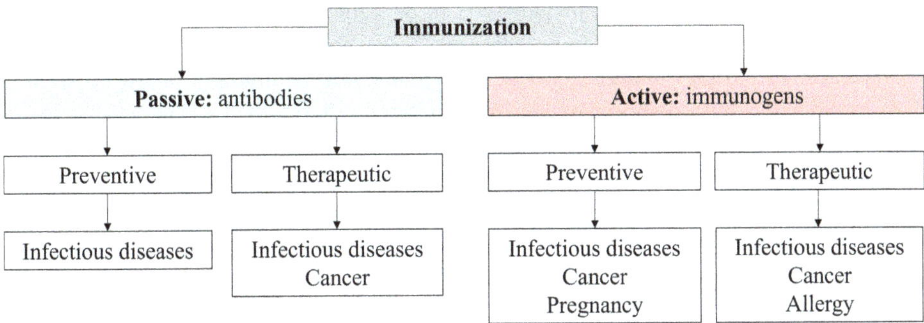

Fig. 4.31   Active immunization (vaccination) and passive immunization: examples of their fields of application

Table 4.8   Five classes of immunoglobulins and their structures

Immunoglobulins (Ig)				
IgM pentamer	IgG monomer	Secretory IgA dimer	IgE monomer	IgD monomer
Heavy chains μ	γ	α	ε	δ
Number of antigen-binding sites 10	2	4	2	2
Molecular weight 900,000	150,000	385,000	200,000	180,000
Percentage in serum 6%	80%	13%	0,002%	1%
Function The main antibody of primary responses, best at fixing complement; the monomer form serves as the B cell receptor	The main blood antibody of secondary responses neutralizes toxins	Secreted into mucus, tears, saliva, colostrum	Antibody of allergy, anti- parasitic activity	B cell receptor

Vaccines are biological tools that create active immunity in humans or animals. Their main component, the immunogen, resembles parts of the disease-causing agent. Vaccines mimic an infection, prompting the immune system to respond. The system then recalls the "invader." If the same microbe enters again, the immune system quickly combats it.

Vaccine antigens require the ability to elicit a response, which is known as immunogenicity. Immunogenicity refers to how an antigen can activate the immune system to combat foreign substances. The strength of the response is often associated with the size of the antigen. Antigens that are particle-based, such as bacteria, exhibit immunogenic properties. In the case of proteins, a higher molecular weight typically leads to a more robust immune response. Therefore, it can generally be said that a larger molecular weight correlates with immunogenicity.

Vaccines can prompt antibody production against one or multiple pathogen strains. Besides the main active component, vaccines contain:

- Sorbents
- Preservatives (for maintaining sterility against bacterial contamination)
- Adjuvants (boosters like aluminum salts, proteins, and amino acids)
- non-specific impurities (such as cultivation proteins, trace antibiotics, and certain animal serum proteins)

Vaccines are categorized based on their composition and methods of production. Different Types of Vaccines for Active Immunization

1. Attenuated vaccines: These vaccines utilize live, weakened pathogens, such as the measles vaccine.
2. Inactivated vaccines: In this category, pathogens are rendered inactive. Examples include;
    - Solid/whole virion vaccines: These vaccines employ killed disease agents like the antiplague vaccine.
    - Subunit vaccines: These vaccines contain specific antigens, such as:
        - Metabolic byproducts of microbes (e.g., tetanus toxoid)
        - Pure capsular polysaccharides (pneumococcal vaccine)
        - polysaccharides combined with protein carriers
3. Genetically engineered vaccines: These vaccines are produced using techniques. Different types include:
    - Recombinants: They contain products derived from microbe genes manufactured in unique cells (e.g., the hepatitis B vaccine).
    - Chimeric/vector vaccines: Non-pathogenic microbes are modified to produce proteins (e.g., herpes, influenza vaccines).
    - Ribosomal: These vaccines employ isolated ribosomes. Their matrix.
4. Synthetic vaccines: Created through chemical synthesis to replicate a protective protein There are more ways to classify vaccines based on various criteria (see Fig. 4.32).

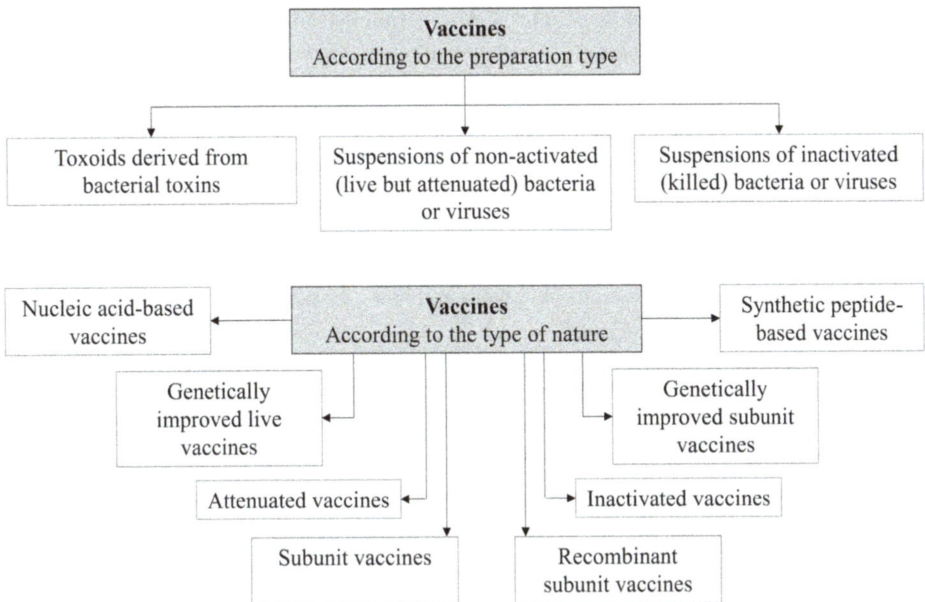

Fig. 4.32 Vaccine classification

4.5.2   Production of Vaccines

The development of vaccines involves several key stages:

1. *Identification of protective antigens.* Scientists begin by identifying antigens capable of eliciting a protective immune response. This is typically done through animal studies. Promising antigen candidates are then tested in animals for both safety and efficacy.

2. *Determination of antigen form and administration method.* Researchers determine the most effective form and delivery method of the antigen. Human volunteers are enrolled in early-phase trials after providing informed consent. A small group receives varying doses of the vaccine to evaluate safety, determine the optimal dosage, and assess the initial immune response.

3. *Clinical trials.* The vaccine's safety and efficacy are further tested in human clinical trials involving participants of various age groups. During these trials, volunteers may be exposed to the target pathogen, either naturally or under controlled conditions, to evaluate immune protection.

4. *Evaluation of safety and efficacy for licensing.* Before a vaccine can be approved for public use, it undergoes large-scale clinical trials to confirm its effectiveness and monitor potential side effects.

Since even large trials may not detect rare adverse reactions, *post-marketing surveillance* is conducted after a vaccine is introduced. This ongoing monitoring helps ensure continued safety and effectiveness in the general population.

Additional information regarding the properties of vaccines can be found in Fig. 4.35.

1. Identifying protective antigens: Through experiments on animals, scientists identify antigens that provide protection. These potential vaccines are then tested for safety and effectiveness in animals.
2. Determining the form of antigen: Researchers determine the best way to administer the antigen. Volunteers are provided with consent forms to participate in the study. A selected few receive varying doses of the vaccine to determine the dosage and evaluate safety and immune response.
3. Conducting clinical trials: The vaccine's safety and effectiveness are tested on humans across age groups. These trials involve a number of volunteers who are exposed to microorganisms that cause diseases.
4. Evaluating safety and efficacy: The vaccine is subjected to large-scale testing on humans before it can be approved for licensing.

However, since clinical trials involve a number of participants, some rare side effects might go unnoticed. Therefore, after a vaccine is introduced, continuous monitoring takes place to ensure its effectiveness and detect any complications.

Further details regarding the characteristics of a "vaccine can be found in Fig. 4.33.

Vaccine production involves multiple steps:

1. Growing organisms or collecting their byproducts
2. Concentrating and purifying antigens
3. Inactivating antigens when needed
4. Adding enhancers like sorbents and adjuvants
5. Freeze-drying (lyophilization)
6. Packaging for distribution

Live vaccines contain genetically modified disease-causing agents known as vaccine strains. These strains are non-virulent but can still stimulate immunity. Typically made from weakened strains of harmful microbes, these vaccines can't cause the actual disease. However, they can multiply in a vaccinated individual, leading to a mild reaction. This reaction, whether local inflammation or a general, symptom-free infection, triggers the immune system to build immunity.

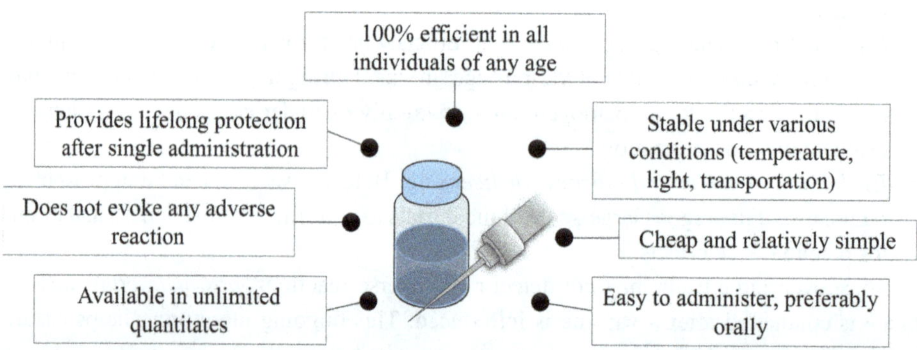

Fig. 4.33 Properties of ideal vaccine

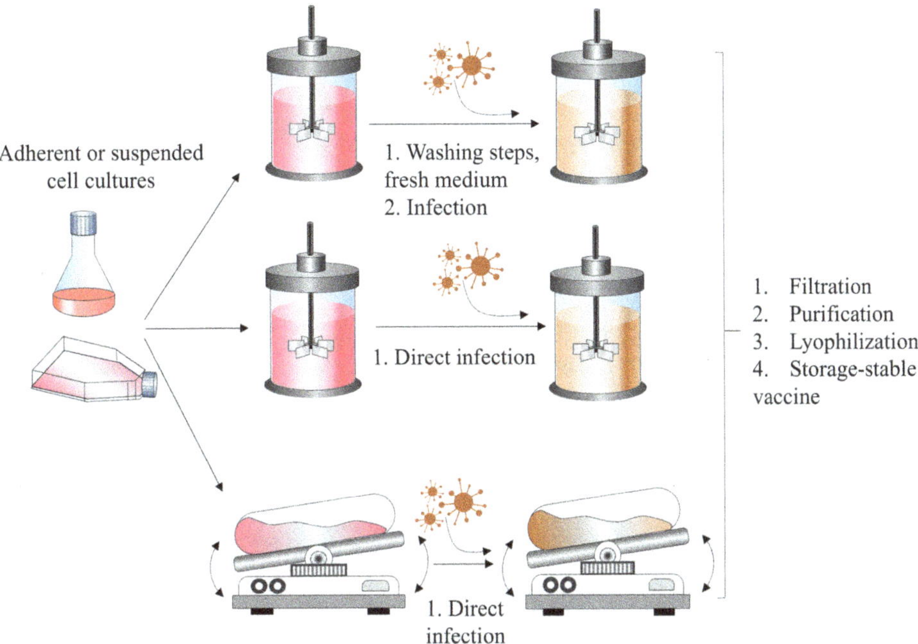

Fig. 4.34 Extended scheme representation for vaccine production using mammalian cells

This vaccine type is simple to produce, as it primarily involves growing a weakened vaccine strain while maintaining purity and preventing contamination from other microbes like mycoplasma or oncoviruses. The process follows with stabilizing and standardizing the final product. Bacterial vaccine strains are cultivated in liquid nutrient solutions like casein hydrolysates, in large bioreactors over 2 cubic meters in capacity. The purified vaccine strain is then freeze-dried, with protectors added, as shown in Fig. 4.34.

Live vaccines for viruses and rickettsia are produced by growing vaccine strains in chicken or quail embryos free of leukemia viruses, or in cell cultures without mycoplasmas (see Fig. 4.35). These strains can also be cultivated in primary animal cells or human diploid cells. Generally, live vaccines are made from naturally occurring bacteria and viruses, which are attenuated through selection or passage through biological systems like animals, embryos, cell cultures, or nutrient media.

Inactivated bacterial vaccines or whole-virion inactivated vaccines are derived from bacteria and virus cultures grown in similar media as live vaccines. These cultures are then deactivated using methods like heat, formalin, ultraviolet light, ionizing radiation, or alcohol, as illustrated in Fig. 4.36.

Developing vaccines is a complex process. It includes extracting antigens from microbial masses, purifying and concentrating them, and incorporating adjuvants. In order to isolate and purify these antigens, traditional methods such as trichloroacetic acid extraction, hydrolysis, or alcohol precipitation are employed. Additionally, advanced techniques

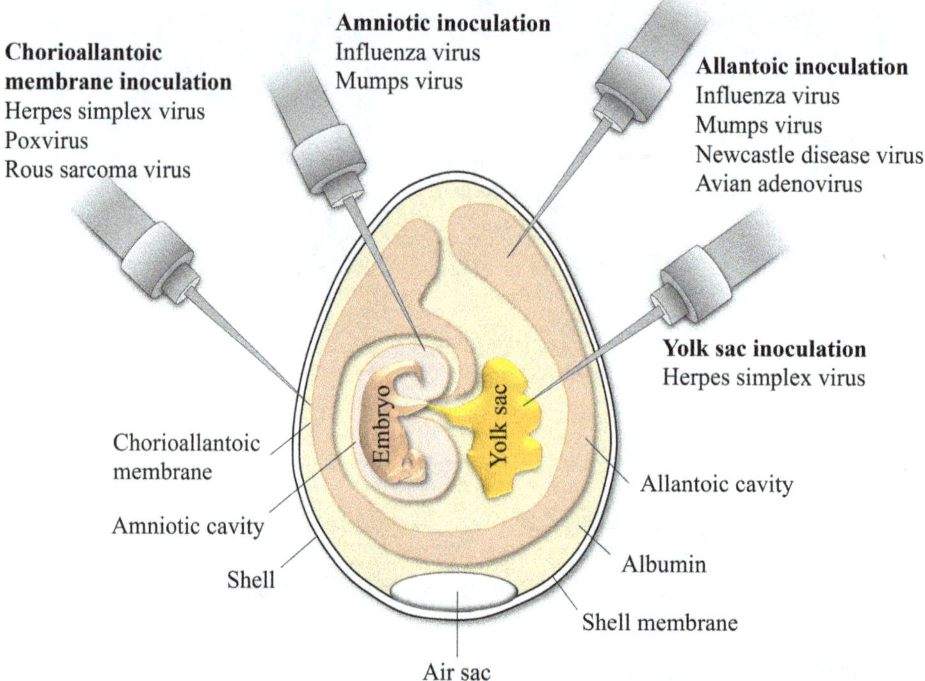

Fig. 4.35 Routes of viral inoculation into chicken embryo

like ultracentrifugation, ultrafiltration, and chromatography are utilized. These approaches result in the production of antigens that're highly pure and concentrated.

Thanks to genetic engineering advancements, vaccine strains can now be specifically designed. As a result, we have recombinant strains of the influenza virus and vaccine strains with integrated genes from protective hepatitis B virus antigens.

4.5.3 Vaccination Effectiveness

Vaccination-induced immunity is termed postvaccinal immunity. Vaccinations aren't always successful. Improper storage can degrade vaccine quality. Plus, even under optimal storage, there's a chance the immune response might not be adequately triggered.

Factors influencing immunity after vaccination:
1. Factors related to the vaccine
   - The quality of the vaccine
   - How long does the antigen remain effective
   - The dosage was given
   - The presence of antigens in the vaccine
   - How often is the vaccine administered

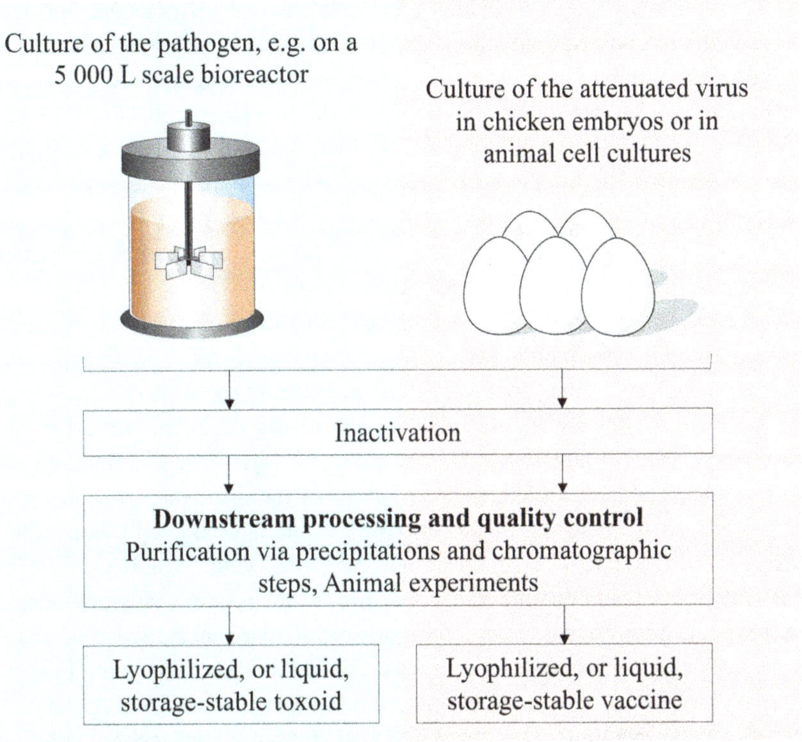

Fig. 4.36 Fermentation and recovery of inactivated vaccines

2. Specific factors:
   - The overall health of a person's system
   - Age
   - Any existing immunodeficiency conditions
   - General health status
   - Genetic factors
3. Environmental factors:
   - Diet and nutrition
   - Working and living conditions
   - Climate and weather patterns
   - Physical and chemical aspects of the environment

In addition to the descriptions mentioned above about vaccine classification, brief information on selected vaccines is provided.

Inactivated (Killed) Vaccines

Inactivated vaccines are obtained by exposure to microorganisms by chemical means or by heating. Such vaccines are quite stable and safe since they cannot cause reversion of virulence. They are usually stable to cold, which is convenient for practical use. However, these vaccines also have several disadvantages; in particular, they stimulate a weaker immune response and require several doses of booster immunization. They contain either killed whole microorganisms (e.g., whole cell pertussis vaccine, inactivated rabies vaccine, viral hepatitis A vaccine) or components of the cell wall or other parts of the pathogen, such as acellular pertussis vaccine, conjugated Haemophilus infection vaccine or vaccine against meningococcal infection. They are killed by physical (temperature, radiation, UV light) or chemical (alcohol, formaldehyde) methods. Such vaccines are reactogenic; few are used against pertussis and hepatitis A.

Inactivated vaccines are also corpuscular. Analyzing the properties of corpuscular vaccines should also distinguish both positive and negative qualities. Positive aspects are that corpuscular slaughter vaccines are easier to dose, better to clean, stored for a long time, and less sensitive to temperature fluctuations. Negative aspects are that the corpuscular vaccine contains 99% of ballast and is, therefore, reactogenic; it also contains an agent used to kill microbial cells. Another disadvantage of the inactivated vaccine is that the microbial strain does not survive; therefore, the vaccine is weak, and vaccination is performed in 2 or 3 doses; it requires frequent revaccinations, which is more complex in terms of organization compared to live vaccines. Inactivated vaccines are produced in dry (lyophilized) and liquid forms. Many microorganisms that cause disease in humans are dangerous because they release exotoxins, which are the main pathogenic factors of the disease (e.g., diphtheria, tetanus). Anatoxins used as vaccines induce a specific immune response. For the preparation of vaccines, toxins are most often neutralized with formalin.

Live Vaccines

They contain a weakened living microorganism. Examples include vaccines against poliomyelitis, measles, mumps, rubella, or tuberculosis. They can multiply in the organism and cause the vaccinal process, forming immunity. The loss of virulence in such strains is genetically fixed; however, in persons with immunodeficiency, serious problems can arise. As a rule, live vaccines are corpuscular. These vaccines are produced by artificial attenuation (strain weakening) or by selecting natural avirulent strains. Currently, it is possible to create live vaccines by genetic engineering at the level of chromosomes using restriction enzymes. When analyzing the properties of living vaccines, one should distinguish between positive and negative qualities. On the positive side, the mechanism of action on the body is reminiscent of a "wild" strain, which can become engorged in the body, and lasts for a long time to remain immune, displacing the "wild" strain.

Furthermore, small vaccination doses are used so that vaccination can be easily conducted organizationally. The latter allows us to recommend this type of vaccine for further use. Negative aspects are that the live vaccine corpuscular contains 99% of ballast and therefore is usually quite reactogenic; in addition, it is capable of causing mutations in host cells, which is especially dangerous for germ cells. Furthermore, live vaccines contain contaminant viruses, hazardous for monkey AIDS and oncoviruses. Unfortunately, live vaccines are difficult to dose and easily sensitive to high temperatures.

Although live vaccines require special storage conditions, they produce sufficiently effective cellular and humoral immunity and usually require only one booster injection. Most live vaccines are administered parenterally (except the poliomyelitis vaccine). Examples of live vaccines are vaccines for rubella prevention (Rudivax), measles (Ruvax), poliomyelitis (Polio Sabin Vero), tuberculosis, and mumps (Imovax Oreion). Live vaccines are produced in lyophilized form.

Corpuscular Vaccines

They are bacteria or viruses inactivated by chemical (formalin, alcohol, phenol) or physical (heat, UV radiation) exposure. Examples of corpuscular vaccines are pertussis, rabies, leptospirosis, vaccine against encephalitis, hepatitis A, and the inactivated polio vaccine.

Chemical Vaccines

Contains components of the cell wall or other parts of the pathogen, such as acellular pertussis vaccine, a conjugated vaccine against hemophilia, or a vaccine against meningococcal infection. These vaccines are created from antigenic components extracted from microbial cells. Such vaccines include polysaccharide vaccines and acellular pertussis vaccines.

Biosynthetic Vaccines

Biosynthetic vaccines are obtained by genetic engineering and artificially created antigenic determinants of microorganisms. An example is a recombinant vaccine against viral hepatitis B and rotavirus infection. To produce them, yeast cells are used in the culture in which a gene encoding the protein production necessary for the vaccine preparation is built in, which is then isolated in its pure form.

At the present stage of the development of immunology as a fundamental medical and biological science, it became evident that it is necessary to create fundamentally new approaches to the design of vaccines based on knowledge of the pathogen's antigenic structure and the immune response of the organism to the pathogen and its components.

Biosynthetic vaccines are produced from peptide fragments that correspond to specific amino acid sequences of the viral or bacterial protein structures recognized by the immune system. Compared to traditional vaccines, an essential advantage of synthetic vaccines is that they do not contain whole bacteria and viruses, nor the products of their activity, and cause a more narrowly targeted immune response. When creating this type of vaccine, several different peptides can be attached to the carrier, allowing the selection of the most immunigenic fragments for coupling. At the same time, synthetic vaccines are generally less effective than traditional ones. However, using one or two immunogenic proteins instead of the entire pathogen enables the formation of immunity while significantly reducing the vaccine reactogenecity and the likelihood of side effects.

Recombinant Vaccines

Vaccines obtained by genetic engineering. The essence of the method is that the genes of a virulent microorganism responsible for synthesizing protective antigens are inserted into the genome of a harmless microorganism that, when cultivated, produces and accumulates the appropriate antigen. An example is a recombinant vaccine against viral hepatitis B and rotavirus infection. The activity of individual components of microbial, viral, and parasitic antigens manifests at different levels and in various parts of the immune system. Their result can be only one—clinical signs of the disease:

- Recovery
- Remission
- Relapse
- Aggravation
- Other conditions of the organism

Thus, recombinant technology is used to produce these vaccines by integrating the genetic material of the microorganism into yeast cells that produce antigens. After yeast cultivation, the necessary antigen is isolated from them; the vaccine is purified and prepared.

Ribosomal Vaccines

Ribosomes available in each cell are used to obtain this type of vaccine. *Ribosomes* are organelles that produce protein by the template mRNA. The isolated ribosomes with the matrix in pure form represent the vaccine. An example is bronchial and dysenteric vaccines.

New Vaccine Generations

Modern technologies have led to the development of second-generation vaccines, offering greater specificity, safety, and efficacy:

- *Conjugated vaccines*:

  Some bacterial pathogens—such as those causing meningitis or pneumonia—possess antigens that are poorly recognized by immature immune systems, particularly in infants. Conjugated vaccines address this by chemically linking these antigens to carrier proteins or toxoids that are easily recognized by the immune system. A notable example is the *Haemophilus influenzae* type B (Hib) vaccine.

- *Subunit vaccines*:

  These vaccines contain only specific antigenic fragments of a pathogen rather than whole cells. The selected components are often produced using genetic engineering techniques. Examples include vaccines against *Streptococcus pneumoniae*, *Neisseria meningitidis* type A, and hepatitis B—where antigens are produced in baker's yeast via recombinant technology.

- *Recombinant vector vaccines*:

  These involve inserting genetic material from a pathogenic organism into a non-pathogenic carrier (vector), such as a weakened virus or bacterium. For instance, cowpox virus has been explored as a vector for HIV vaccines, while *Salmonella* is used experimentally to deliver hepatitis B virus antigens. Although promising, these vaccines are not yet widely used in clinical practice.

Vaccine Quality Assurance

Modern vaccine development is supported by stringent quality control measures to ensure safety and efficacy. These standards are often based on guidelines from the World Health Organization (WHO), drawing from international expert consensus. The complexity of contemporary vaccine production integrates knowledge from molecular biology, immunology, genetic engineering, and industrial biotechnology.

Overview of Antibodies and Their Production

Antibodies play a key role in diagnosing genetic and infectious diseases, as their presence in blood serum reflects the immune response to specific antigens.

- *Polyclonal antisera*: When animals are immunized with antigen mixtures (e.g., biopolymer combinations), the resulting antisera typically contain a range of antibodies, only a portion of which are specific to the target antigen (see Fig. 4.37).
- Monoclonal antibodies (mAbs):

  Each β-lymphocyte clone produces one specific antibody. To obtain monoclonal antibodies, individual lymphocytes are fused with myeloma cells—immortal cell lines derived from lymphoid tumors—to create hybridomas. These hybrid cells combine the desired antibody specificity with the capacity for continuous division in vitro.

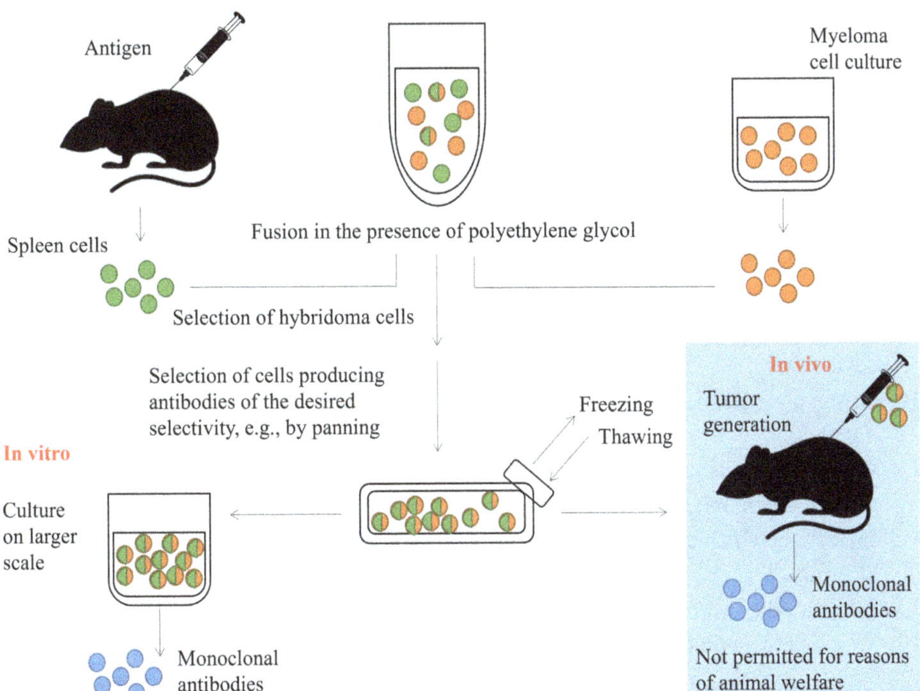

Fig. 4.37 Overview of monoclonal antibody preparation. (Adapted from Rolf D. Schmid)

4.5.4   Cell Fusion Techniques

Mammalian cells, which lack rigid cell walls, are amenable to fusion using various agents:

- *Sendai virus method*: The inactivated Sendai virus (a paramyxovirus) is a lipid-enveloped virus that fuses with cell membranes, facilitating the fusion of two mammalian cells.
- *Chemical fusion*: Agents such as polyethylene glycol (PEG), calcium ions, and lysolecithin can also induce cell fusion. After fusion, hybrid cells are selected based on their growth characteristics and antibody production.

Monoclonal antibodies generated through such methods are widely used in diagnostics, especially in assessing tissue compatibility for organ transplantation.

A key feature of modern immunopharmacology is its integration with a wide range of life sciences, particularly biotechnology. Immunobiotechnology holds great potential for the development of advanced medical and therapeutic products, blending immunology, molecular biology, industrial biotechnology, and pharmacology.

4.5.5   Brainstorming

1. What are monoclonal antibodies, and where are they used?
2. Formulate the main stages of hybrid technology. How is it carried out?
3. What types of vaccines are there?
4. What are the components of vaccines? Identify the requirements for vaccine preparations.
5. Formulate general principles and techniques to produce live vaccines.
6. Discuss the merits and demerits of attenuated live vaccines.
7. Describe the technology for producing killed (inactivated) vaccines. Discuss their advantages and disadvantages, as well as ways to eliminate them.
8. Describe modern ways and directions of development of subunit vaccines.
9. Describe the principle of creating vaccines with artificial adjuvants. What are their advantages?
10. Give a classification of adjuvants for subunit vaccines and clarify the mechanisms of their action.
11. Describe the principles of the creation and technology of producing peptide (ribosomal), chemical, and complex vaccines.
12. Indicate in which cases the sera are used. What is the basis of the industrial production of serums?
13. What are the technological features of obtaining serums?
14. What is the name of immunity that develops when serums are introduced?

Take-Home Messages

- Immunobiotechnology combines immunology with modern biotechnology to develop tools for disease prevention, diagnosis, and treatment, including vaccines, monoclonal antibodies, and immunomodulators.
- Immunomodulatory products (IMPs) include:
  - Microbial-based vaccines (live, inactivated, subcellular).
  - Antibody-based products (sera, immunoglobulins, monoclonals).
  - Immune-modulating cytokines (interleukins, interferons).
  - Natural adaptogens (e.g., ginseng, eleutherococcus).
- Vaccines are categorized into several types, each with advantages and limitations:
  - Inactivated (killed) vaccines are safe but require boosters.
  - Live attenuated vaccines induce strong immunity but may pose risks for immunocompromised individuals.
  - Subunit, chemical, and corpuscular vaccines use selected antigenic parts, offering safety with lower immunogenicity.
  - Biosynthetic and recombinant vaccines are engineered to include specific genes or proteins for enhanced safety and targeted response.
  - Ribosomal vaccines utilize cell organelles like ribosomes as antigens.
  - Second-generation vaccines, such as conjugated, subunit, and recombinant vector vaccines, improve immune system recognition and reduce reactogenicity—especially in infants and high-risk groups.
- Monoclonal antibodies (mAbs) are produced by fusing antibody-producing lymphocytes with myeloma cells, resulting in immortal hybridomas that secrete specific antibodies used in diagnostics, therapy, and transplant medicine.
- Cell fusion techniques, including the use of PEG and Sendai virus, are critical for creating hybridomas used in monoclonal antibody production.

Technological Bioenergy: Bioproduction of Renewable Energy Source   5

Modern society urgently needs new methods to generate sustainable energy and replace raw materials due to global shortages. The Sun is Earth's inexhaustible energy source. Each year, Earth receives approximately 3,850,000 EJ of solar energy. Through photosynthesis, over 170 billion tons of dry biomass are produced annually—yielding more than 20 times the energy currently consumed worldwide. Therefore, solar energy can sustainably meet both present and future human energy demands.

Fuel production via the "biomass-biotechnology" approach integrates solar energy through photosynthesis with livestock, fodder cultivation, and fermentation using specific biological agents (see Fig. 5.1).

5.1 Biogas

Biogas mainly consists of 60% methane (CH_4) and 40% carbon dioxide (CO_2), with trace amounts of hydrogen sulfide and other gases. It is generated through methane fermentation, which converts biomass into energy.

When biodegradable waste undergoes fermentation in an oxygen-deprived environment, it breaks down into CO_2 and CH_4 through an anaerobic process. This breakdown can occur for all natural compounds and many synthetic organic substances. Various microorganisms play roles in different stages of this process, as mentioned in Table 5.1.

5.1.1 Hydrolysis

Extracellular enzymes like amylases, proteases, and lipases break down complex compounds like proteins, lipids, and polysaccharides. However, only half of these organic compounds undergo biodegradation due to the absence of specific enzymes for full decomposition.

© The Author(s), under exclusive license to Springer Nature Switzerland AG 2025   183
I. Digel et al., *Introduction to Industrial Biotechnology*, Learning Materials in Biosciences,
https://doi.org/10.1007/978-3-032-07918-3_5

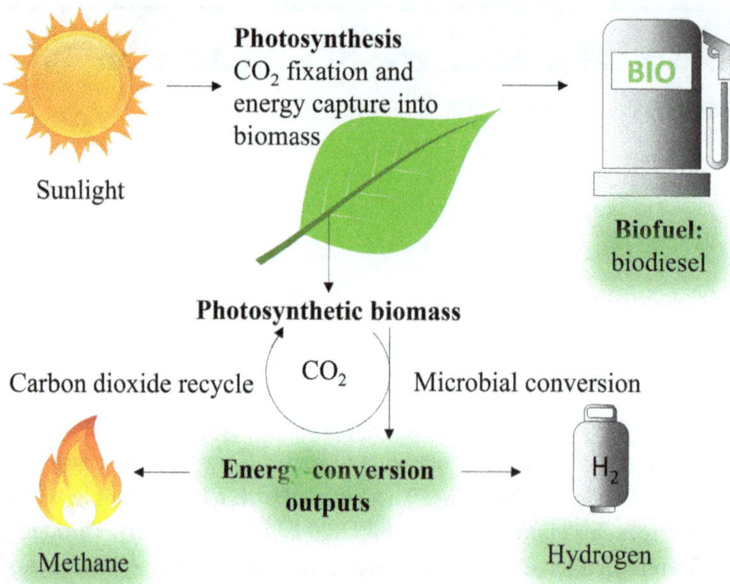

Fig. 5.1 Photosynthesis is the fundamental biomass that can be converted to renewable bio-based energy

Table 5.1 Biological background of biogas formation

#	Reaction type	Groups of bacteria	Electron donor	Electron acceptor	Product
1	Fermentation	Fermentative bacteria	Organic carbon	Organic carbon	CO_2
2	Acidogenesis	Syntrophic bacteria	Organic carbon	Organic carbon	H_2
3	Acetogenesis	Acetogenic bacteria	Organic carbon/H_2	CO_2	CH_3COOH
4	Methanogenesis	Methogenic bacteria	Organic carbon/H_2	CO_2	CH_4

5.1.2 Acidogenesis

Water-soluble compounds, including those from hydrolysis, are turned into short-chain organic acids, alcohols, aldehydes, CO_2, and hydrogen by acidifying microorganisms. The process varies due to different microbes and can be split into hydrogenation and dehydrogenation stages. Essential changes occur through CO_2, H_2, and acetates. Methanogens use these new products for energy. An increase in hydrogen concentration in the environment leads to the accumulation of compounds like propionate and butyrate. Obligatory bacteria, known as acetogenesis, must first convert these products. Facultative anaerobes, which need oxygen, dominate this phase, paving the way for obligatory anaerobes like *Flavobacterium, Pseudomonas, Clostridium, Micrococcus,* and *Bacillus.*

5.1.3 Acetogenesis

In this phase, acetate microorganisms transform acid phase products into hydrogen and acetates. These are later used by methanogenic bacteria. The hydrogen released can be toxic to the bacteria, leading to a symbiotic relationship between acetogenic organisms and methane bacteria that consume the hydrogen. Acetates are vital in this phase, with about 25% of acetates and 10% of hydrogen produced during biodegradation.

5.1.4 Methanogenesis

Methanogenic bacteria play a role in the production of methane by transforming different substances such as acetic acid and CO_2 into methane. While a few bacteria can directly convert acetic acid into methane, a significant portion of the conversions are carried out by heterotrophic methane organisms. Autotrophic methane bacteria are responsible for 30% of methane production through the reduction of CO_2. This process involves the use of hydrogen, which promotes the growth of bacteria and leads to the production of short-chain organic acids. As a result a gas rich in CO_2 is formed, with only a small fraction being converted into methane (refer to Fig. 5.2).

$$4H_2 + CO_2 \rightarrow CH_4 + 2H_2O$$

$$3H_2 + CO \rightarrow CH_4 + H_2O$$

$$2H_2O + 4CO \rightarrow CH_4 + 3CO_2$$

$$4HCOOH \rightarrow CH_4 + 3CO_2 + 2H_2O$$

$$4CH_3CH_2OH \rightarrow 3CH_4 + CO_2 + 2H_2O$$

$$CH_3COOH \rightarrow CH_4 + CO_2$$

Methane production contributes to sustainable organic waste management, offering benefits such as local energy generation, improved organic fertilizer quality, and environmental protection. Clean energy sources like biogas are eco-friendly alternatives to fossil fuels.

Methanogenic bacteria can be grouped into two categories based on temperature: mesophilic and thermophilic. Mesophilic bacteria thrive in temperatures around 35 °C while thermophilic bacteria prefer temperatures around 55 °C. Methanogenesis is more efficient under thermophilic conditions and can also help deactivate pathogenic microorganisms and pests. When processing livestock waste, microbial communities in methane tanks originate mainly from animal gut microbiota and the environment. Common bacteria include *Lactobacillus* spp. and *Bacteroides* spp., along with cellulose-degrading strains such as *Bacteroides succinogenes* and *Ruminococcus flavefaciens*. Over time, stable microbial communities form, adapted to specific temperature and substrate conditions.

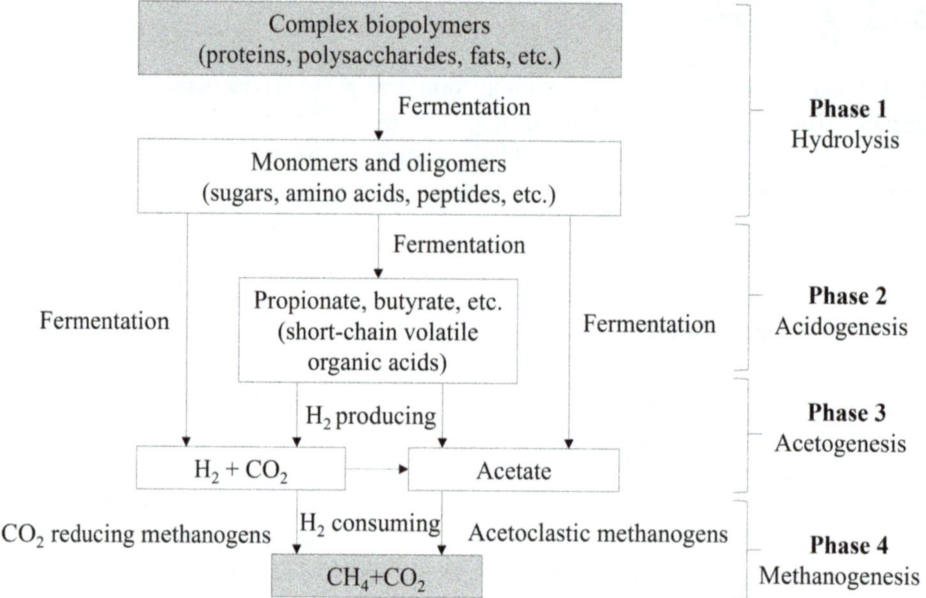

Fig. 5.2 The phases of biogas production

The initial breakdown of complex organic polymers involves bacteria like *Clostridium*, *Bacteroides*, and *Ruminococcus*.

During the acetogenic phase, bacteria such as *Syntrophobacter, Syntrophomonas* and *Desulfovibrio* help break down compounds into acetate, H_2, and CO_2—all precursors for methane. Some microbes, like *Clostridium* sp., *Methanothrix*, *Methanosarcina*, *Methanococcus*, *Methanogenium*, *Methanospirillum* and *Methanosarcina*, produce acetate from CO_2 under hot conditions.

Biogas production mainly uses waste from farms, expired goods, household trash, and city sewage. The procedure occurs at about 40 °C with a pH of 6–8. Figure 5.3 illustrates a methane tank designed for agricultural waste processing, like manure and crop leftovers.

Waste is supplied to the reactor's lower part, where treated runoff is also selected. The reactor can operate either intermittently or semi-continuously. It usually features two sections to differentiate the stages of the process.

5.1.5 Brainstorming

1. What is the composition of biogas?
2. What is methanogenesis? What stages does it consist of?
3. What microorganisms are involved in the process of biogas formation?
4. Describe microorganisms-methanogens.
5. What raw materials can be processed using methanogenesis?
6. List the plants/installations used for methanogenesis.
7. Where can methanogenesis products be used?

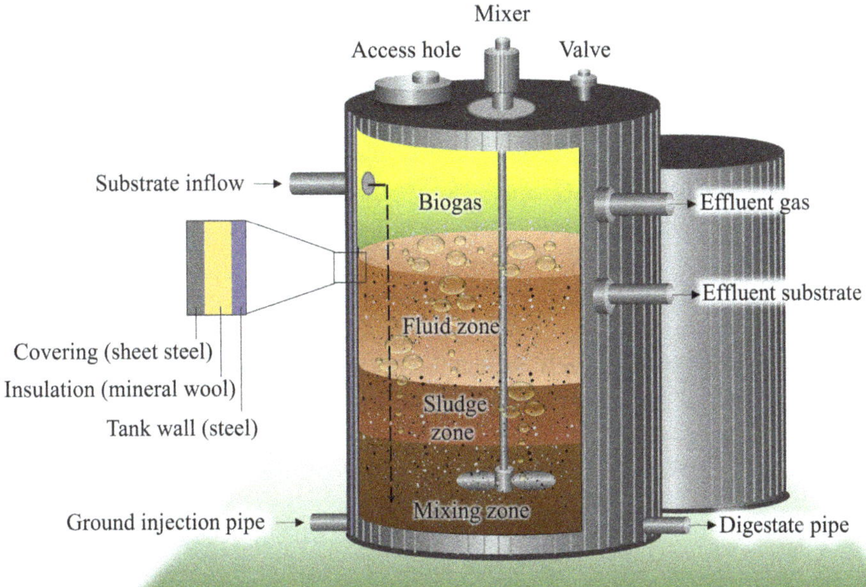

Fig. 5.3 Outline of biogas power installation

8. What is the efficiency of using biogas?
9. In what ways can the processes of obtaining bioethanol be intensified?

Take-Home Messages

- Biogas is a valuable renewable energy source derived from the anaerobic digestion of organic biomass, primarily producing methane (CH_4) and carbon dioxide (CO_2).
- Methane fermentation is a multi-stage biological process involving hydrolysis, acidogenesis, acetogenesis, and methanogenesis—each catalyzed by distinct microbial communities.
- Methanogenic bacteria are sensitive to environmental conditions, especially temperature. Thermophilic methanogenesis (at ~55 °C) is more effective than mesophilic (at ~35 °C) and also contributes to pathogen reduction. Hydrogen levels must be regulated, as excess H_2 can inhibit key metabolic steps. Methanogens help by consuming hydrogen, maintaining a favorable microbial balance.
- Only a portion of biomass is degradable due to limitations in enzymatic capabilities, emphasizing the importance of optimizing hydrolytic conditions and microbial consortia.
- The microbial diversity and symbiosis are critical: Fermentative, syntrophic, acetogenic, and methanogenic bacteria each perform specific roles, ensuring efficient biomass conversion. Stable microbial communities develop over time in methane digesters, adapting to substrate composition and operating conditions—highlighting the need for consistent input management.

5.2   Ethanol Production

In recent decades, ethanol has gained considerable attention as a renewable fuel. Its appeal lies in several key advantages:

- It is a liquid and thus easily transported.
- It has a calorific value approximately two-thirds that of gasoline.
- It increases the octane rating in unleaded gasoline blends.

Ethanol can be produced biochemically from a wide range of sources, including sugarcane juice, cassava, molasses, cereals, agricultural waste, and industrial by-products.

Biologically, ethanol is most commonly produced through the fermentation of hexoses by bacteria such as *Zymomonas mobilis*, *Z. anaerobica*, *Sarcina ventriculi*, and *Clostridium thermocellum*, as well as by the yeast *Saccharomyces cerevisiae*:

$$C_6H_{12}O_6 \rightarrow 2CH_3CH_2OH + 2CO_2$$

Yeast produces ethanol anaerobically, but a small amount of oxygen is needed for its growth and reproduction. The transformation of sugars into alcohol is coupled with metabolic activity and cell proliferation. The efficiency of this process depends on several factors, including substrate concentration, oxygen levels, and the accumulation of ethanol, which can inhibit microbial activity. Therefore, industrial strains with high alcohol tolerance are essential.

Industrial ethanol fermentation can be conducted in three main modes: batch, continuous, and semi-continuous processes.

In *batch fermentation*, microorganisms are inoculated into a prepared medium and allowed to ferment over several days at approximately 30 °C and pH 4.5. Once fermentation is complete, the cells can be reused. The typical ethanol yield is around 45% of the substrate's weight (see Fig. 5.4).

In *continuous fermentation*, process intensity—especially the flow rate—affects efficiency. The goal is to optimize alcohol yield while ensuring maximum sugar utilization. By the end of fermentation, the mash may contain approximately 10% ethanol. Higher ethanol concentrations reduce energy requirements during distillation. For instance, producing 96% ethanol from a 10% mash typically requires over 4 kg of steam per liter of ethanol. Valuable by-products of this process include carbon dioxide, fusel oils, residual mash, and yeast biomass, each with potential commercial uses.

Ethanol recovery and purification is a major technical challenge. The choice of separation technique depends on multiple factors: the viscosity and composition of the culture liquid, the concentration of ethanol, the presence of impurities, and the required final purity and physical state of the product.

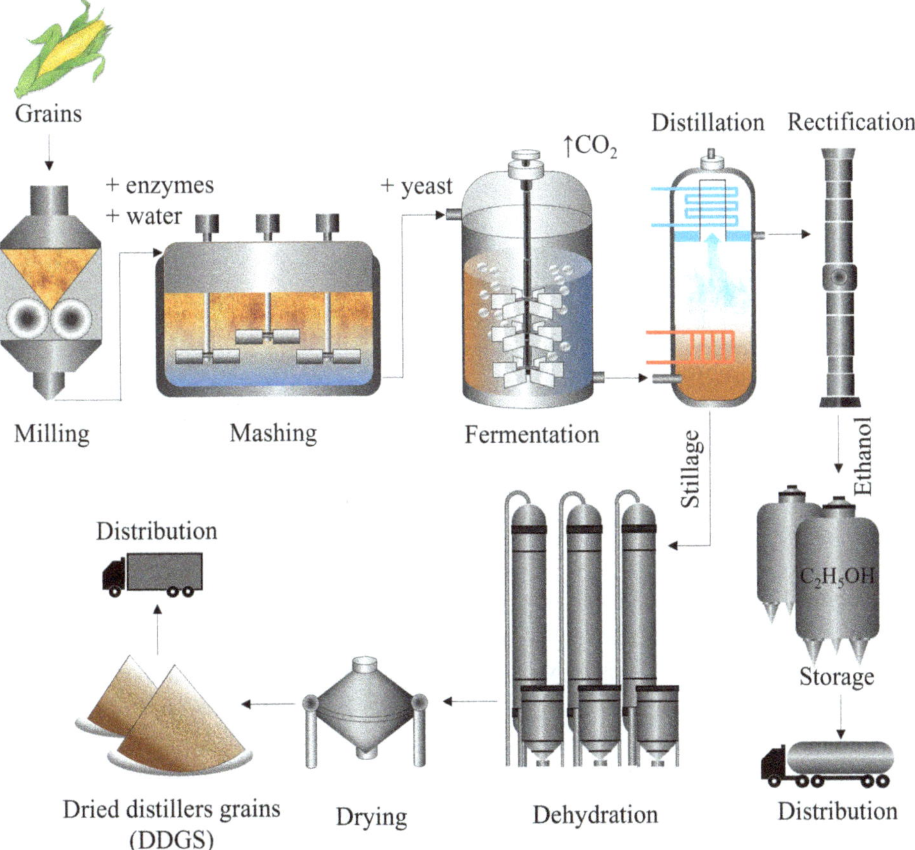

Fig. 5.4 Schematic diagram of ethanol production

Product isolation and purification typically proceed in three stages

1. Removal of insoluble substances using filtration, centrifugation, or sedimentation.
2. Ethanol isolation via solvent extraction, sorption, precipitation, or ultrafiltration.
3. Final purification, which eliminates residual impurities to achieve the desired product quality.

Recent technological improvements in ethanol production include

- Vacuum distillation during fermentation, which improves sugar conversion and fermentation efficiency.
- Processes involving biomass recirculation and low-pressure conditions, which can increase bioreactor productivity up to 12 times compared to traditional batch methods.

- Utilization of *Zymomonas mobilis*, which surpasses yeast in both glucose consumption and alcohol production rates.
- Application of immobilized cells, which significantly enhances productivity compared to conventional suspension cultures.

5.2.1 Brainstorming

1. What microorganisms are used to produce alcohol?
2. What is bioethanol?
3. From which substrates can bioethanol be obtained?
4. What are the areas of application of bioethanol?
5. Draw the schematic diagram of ethanol production.
6. What are the ways of concentrating alcohol from the culture liquid?
7. In what ways can the processes of obtaining bioethanol be intensified?

Take-Home Messages
- Ethanol is a sustainable fuel alternative with favorable physical properties and broad applicability in energy and industry. Its production relies on microbial fermentation, particularly using yeast and ethanol-tolerant bacteria, under carefully controlled conditions.
- Ethanol purification is complex, requiring multi-step separation and purification methods tailored to product quality and process conditions. Continuous fermentation and cell immobilization significantly improve productivity, while technological innovations reduce energy demands.

5.3   Hydrogen Biotechnology

Hydrogen is a clean fuel with an exceptionally high calorific value (per mass). Current research is focused on developing sustainable methods to obtain hydrogen, particularly through the photochemical decomposition of water. In this process, chloroplasts, in the presence of an artificial electron donor and a bacterial extract containing the enzyme hydrogenase, can produce hydrogen:

$$\text{Electron donor} \xrightarrow{} \text{Photosystem I} \xrightarrow{e^-} \text{e} - \text{carrier} \xrightarrow{e^-} \text{H}_2 \uparrow$$
$$\text{H}^+ \uparrow$$

Hydrogenase receives electrons from ferredoxin, and various organic compounds may serve as electron donors. This light-driven process requires visible light for activation (see Fig. 5.5).

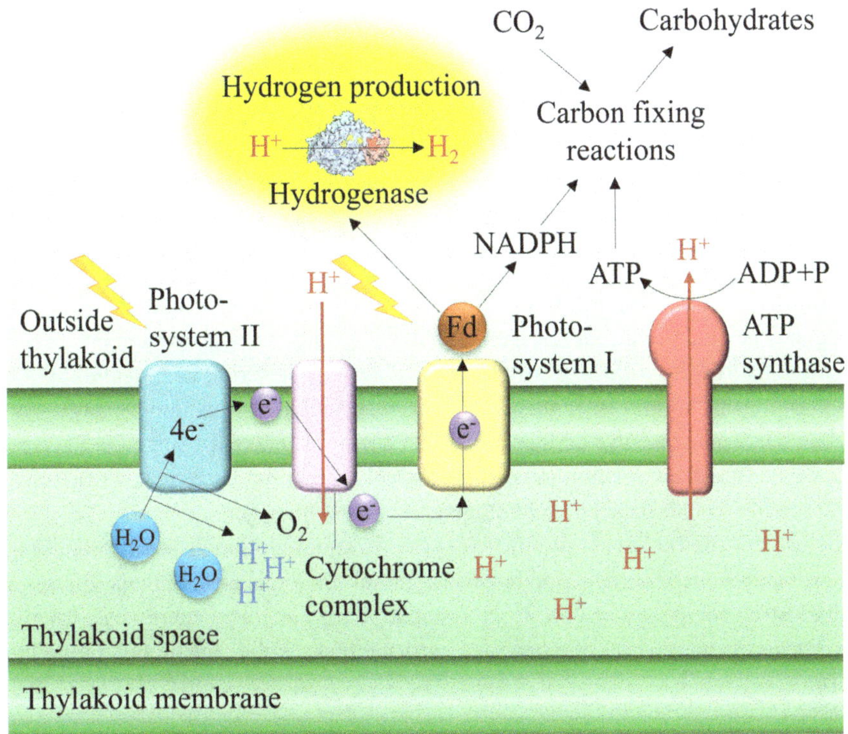

Fig. 5.5 Photosynthesis production of hydrogen

Hydrogen has a much higher calorific value (per mass) than hydrocarbons, and the process of its bioproduction is renewable, depending primarily on the stability of the isolated chloroplasts. Hydrogen can also be generated using artificial electron donors and light-absorbing pigments, without requiring chloroplast membranes. Some microorganisms, such as cyanobacteria, are also capable of hydrogen release.

Microbial hydrogen formation can proceed from carbohydrate-rich compounds, including starch, cellulose, amino acids, and keto acids.

Efforts to develop biophotolysis systems for water have resulted in the creation of various configurations. Regardless of their specific components, these systems must include two critical elements:

1. The electron-transport chain of photosynthesis, including the water-splitting complex.
2. Catalysts for hydrogen formation.

Catalysts may be inorganic (e.g., metallic platinum) or enzymatic (e.g., hydrogenases). Enzymatic hydrogenases can function in both soluble and immobilized forms:

$$H_2 \xrightarrow{\textit{Microalgae}} \text{Photosynthesis in light}$$

$$O_2 \longrightarrow H_2 \left(\text{in the dark}\right)$$

As well as with bacterial immobilized cells:

$$H_2O \xrightarrow{\textit{Anabaema nidulans}} NADH + O_2$$

$$NADH \xrightarrow{\textit{Rhodospirillum sp.}} H_2 + NAD^+$$

Using hydrogenases, almost any plant photosystem can be engineered to produce hydrogen. The goal of current research is to develop fully artificial systems that replicate the natural hydrogen-producing mechanisms of algae or bacteria-plant symbioses.

A comprehensive and integrated approach is essential for planning energy systems based on renewable resources. Since renewable energy is inherently linked to the environment, its study and application must span multiple scientific disciplines—from environmental science to industrial biotechnology and process management.

The use of renewable energy must be systemic and multidimensional, promoting not only sustainable energy production but also regional economic development. For instance, agro-industrial complexes offer a strong foundation for integrated energy use, where livestock and crop waste can be converted into methane, solid and liquid biofuels, and fertilizers.

5.3.1   Brainstorming

1. What are the advantages of hydrogen as an energy carrier?
2. List ways of obtaining hydrogen from water.
3. Name and characterize the biocatalytic system's elements for water photolysis.
4. What are the advantages of producing hydrogen using biocatalysts?
5. How can hydrogen be obtained based on growing microbial populations (chemotrophs, phototrophs, halobacteria)?
6. What are biofuel elements and bioelectrocatalysis?
7. What are the prospects for obtaining molecular hydrogen?
8. What types of renewable fuels are intended to replace fossil resources?
9. What are the types of renewable fuels? Give examples.

5.4   Acetone and Butanol Production via Fermentation

Acetone and butanol are essential industrial chemicals that can be produced on a large scale via bacterial fermentation. This process, known as acetone-butanol fermentation, is primarily carried out by bacteria of the genus *Clostridium*, including *C. acetobutylicum*,

Take-Home Messages
- Hydrogen is a high-energy, clean fuel that can be produced biologically through light-driven photosynthetic and enzymatic processes. Integrating hydrogen biotechnology into agro-industrial systems enhances environmental sustainability while supporting local economies.
- Hydrogenase enzymes and photosystems, whether from plants or microbes, are key components in biohydrogen production systems. Artificial and immobilized systems mimicking natural photolytic processes are a promising direction for sustainable hydrogen generation.

C. beijerinckii, *C. chauvoei*, *C. roseum*, and others. While most *Clostridium* species are capable of this fermentation, only *C. acetobutylicum* and *C. beijerinckii* are commonly used in industrial applications.

Most industrial strains primarily produce butanol, with smaller amounts of acetone, ethanol, and trace levels of acetic and butyric acids (see Fig. 5.6).

Clostridium spores are widespread in soil, and many strains are also isolated from the rumen of animals. These bacteria can ferment substrates such as glucose, cellobiose, and starch.

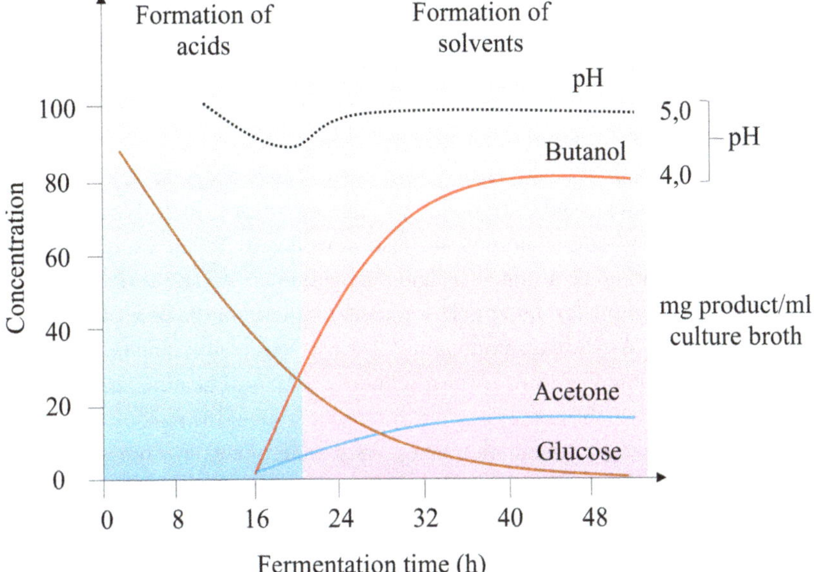

Fig. 5.6 Fermentation curve for *Clostridium acetobutylicum*

5.4.1 Inoculum Preparation

There are several methods for preparing commercial inoculum, many involving thermal shock to induce spore germination. The process includes:

- Destroying vegetative cells and weak spores through heat treatments, such as:
  - 50 s or 2 min at 100 °C
  - 30 min at 95 °C
  - 5 min at 80 °C.

Repeated heating during sub-cultivation selects for the most thermally resistant and active spores. Industrially, bacteria are stored as spores, and highly motile cells are associated with higher solvent yields.

Initial inoculum generations are grown in nutrient medium at 37 °C for 24 h, then transferred to a molasses or starch-based medium in 500 mL flasks. These are incubated again, and the culture is scaled up into 4-liter flasks. During this stage, inert gas is bubbled through the medium, and the pH drops to 3.9–4.5.

In batch cultivation, carbon source concentration and pH are continuously monitored. High solvent concentrations, particularly butanol, can inhibit fermentation, thus limiting solvent production to around 20 kg/m^3. Typical solvent yields reach up to 0.6 kg/m^3/h, depending on the substrate and conditions.

To facilitate product removal, non-soluble organic solvents are introduced into the bioreactor as extractants. The best results have been achieved with oleyl alcohol and its mixtures in benzyl benzoate. When glucose is supplied periodically (about 80 kg/m^3), the removal of butanol via extraction significantly boosts productivity—up to 67% more butanol compared to standard batch fermentation (see Fig. 5.7).

Different raw materials are used as sources of carbon and energy, such as grain (starch), molasses (crude molasses), and pure molasses. Various wastes, hydrolyzed wood, waste hydrolysates, whey, and sulfite liquids are also considered. When using molasses, the sugar concentration should be around 7%, and the medium should be enriched with superphosphate and ammonium. The latter is introduced gradually in the case of pure molasses to maintain the pH at 5,6–6,0 in an amount of 2,2–1,3% NH$_3$ concerning the sugar concentration. This increases the yield of solvents, reduces the cost of heating, and gives significant savings in steam, which is necessary for the evaporation of the residue. The bioreactor and medium are sterilized. Fermentation takes place under the excess pressure of gases to create anaerobic conditions. The initial pH is 6,0, which decreases to 5,4 after 24 h.

Distillation is a traditional way of separating ethanol, acetone, and butanol from diluted aqueous solutions. When distilling the products of acetone-butyl fermentation, the heat generated by cooling the heat-sterilized fermenter can be used. *Distillation* is a flexible process in which fractionation and isolation of each product are possible. Novel methods of product isolation are membrane processes, for example, ultrafiltration and reverse osmosis.

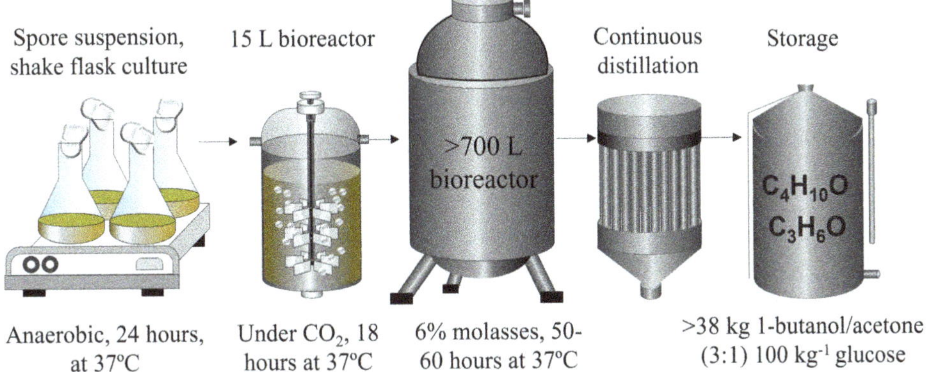

Fig. 5.7 Fermentation and recovery of solvents

5.4.2 Brainstorming

1. Application areas of acetone.
2. Name the raw materials for the production of acetone and butanol.
3. What wastes are produced from the manufacturing process of acetone and butanol.
4. Characteristics of acetone-butyl fermentation, pathogens.
5. Alternative technologies for the production of acetone and butanol.
6. Ways to intensify the production of acetone and butanol.

Take-Home Messages
- Acetone-butanol fermentation, primarily driven by *Clostridium acetobutylicum* and *C. beijerinckii*, remains a viable microbial method for solvent production.
- Butanol inhibition is a limiting factor in batch fermentation, but in situ extraction significantly improves productivity.
- The choice of substrate, nutrient enrichment, and controlled pH play a key role in optimizing solvent yields.

6.1 Microbiological Production of Organic Acids

Organic acids are widely used in the food and pharmaceutical industries, engineering, (e.g., as buffer solutions, see Table 6.1).

They can be obtained from natural raw materials or via organic synthesis. More than 50 organic acids are currently produced through microbiological synthesis. Some acids—such as citric, gluconic, ketogluconic, and itaconic acids—are exclusively obtained through microbial processes. Others, like lactic, salicylic, and acetic acids, can be synthesized both chemically and microbiologically. Malic acid can be produced through chemical and enzymatic methods.

From the standpoint of microbial metabolism, organic acids are intermediates or end-products of carbon and energy source degradation:

- Citric, isocitric, α-ketoglutaric, succinic, fumaric, and malic acids are intermediates of the tricarboxylic acid (TCA) cycle.
- Gluconic, ketogluconic, and tartaric acids are intermediates of the direct glucose oxidation pathway.
- Lactic, butyric, and propionic acids are end-products of carbohydrate metabolism.
- Acetic acid is a product of ethanol oxidation, and aliphatic mono- and dicarboxylic acids arise during alkane oxidation (Fig. 6.1).

In general, the production of organic acids can be divided into two groups: *anaerobic* (lactic, propionic, etc.) and *aerobic* (acetic, citric, itaconic, gluconic). Aerobic acid production is typically carried out through both surface and submerged fermentation, while anaerobic acids are usually obtained via submerged fermentation.

I. Digel et al., *Introduction to Industrial Biotechnology*, Learning Materials in Biosciences, https://doi.org/10.1007/978-3-032-07918-3_6

Table 6.1 Types of organic acids and their application areas

C atom number	Organic acid	Application
C_2	Acetic acid	"Green" solvent for the production of chemical compounds
C_2	Oxalic acid	Mordant in dyeing processes, in bleaches, in baking powder. As a reagent in silica analysis instruments
C_3	3-hydroxypropionic acid	Potential substitute for biodegradable polymer production
C_3	Lactic acid	Pharmaceuticals, cosmetics, food and beverages, production of biodegradable polymer, etc.
C_3	Propionic acid	Pharmaceuticals, preservatives, food additives, solvent, artificial flavoring, pesticides, etc.
C_4	Butyric acid	Adjuvant in animal feed, synthetic flavoring, therapeutics, etc.
C_4	Fumaric acid	Souring agent in food and feed, polyester resins
C_4	Malic acid	Additive in food and feed, acidulant and flavoring agent
C_4	Succinic acid	Additive in food and feed, effective in product packaging, coating, etc.
C_5	Itaconic acid	Polymer industry, production of synthetic fibers, coatings, thickeners, adhesives, etc.
C_5	Levulinic acid	Anti-inflammatory drugs, anti-allergy agents, supplements, etc.
C_6	Adipic acid	Gelling aid, acidulant, leavening, and buffering agent
C_6	Ascorbic acid	Food additives, pharmaceutical
C_6	Citric acid	Food additive
C_6	Glucaric acid	Production of cleaners and detergents, concrete formulations
C_6	Gluconic acid	Food and beverages, chemical analysis, buffering agent

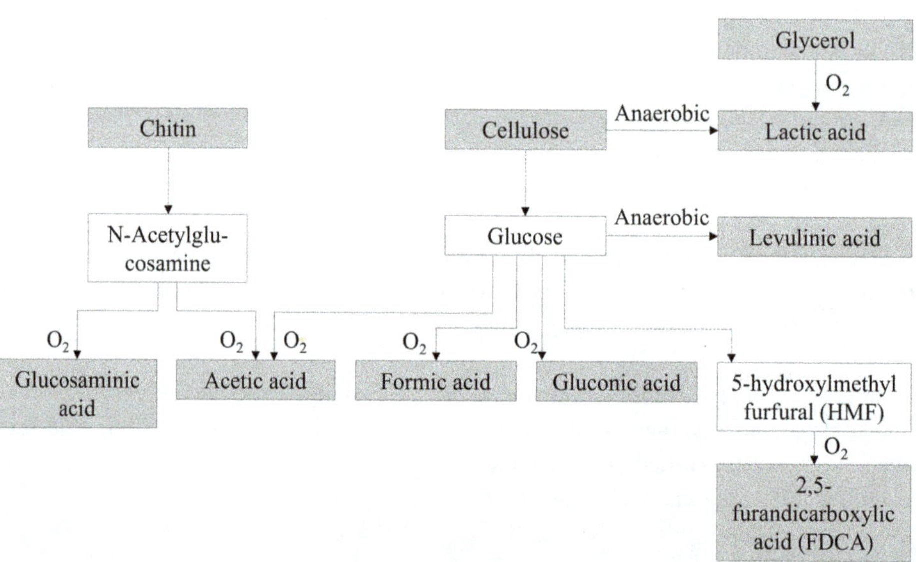

Fig. 6.1 Selected examples of biomass conversion to various organic acids

Specialized high-yield microbial strains are used in industry to ensure efficient carbon substrate utilization and mono-product formation. Substrates vary widely: initially glucose and sucrose were most common, later replaced by molasses, hydrolyzed starch, and other complex media.

6.1.1   Lactic Acid

Numerous bacteria can produce lactic acid as a primary or secondary end-product of fermentation. These are collectively known as *lactic acid bacteria (LAB)*, and include genera such as:

- Lactobacillus, Leuconostoc, Pediococcus, Lactococcus, and Streptococcus.
- Also Carnobacterium, Enterococcus, Vagococcus, Oenococcus, Weisella, and Tetragenococcus.

The genus *Lactobacillus* contains three main subgroups:

- *Thermobacterium:* cannot grow below 15 °C but tolerates over 50 °C.
- *Streptobacterium:* not thermophilic.
- *Betabacterium:* produces D- and L-lactic acid from glucose.

Thermostreptobacterium, *Streptococcus*, and *Pediococcus* are *homofermentative*, meaning they produce predominantly *lactic acid* from hexoses. *Betabacterium* and *Leuconostoc* are *heterofermentative*, producing lactic acid along with *acetic acid, CO_2,* and *ethanol*.

Glucose is converted to lactic acid via:

1. Homofermentative fermentation.
2. *Heterofermentative fermentation*, which includes glucose breakdown to *pyruvate*, followed by its reduction to *lactate* (see Fig. 6.2).

Lactic acid bacteria consume maltose, lactose, glucose, saccharified starch, etc., as a substrate. In industrial-scale natural raw materials, such as corn stover, raw sweet potato, molasses, fruit and vegetable waste, waste-derived solids and algal biomass, a mixture of glucose and xylose, lignocellulose-derived sugars, oil palm trunk, cellulosic feedstock, etc., are widely used. The medium for LAB must also contain B vitamins, amino acids, purines, pyrimidines, organic acids, etc.

Thermophilic homofermentative species are promising in industrial production. For example, *L. delbrueckii* yields lactic acid up to 95–98% from consumed sucrose.

Some developed processes for biotechnological production of lactic acid are based on the use of highly productive strains of microorganisms, such as *Lactobacillus casei* M-15,

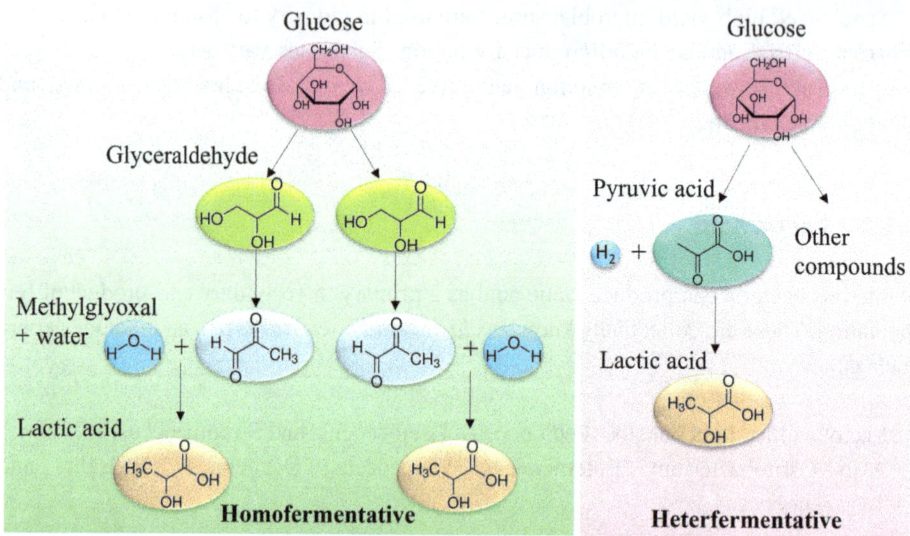

Fig. 6.2 Two pathways of lactic acid formation

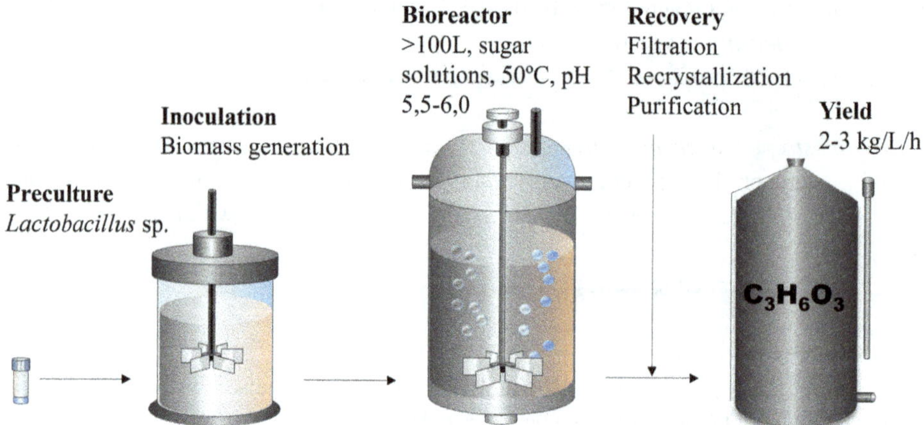

Fig. 6.3 Schematic representation of lactic acid microbial production

Bacillus coagulans, Enterococcus mundtii QU 25, *L. brevis* ATCC 367, *Lactobacillus plantarum* ATCC 21028, *L. paracasei, Rhizopus oryzae* and mixed cultures of *Lactobacillus rhamnosus* and *Lactobacillus brevis*; *Lactobacillus paracasei* and *Lactobacillus coryniformis*.

The manufacturing scheme of obtaining L-lactic acid starts from upstream processing; molasses medium containing a high amount of sugar, malt, and yeast extracts, vitamins, and ammonium phosphate is inoculated with *Lactobacillus* sp. (Figure 6.3). Fermentation takes place at around 50 °C with a pH of 6,0. As the lactic acid begins to express in the medium, it is periodically neutralized with chalk. The entire fermentation cycle is completed in 5–10 days, depending on the microbial strain types used. During downstream processing, culture liquid is separated by filtration, then the filtrate is evaporated, cooled down and applied to crystallization that lasts two days. The formulation phase includes several steps: crystals of calcium lactate are treated with sulfuric acid at around 65 °C, which causes them to precipitate, and potassium ferrocyanide is added to the supernatant in order to remove iron ions, then exposed to sodium sulfate to recover from heavy metals. The colorants are removed with activated charcoal. After that, the lactic acid solution used for technical purposes is vacuum evaporated. High-purity lactic acid can be obtained by distilling with methyl esters.

Lactate is widely used as an acidifier in producing jam, jelly, confectionery, liquors, extracts, canning vegetables, etc. It is also an essential ingredient in leather and textile manufacturing, the pharmaceutical industry, the production of solvents and detergents, etc.

Homo- and heterofermentative LAB have long been used in the food industry, especially in the bakery. The starter culture, based on a consortium of LAB and yeast, provides favorable fragrance, taste, porosity, color, and freshness for baked products. Lactic acid fermentation is the basis for ensilage in livestock and vegetable and fruit souring. They are also well-known for the formulation of dairy products and cheeses. Finally, LAB is part of preventive and curative medicines.

6.1.2 Citric Acid

Citric acid was originally obtained from lemons and limes. Today, it is produced through submerged fermentation (SmF) and solid-state fermentation (SSF) using specially selected microbial strains (see Fig. 6.4).

Currently, selected strains of *Aspergillus*, including *A. niger*, *A. aculeatus*, *A. awamori*, *A. carbonarius*, *A. wentii*, and *A. foetidus*, are widely used as citric acid-secreting cell factories. The main raw material for citric acid production is molasses, which contains iron that must be removed via precipitation with potassium ferrocyanide.

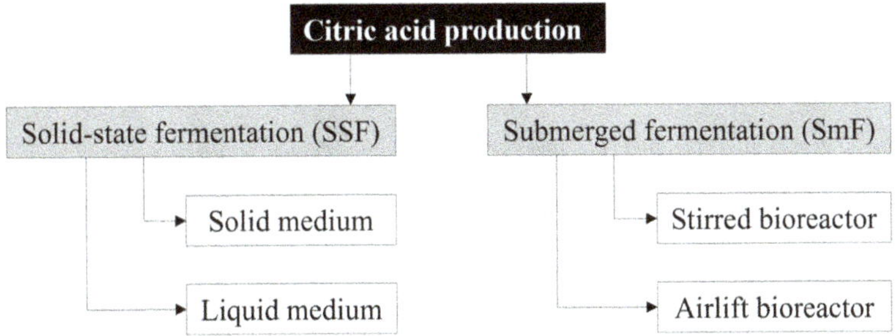

Fig. 6.4 Industrial process for the citric acid production

The fungal spores, or conidia, are produced separately in multiple steps. First, *Aspergillus* species are grown on an agar medium. After initial growth, the culture is scaled up on either a solid or liquid medium at 32 °C for approximately 10 days. As conidia form and mature, the mycelium changes from colorless to black. Conidia are then harvested by vacuum aspiration, dried at 30 °C in a thermal chamber, mixed with sterile sorbents, packed into sterile containers, and stored for up to two years.

There are various types of citric acid production. First, solid-state fermentation is achieved on a free-flowing medium and in the liquid phase. Here, the nutrient medium is poured into the cuvettes, then placed on the shelves in a sterilized fermentation chamber. Next, the surface of the medium is inoculated with starter culture or spores (conidia) in the mass of 60–70 mg per 1 m² of the cuvette area.

The mycelium grows intensively during the first 2–3 days of mid-stream processing. The medium temperature of this period stabilizes at around 33 °C, and the aeration intensity is also adjusted. The air supply is increased 5 times during the active acid formation period. As a result of more intensive thermogenesis, the temperature drops to 31 °C. As the acidification process decreases, the aeration regime becomes less intense. The process is controlled by the indicators of the titratable acidity of the medium; when its level reaches 12–20%, and the residual sugar concentration is about 1–2%, the process is considered to be completed. The citric acid content from the level of all acids reaches around 96%. The discarded solution is poured into a collection container and transported to the treatment, where purified mycelium is used in fodder production.

The submerged technique of citric acid production is economically advantageous (Fig. 6.5). The active strains of Aspergillus are used for its realization. Conidia are revived

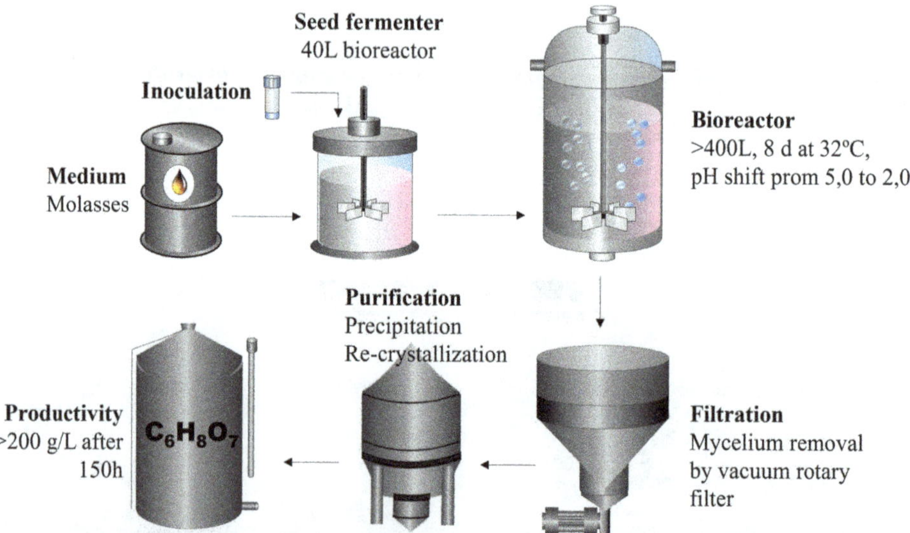

Fig. 6.5 Schematic representation of citric acid microbial production

first and then scaled up in the inoculator-containing medium with 3–4% sugar content. After a couple of days, the inoculum is transferred from the seed reactor to the main fermenter, where the process is carried out for 8–10 days, under 32 °C, supplied with molasses solution. When the fermentation process is completed, the mycelia are filtered out, and the culture liquid is treated according to the set regime.

The finished product, highly purified crystalline citric acid, is obtained during the downstream processing. The obtained solution contains the remains of unfermented sugars and mineral salts, in addition to the target acid, as well as gluconic and oxalic acids. It is bound with calcium hydroxide to isolate citric acid from this solution to form a hardly soluble calcium citrate. Calcium salts of citric and oxalic acid precipitate, calcium gluconate, and most of the molasses' organic and mineral components remain in the solution. The precipitate is separated on a vacuum filter, washed, and dried. Further, the precipitate is treated with sulfuric acid to convert the citric acid to its free state and remove calcium oxalate. Finally, the citric acid solution is filtered, concentrated by vacuum evaporation, and then subjected to crystallization under slow cooling.

Citric acid enhances the pancreas's activity, stimulating appetite and food digestion. It is used in cooking, dairy, food preservation, producing non-alcoholic beverages, winemaking, refining vegetable oils, marmalade, waffles, etc. It retains its natural flavor when stored for a long time in the frozen state of meat and fish.

6.1.3   Acetic Acid

Acetic acid fermentation is primarily carried out by *Acetobacter* species—especially *A. oxidans*, *A. aceti*, and *A. xylinum*—collectively referred to as acetic acid bacteria (AAB). This process follows alcoholic fermentation: the sugar in the substrate is first converted to ethanol, which is then oxidized to acetic acid under aerobic conditions.

Alcoholic fermentation is most efficiently performed by selected wine yeast strains, which not only produce ethanol but also release metabolic by-products that enhance the flavor and aroma of vinegar.

Vinegar is a valuable liquid containing approximately 15% acetic acid, along with various aromatic and flavor compounds. Its characteristics vary depending on the fermented substrate—e.g., apple, grape, or pear vinegar. The specific flavor of vinegar is attributed to the presence of esters, higher alcohols, and organic acids.

In addition to AAB, certain acetogenic anaerobic bacteria, such as species from the genera *Acetobacterium* and *Clostridium*, can directly convert sugars into acetic acid without ethanol as an intermediate. The reaction may be represented as:

$$C_6H_{12}O_6 \rightarrow 3CH_3COOH$$

The efficiency of acetic acid production can be greatly improved by immobilizing microbial producers on natural matrices or sorbents. When an ethanol solution is passed through such a system, a high-concentration acetic acid solution (>20%) can be obtained.

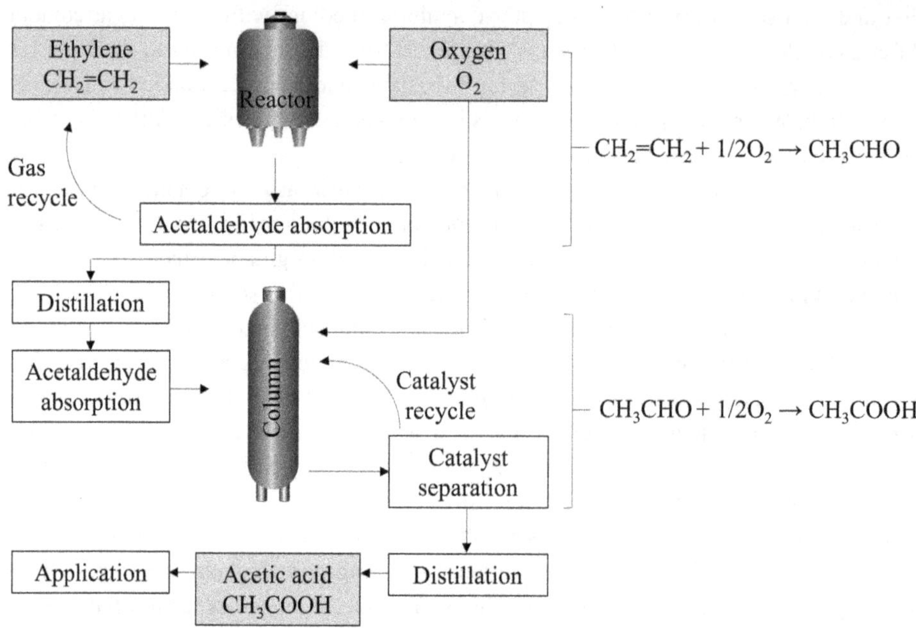

Fig. 6.6 Conventionalized outline of acetic acid production

The general production scheme for acetic acid using *Acetobacter* species is illustrated in Fig. 6.6.

Globally, over 10 megatons of acetic acid are produced each year. Approximately 65% of this amount is obtained via chemical synthesis, resulting in technical-grade acetic acid. It is extensively used to produce acetone, acetylene, synthetic dyes, medicines, fragrances, etc. In addition, bio-based acetic acid is well applied in the food and beverage industries, especially in vinegar and flavoring production.

6.1.4 Propionic Acid

Propionic acid is synthesized by Gram-positive, non-motile, non-spore-forming bacteria of the genus *Propionibacterium*. It is widely used in the food and pharmaceutical industries as both an antimicrobial and flavoring agent (see Table 6.2).

The biochemistry of propionic acid formation is relatively complex. It begins with pyruvic acid, which is carboxylated to oxaloacetic acid with the help of carbon dioxide (CO_2) and the coenzyme biotin. Oxaloacetic acid is then reduced to succinic acid via malic and fumaric acid intermediates in the next steps:

- Succinic acid, with the assistance of ATP and Coenzyme A (CoA), is converted to succinyl-CoA.

Table 6.2 Industrial uses of propionic acid

The industrial application	Remarks
Herbicides	Synthesis of its constituent
Chemical intermediates	Production of plastics, textiles, rubber, and dye intermediates
Flavors and fragrances	Precursor
Pharmaceuticals	Synthesis of propionic anhydride
Cellulose acetate propionate	Precursor
Food preservatives	Aids to prevent the food from molding
Feed and fodder	Feed preservation, such as hay, silage, and grains, as it inhibits undesirable microbial populations

- Succinyl-CoA is then transformed into methylmalonyl-CoA under the influence of methylmalonyl-CoA isomerase, with vitamin B_{12} as a cofactor.
- Finally, methylmalonyl-CoA undergoes decarboxylation, producing free CoA and propionic acid as the end product (see Fig. 6.7).

Important starter cultures for commercial propionic acid production are P. acidipropionici, P. freudenreichii, P. zeae, P. jensenii, P. shermanii, P. beijingense, etc.

Different raw materials and compounds are used as a substrate for fermentation, including flour hydrolysate, glycerol, lactate, sugarcane molasses, cheese whey, etc. Propionic acid is obtained based on various culture systems, such as batch, bed-batch, fibrous bed reactor, semi-continuous, batch with extractive fermentation, and continuous in situ cell retention. According to the strain type, fermentation can take 2 days to 4 months.

6.1.5 Gluconic Acid

Gluconic acid, also known as pentahydroxycaproic acid, is produced via enzymatic oxidation of glucose, catalyzed by the enzyme glucose oxidase. Its alkaline salts are widely used in the chemical, pharmaceutical, food, and beverage industries.

Producers of gluconic acid are mainly fungi of the genera *Aspergillus, Penicillium, Endomycopsis, Pillularia,* and *Scopulariopsis.* Some bacterial species can also produce gluconic acid, including the members of the genera *Acetobacter, Gluconomacter, Acinetobacter, Enterobacter, Micrococcus,* etc.

Fermentation on an industrial scale is carried out using the SmF and SSF methods. The medium with a standard mixture of high glucose content (up to 35%), magnesium sulfate, potassium phosphate, nitrogen source, and calcium carbonate is used. The process is complete with a residual sugar concentration of about 1–3%. The finished product is crystallized and evaporated gluconic acid (Fig. 6.8).

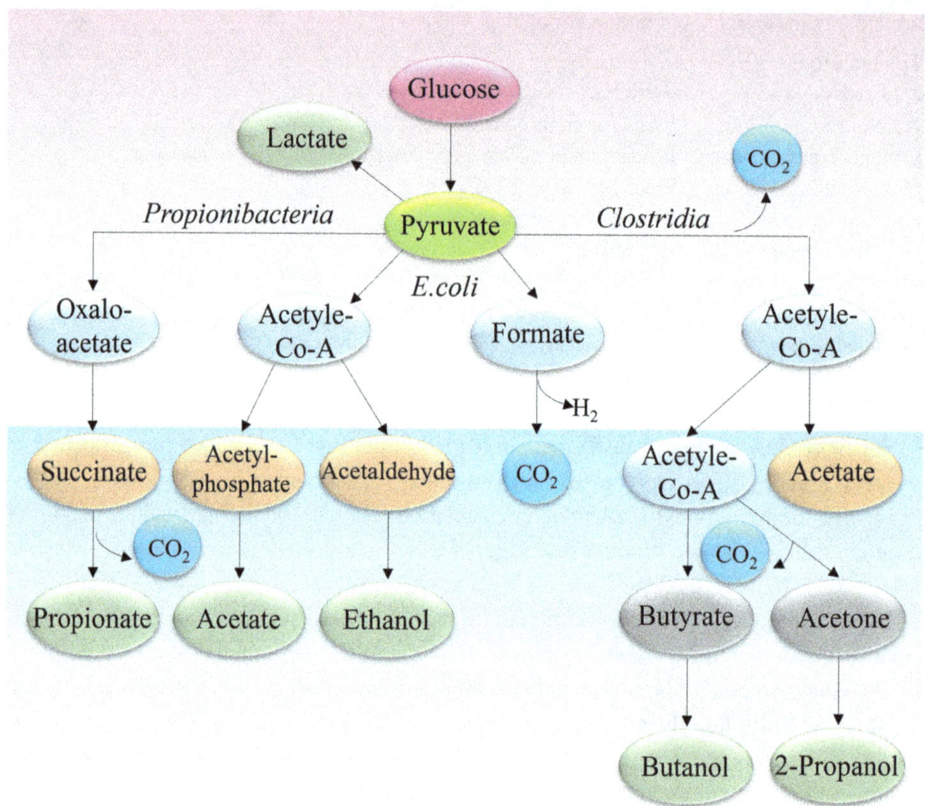

Fig. 6.7 Various pathways involved in propionic acid production

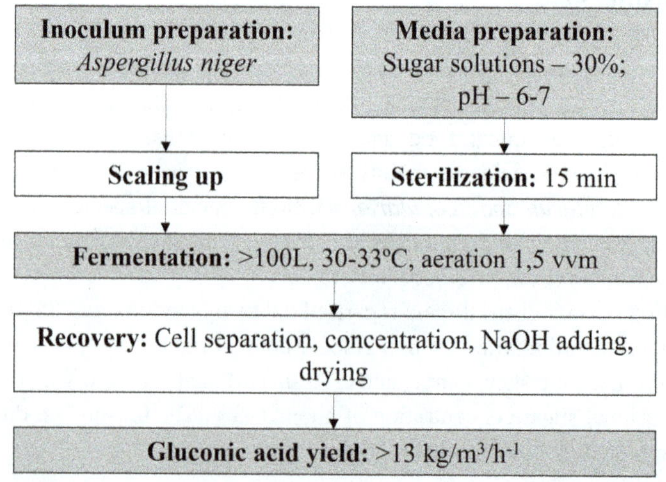

Fig. 6.8 Flowsheet of conventional production of gluconic acid

In recent years, a modern multi-step, membrane-based production method has emerged. This system involves fewer fermentation stages than traditional processes and is gaining attention as a versatile and efficient approach to gluconic acid manufacturing.

6.1.6 Itaconic Acid

Itaconic acid is a key intermediate in the production of polymers. It forms copolymers with esters and other monomers, serving as a valuable component in the manufacture of synthetic fibers, resins, adhesives, surfactants, dyes, and other complex organic compounds.

Genetically engineered strains such as *E. coli* pET-9971, *E. coli* BW25133, *Aspergillus niger* AB 1.13, and *Aspergillus terreus* are commonly employed for itaconic acid production.

The production process is generally similar to that used for citric acid. The culture media contain high concentrations of sugars, as well as zinc, magnesium, and copper salts. In solid-state fermentation (SSF), the process takes about 10 days, yielding a product mixture (containing amber, oxalic, and fumaric acids), with itaconic acid comprising around 25% of the total. The overall yield can reach 65% of the initial sugar content. In contrast, submerged fermentation (SmF) typically results in a lower itaconic acid concentration—around 8%.

Recent studies highlight the use of *Ustilago maydis* as a promising fungal producer for itaconic acid. A novel continuous fermentation technique called reverse-flow diafiltration utilizes membranes to enhance process efficiency (Fig. 6.9).

6.1.7 Fumaric Acid

Fumaric acid is a vital chemical substance with wide application in different fields; around 35% of annual production is used to produce paper resins, 6%—alkyl resins, 15%—polyester resins, and 5%—plasticizers. It is also used as an acidulant in many food products (22%) and for miscellaneous (17%) purposes.

Fumaric acid is a metabolite of the citric acid cycle and is present in all living cells. The producer of this acid is various fungi, such as *Rhizopus, Mucor, Cunninghamella,* and *Circinella.* Among them, some *Rhizopus* species (*R. nigricans, R. arrhizus, R. oryzae* and *R. formosa*) are the best producers in industrial-scale fermentation.

The fermentation medium contains about 10% glucose, with nitrogen and zinc as limiting factors. Fermentation is conducted under intensive aeration, using either submerged (SmF) or solid-state fermentation (SSF) methods (Fig. 6.10). During the process, the medium is neutralized with calcium carbonate or an alkaline solution. The maximum yield of fumaric acid reaches about 60% of the consumed glucose.

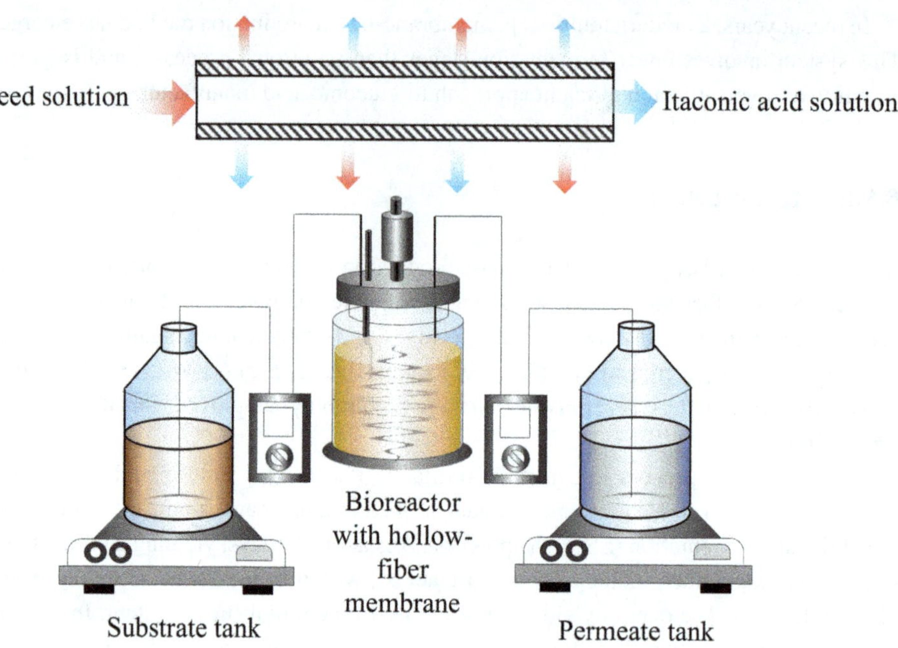

Fig. 6.9 Outline of reverse-flow diafiltration for itoconic acid production

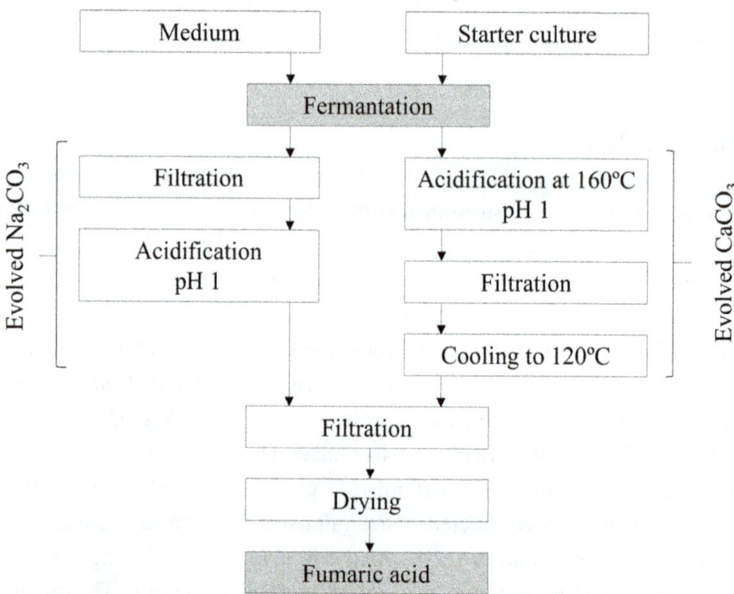

Fig. 6.10 Schematic representation for fumaric acid production via fermentation

Brainstorming

1. What main principles are the basis for the microbiological production of organic acids?
2. Where are organic acids used in industry? Give examples.
3. What substrates are used in the production of organic acids?
4. Name the producers of lactic acid.
5. Give the schematic representation of lactic acid microbial production.
6. Name the producers of citric acid. What are their technological criteria, and how can they be improved?
7. What methods of cultivation of citric acid producers are used in production?
8. Describe the industrial process for citric acid production.
9. Give the schematic representation of citric acid microbial production.
10. Describe the characteristics of acetic acid bacteria in terms of their importance in biotechnology.
11. Describe industrial processes to produce acetic acid.
12. Give the conventionalized outline of acetic acid production.
13. Give characteristics of propionic acid bacteria in terms of their importance in biotechnology.
14. How is propionic acid produced?
15. What are the ways to obtain gluconic acid?
16. What is the industrial biotechnology of itaconic acid?
17. Give the outline of reverse-flow diafiltration for itaconic acid production.
18. Give the schematic representation for fumaric acid production via fermentation.

> **Take-Home Messages**
> - Lactic acid is widely used in the food, pharmaceutical, textile, and chemical industries. It is produced by homo- and heterofermentative lactic acid bacteria (LAB), such as *Lactobacillus* spp., using molasses, starchy biomass, or agro-wastes. Industrial strains like *L. delbrueckii* can yield up to 95–98% conversion efficiency.
> - Citric acid production is dominated by *Aspergillus niger* via solid-state (SSF) or submerged (SmF) fermentation of sugar-rich media like molasses. Yield optimization requires aeration control, temperature management, and acid precipitation with calcium hydroxide.
> - Propionic acid is a valuable antimicrobial and flavoring agent, produced by *Propionibacterium* spp., through complex biochemical pathways involving methylmalonyl-CoA. Substrates include cheese whey, glycerol, and molasses, with fermentation durations ranging from 2 days to 4 months.
> - Gluconic acid, used in pharmaceuticals, foods, and detergents, is produced by fungi (*Aspergillus*, *Penicillium*) and bacteria (*Acetobacter*, *Gluconobacter*)

through glucose oxidation. Modern production incorporates membrane-based bioprocessing for higher efficiency and fewer steps.

- Itaconic acid is used in the polymer and resin industries and is biosynthesized by *Aspergillus terreus* or genetically modified microbes using sugar-rich media. SSF yields are higher than SmF, and innovative membrane-coupled fermentation systems (e.g., with *Ustilago maydis*) show promise.
- Fumaric acid, a key chemical and food additive, is a TCA cycle intermediate produced by fungi like *Rhizopus* spp.. Glucose is the primary substrate, and aerated fermentation with neutralization agents (e.g., calcium carbonate) yields up to 60% of the input glucose as product.

6.2   Production of Amino Acids

6.2.1   General Characteristics of Amin.o Acids Production Methods

Amino acids hold a leading position in terms of production volume among compounds obtained through biotechnological methods. The 20 standard amino acids are the building blocks of proteins and peptides. They exist as optical isomers, but only the L-forms are incorporated into proteins. Eight of these amino acids (isoleucine, leucine, lysine, methionine, threonine, tryptophan, valine, and phenylalanine) are essential for humans. This list is supplemented by histidine and arginine for farm animals, and proline for young poultry. Therefore, amino acids are widely used as additives in animal feed. Their inclusion in compound feeds reduces the need for protein sources of animal origin.

Essential amino acids can also be added to human food, either for medical reasons or to improve the nutritional quality of plant-based diets. Plant-derived proteins can be optimized by balancing their amino acid composition with lysine, threonine, and methionine (see Table 6.3).

Approximately 30% of all produced amino acids are used in the food industry. For example, cysteine prevents scorching during cooking and improves bread quality, while glycine, with its mildly sweet taste, is commonly used in beverages. L-glutamate is known for enhancing food flavor, and the dipeptide Aspartame™ serves as a popular low-calorie sweetener.

Several amino acids (e.g., arginine, aspartate, cysteine, phenylalanine) are also used in medicine, especially in infusion therapy. The list of amino-acid-based pharmaceuticals continues to grow. Parenteral nutrition formulations containing amino acid complexes are of increasing medical importance, especially when oral intake is contraindicated.

Non-proteinogenic amino acids, e.g., having the D configuration at their Cα atom, occur in natural compounds. They serve as chiral synthons in organic synthesis, e.g., for manufacturing semi-synthetic antibiotics.

Table 6.3 Manufacturing procedures and application of amino acids

Amino acid	Production type	Application
L-glutamate	Fermentation	Flavor enhancer
L-lysine	Fermentation	Feed additive
D,L-methionine	Chemical synthesis	Feed additive
L-threonine	Fermentation	Feed additive
L-aspartic acid	Cell reactor, chiral pool	AspartamePP™
L-phenylalanine	Fermentation	Medication, AspartamePP™
Glycine	Chemical synthesis	Sweetener
L-arginine	Fermentation	Medication, personal care items
L-tryptophan	Fermentation	Feed additive
All others	Fermentation, enzyme and cell reactors, chiral pool	Various applications

Table 6.4 Methods of producing amino acids

Method	Process	Disadvantages
Biological	Hydrolysis of protein-containing substrates	Racemization (mixture) of amino acids, destruction of amino acids
Chemical	Fine organic synthesis	Racemates of D and L isomers, expensive equipment and catalysts, aggressive, toxic compounds in the feed, environmentally polluting by-products
Chemical-enzymatic	Biotransformation of D-amino acids into L-amino acids	Economic inefficiency, the complexity of obtaining the amino acid precursors
Microbiological	Obtaining L-amino acids	Complex fermentation, production of mutant producers

Non-proteinogenic amino acids—such as D-isomers—occur naturally and are valuable in organic synthesis, particularly in the semi-synthesis of antibiotics.

Amino acids are also widely used in chemical and industrial sectors as precursors for producing detergents, polyamino acids, polyurethanes, and agrochemicals. Their growing use as feed and food additives, seasonings, and raw materials for pharmaceutical and cosmetic products has stimulated the industrial-scale production of amino acids.

Methods of Producing Amino Acids

Four different methods are used for manufacturing amino acids:

- Biological, i.e., extraction from protein hydrolysates.
- Chemical synthesis.
- Biotransformation of chemical precursors using enzyme or cell reactors.
- Microbial production via fermentation (Table 6.4).

Hydrolysis of protein substrates using acid, alkali, or enzymes is the oldest technique. Economically, it is viable for producing L-cysteine, L-cystine, L-leucine, L-asparagine, L-arginine, and L-tyrosine. Raw materials include plant proteins and protein-rich animal waste products (e.g., horns, hooves, feathers), which are rich in keratin. Hydrolysis typically involves heating these materials with 20% hydrochloric acid at 100–105 °C for 20–48 h. To accelerate the reaction, immobilized proteolytic enzymes or ion exchange resins may be used. The resulting mixture contains amino acids and peptides.

However, preparing optically pure L-amino acids from hydrolysates is expensive and involves multiple purification steps. After acid hydrolysis, hydrophobic amino acids like L-phenylalanine, L-leucine, and L-isoleucine are first extracted with ethanol. Subsequent steps include ion exchange chromatography to separate amino acids into acidic, basic, and neutral fractions, followed by crystallization.

During acid hydrolysis, some amino acids degrade or racemize. Tryptophan, for instance, breaks down completely, while cysteine, methionine, and tyrosine suffer 10–30% losses. These effects can be reduced by conducting hydrolysis in a vacuum and using a high acid-to-protein ratio (e.g., 200:1). Although hydrolysis was historically used mainly for pharmaceutical and research purposes, its application has now expanded significantly.

Chemical synthesis usually results in racemic mixtures (DL-amino acids). These mixtures can be used in animal feed if the absence of chirality does not affect functionality (e.g., DL-alanine, DL-methionine). Chemical synthesis is generally cost-effective if the precursors are inexpensive. However, racemate separation is difficult and costly. Furthermore, chemical synthesis requires high temperatures, often involves toxic reagents, generates hazardous by-products, and poses safety and environmental risks.

Biotechnological production includes microbial fermentation and enzymatic or microbial biotransformation of chemical precursors. Today, the most widely used approaches are chemical-enzymatic and microbiological methods (see Fig. 6.11).

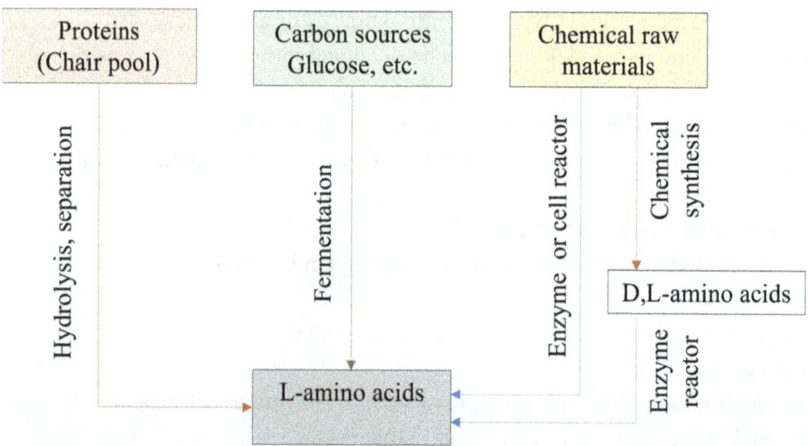

Fig. 6.11 Manufacturing types of amino acids

Brainstorming

1. Where are the amino acids used?
2. Which amino acids are irreplaceable for humans, and which are for farm animals?
3. Why should some amino acids be added to the diet of humans and animals? List them.
4. What methods are used to produce amino acids?
5. How is the biotechnological production of amino acids implemented?
6. How can the level of amino acid output be increased in microbiological synthesis?

Take-Home Messages
- Amino acids are essential compounds with wide applications in animal feed, food, medicine, and industry, and are produced via several methods—including protein hydrolysis, chemical synthesis, biotransformation, and especially microbial fermentation, which is the most sustainable and widely used today.
- While traditional hydrolysis and chemical synthesis remain useful, modern biotechnological methods—such as microbial fermentation and enzymatic synthesis—are preferred due to their efficiency, safety, and ability to produce optically pure L-amino acids needed for nutritional and pharmaceutical uses.

6.2.2   Chemical-Enzymatic Method to Produce Amino Acids. Obtaining Optical Isomers of Amino Acids via Microbial Enzymes

The chemical-enzymatic method consists of two main stages. First, a precursor—typically the corresponding carboxylic acid—is synthesized through chemical means. In the second stage, this acid (usually in the presence of ammonia) is converted into the corresponding amino acid. These transformations can be either single-step or multi-step and are carried out by enzymes of living cells. Since this process involves both chemical synthesis and biological catalysis, it is considered semi-biotechnological and is often referred to as a chemical-microbiological method.

A wide variety of enzymes are involved in amino acid production, including hydrolases, dehydrogenases, lyases, and isomerases. For example:

- *Amino acid dehydrogenases* (such as leucine and alanine dehydrogenases) catalyze reversible deamination and are used for continuous synthesis of amino acids from their corresponding keto analogs.
- *Glutamate synthetase* catalyzes the ATP-dependent amination of glutamate, yielding glutamine with up to 92% efficiency.
- *L-tyrosine-phenol-lyase* catalyzes the elimination of phenol, ammonia, and pyruvate from tyrosine and is also used for pyruvate synthesis.

Table 6.5 Types of amino acids obtained by the chemical-enzymatic method

Amino acid	Enzyme	Producer
Aspartic acid—From fumarate	Aspartase	*Escherichia coli, Serratia marcescens*
Phenylalanine—From brown acid; tryptophan—From anthranilic acid	Aminoacylase	*Candida utilis*
Glutamic acid—From ketoglutarate	Glutamate dehydrogenase, transamidase	*Pseudomonas putida, Escherichia coli*
Lysine—From amino-caprolactam	Lactamase (hydrolase), racemase	*Candida laurentii, Alcaligenes obae*

- *L-tryptophan-indole-lyase* enables the biosynthesis of L-tryptophan from indole, pyruvate, and ammonia.

Microbial enzymes—whether isolated, naturally mixed, or present in intact, dried, lysed cells or cell extracts—serve as the principal catalysts. Enzyme preparations can also include *immobilized cells or enzymes*, which offer advantages for industrial use.

Enzyme immobilization has greatly advanced amino acid biotechnology. Immobilized biocatalysts—whether purified enzymes or whole cells—are commonly used in commercial processes, allowing continuous operation and longer catalyst lifetimes.

A major industrial application of immobilized enzymes was developed in Japan: the separation of racemic mixtures of chemically synthesized D- and L-amino acids. A solution containing acyl derivatives of both forms is passed through a column with immobilized L-aminoacylase, which selectively hydrolyzes only the acyl-L-isomers. After cleavage of the acyl group, the L-amino acids are removed through a membrane, leaving behind the acyl-D-amino acids. These remaining D-isomers can be racemized by heating, forming a new D/L mixture, which can again be separated.

This cycle allows for efficient recovery of L-amino acids such as methionine, valine, phenylalanine, and tryptophan, and has been successfully implemented in large-scale production.

One example of the chemical-enzymatic method is the synthesis of aspartic acid from fumaric acid and ammonia (see Table 6.5). In this process, a fumaric acid solution is passed through columns containing immobilized enzymes or microbial cells (e.g., *Escherichia coli* or *Serratia marcescens*) with high aspartase activity. Ammonia is co-supplied to enable the biotransformation into aspartic acid.

The enzyme, in one step, links the ammonia molecule to the double bond of fumaric acid with the formation of optically active L-aspartic acid. The product yield is 99%, and the process is carried out continuously in a column with a volume of 1 m^3 (Fig. 6.12).

L-tryptophan can also be synthesized from anthranilic acid as a precursor. In the first stage, a conventional microbiological scheme is employed using *Candida utilis* yeast over 20–24 h under intensive aeration. Subsequently, the aeration rate is reduced by half, and urea, molasses, and anthranilic acid are periodically added to the culture. Over the next

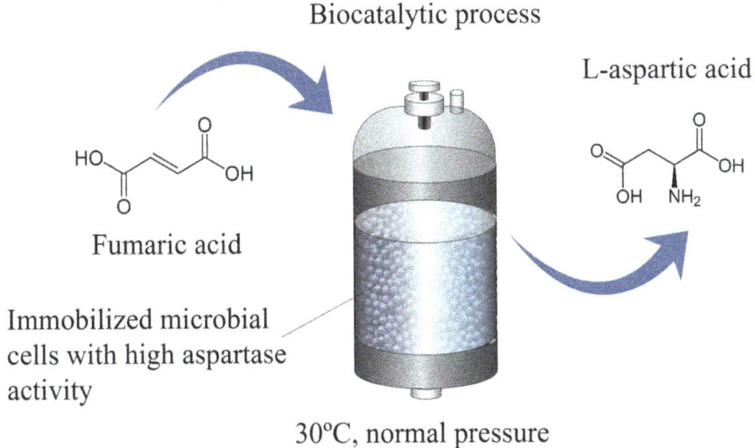

Fig. 6.12 Outline of aspartic acid production processes

22–24 h, microbial biomass increases, serving as the source of enzymes for the biotransformation. The conversion of anthranilic acid into L-tryptophan then proceeds over the following 120 h. The total process time is approximately 140 h, and the yield reaches 60 g/L.

L-phenylalanine can also be synthesized from cinnamic acid using *Candida utilis* yeast cells.

A two-step process is used for the production of L-glutamic acid from α-ketoglutarate, a key precursor. In stage I, α-ketoglutarate is produced microbiologically using *Pseudomonas*, *Escherichia*, or *Candida* cultures. In stage II, L-glutamic acid is obtained via a reductive amination reaction, catalyzed by *Pseudomonas* strains exhibiting strong glutamate dehydrogenase activity. Alternatively, the conversion can occur through transamination, with *E. coli* as the biocatalyst and alanine or aspartic acid as amino group donors.

An alternative and widely adopted approach for glutamic acid production utilizes *Corynebacterium glutamicum* strains. The process follows conventional fermentation conditions using high-yield mutant strains (see Fig. 6.13).

A combined chemical-enzymatic process for producing L-lysine was developed by the Japanese company Toyo Rayon. The process is based on stereospecific enzymatic hydrolysis of D,L-α-amino-ε-caprolactam, which is chemically synthesized from cyclohexane. In the first stage, cyclohexane undergoes chemical transformation to form cyclic lysine anhydride. In the second stage, enzymatic hydrolysis with L-α-amino-ε-caprolactam hydrolase selectively cleaves the L-isomer.

This hydrolase is synthesized by yeast species such as *Candida*, *Trichospora*, and *Cryptococcus*, and its activity is enhanced by manganese, magnesium, and zinc ions. The racemase, which enables conversion between D- and L-forms, is derived from bacteria such as *Flavobacterium* and *Achromobacter*. These two enzymes—hydrolase and racemase—can be introduced as part of the microbial biomass or as immobilized enzymes in a column reactor. The precursor solution is passed through this column, where L-lysine is generated with a purity exceeding 99%, and yields can reach up to 95%.

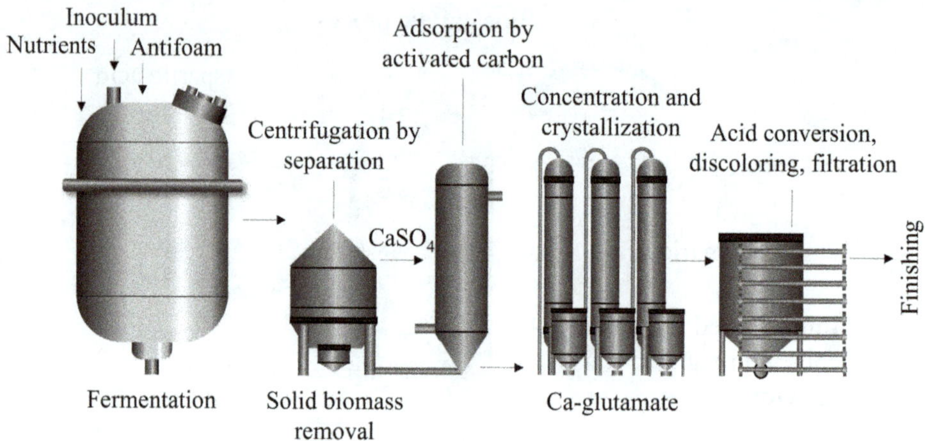

Fig. 6.13 Fermentation and recovery of glutamic acid

This two-step synthesis from biosynthetic precursors is especially advantageous when the precursor is inexpensive and when direct microbial fermentation is less efficient or economically viable. Moreover, using precursors allows the process to bypass metabolic regulation, such as genetic repression or feedback inhibition, common in microbial amino acid biosynthesis.

While the chemical-enzymatic approach does not require extensive purification from by-products and waste streams and can, in principle, be used to produce nearly all amino acids, it is not always economically favorable. The high cost of raw materials and enzymes, along with the complexity of precursor synthesis, often makes this method less attractive than direct microbial fermentation for large-scale amino acid production.

Brainstorming

1. What are the two stages of the chemical-enzymatic method for amino acid production?
2. In which cases do biotechnologists prefer a two-step synthesis of amino acids from biosynthetic precursors?
3. Which amino acids can be obtained by both chemical and enzymatic methods?
4. Describe the process of producing amino acids using immobilized cells or enzymes.
5. List the main advantages and disadvantages of the chemical-enzymatic approach to amino acid production.

Take-Home Messages
- Four main methods are used to produce amino acids: protein hydrolysis, chemical synthesis, enzymatic biotransformation, and microbial fermentation. Among these, fermentation and biotechnological methods are the most economically and environmentally viable.

- Chemical-enzymatic production combines chemical precursor synthesis with biocatalysis using free or immobilized enzymes or microbial cells. This method allows for high stereo-specificity and purity, which is particularly important for L-isomers used in biological systems.
- Immobilized enzyme technology has revolutionized industrial amino acid production by enabling continuous processing and high product yields, though economic feasibility still depends on substrate cost, catalyst stability, and downstream processing efficiency.

6.2.3 Microbiological Method to Produce Amino Acids. Intensification Ways of the Process by Optimizing the Cultivation Conditions

The fourth method for obtaining amino acids is direct microbiological synthesis, which relies entirely on living cells and is therefore considered a fully biotechnological process. This method, currently the most widely used, exploits the natural ability of microorganisms to synthesize all amino acids—and under certain conditions, to overproduce them. Unlike chemical synthesis, microbiological production yields only L-forms of amino acids—the biologically active isomers with therapeutic efficacy—avoiding the formation of racemic mixtures. This advantage makes biotechnological production the method of choice for industrial-scale amino acid manufacturing.

In microbial cells, amino acids are biosynthesized as part of the so-called "free amino acid pool," from which proteins are built during constructive metabolism. For protein synthesis, all 20 standard amino acids are required. Most prokaryotes are capable of synthesizing these amino acids from intermediates such as pyruvate (from glycolysis), and α-ketoglutarate and fumarate (from the tricarboxylic acid cycle). ATP serves as the energy source. The introduction of nitrogen into the carbon backbone typically occurs at the final biosynthesis stages via amination and transamination reactions (see Fig. 6.14).

The synthesis routes of most amino acids are interrelated. In this case, some amino acids are precursors for the biosynthesis of others. For example, Pyruvate is a precursor of alanine, valine, and leucine; 3-phosphoglycerate-serine, glycine, cysteine; fumaric acid-aspartate, asparagine, methionine, lysine, threonine, isoleucine; α-ketoglutaric acid—glutamate, glutamine, arginine, proline; phosphoenolpyruvate + erythrose-4-phosphate-phenylalanine, tyrosine, tryptophan; 5-phosphoribosyl-1-pyrophosphate + ATP-histidine.

Among the producers of amino acids are various microorganisms, representatives of the genera *Corynebacterium, Brevibacterium, Bacillus, Aerobacter, Microbacterium, and Escherichia* are well studied. The producer strains perform hyperexpression of amino acids. As a result, excess amounts of amino acids, for example, L-lysine, L-glutamic acid,

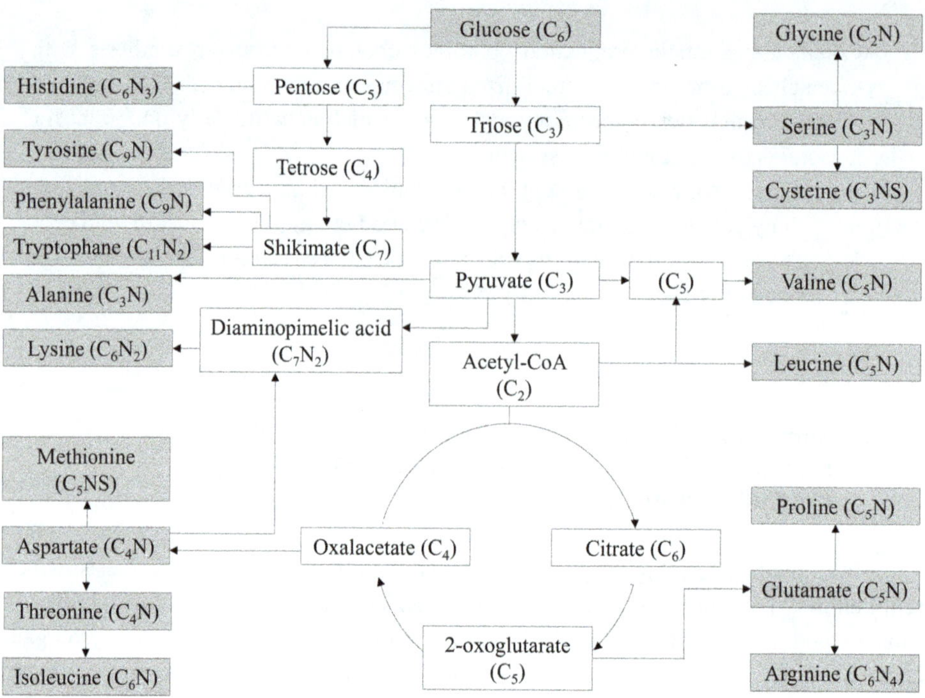

Fig. 6.14 Biosynthesis of amino acids

Table 6.6 Microorganism producers of amino acids

Amino acid	Organism	Producer
Glutamate, valine, alanine, glutamine, proline	Natural (wild) strains	*Corynebacterium glutamicium, Escherichia coli, Bacillus subtilis, Brevibacterium flavum*
Lysine, threonine, methionine, isoleucine	Auxotrophic mutants	*Corynebacterium glutamicium*
Arginine	Regulatory mutants	*Brevibacterium flavum*
Threonine	Recombinant strains	*Escherichia coli*

L-threonine, and L-tryptophan, are excreted into the culture medium. The culture medium, in this case, can contain from four to five and up to 100 grams of the desired amino acid per liter in the liquid phase.

Microorganisms used in industry can be divided into wild strains, auxotrophic mutants, regulatory mutants, and auxotrophic regulatory mutants—recombinant strains (Table 6.6).

The use of both wild-type and mutant microbial strains in amino acid production arises from the fact that microbial cells synthesize amino acids in precisely regulated amounts under strict genetic and enzymatic control. Regulation occurs through feedback mechanisms at two levels:

1. *Genetic (transcriptional) regulation*—At the gene level, the synthesis of enzymes involved in amino acid biosynthesis is repressed. In *E. coli*, for example, structural genes are clustered and controlled by a common regulatory region, which includes a *promoter* (RNA polymerase binding site) and an *attenuator* (a regulatory element sensing feedback signals). When the concentration of a particular amino acid exceeds the cellular requirement, it binds to the attenuator, *terminating transcription*. This halts the expression of enzymes in the corresponding biosynthetic pathway.
2. *Enzymatic (allosteric) regulation*—At the enzymatic level, retroinhibition (feedback inhibition) occurs. Here, the final product—an amino acid—binds to and inhibits an early enzyme in its own biosynthetic pathway. This mechanism prevents the overproduction and secretion of amino acids, maintaining internal metabolic balance.

To achieve overproduction (hyperexpression) of specific amino acids for industrial purposes, these regulatory mechanisms must be disrupted or bypassed.

For certain amino acids—such as L-glutamate, L-valine, L-alanine, L-glutamine, and L-proline—high-yield production is possible using wild-type strains. Productivity can be improved by modifying fermentation conditions or by increasing membrane permeability.

For example, high L-glutamate yields (up to 30 g/L) are obtained by partially inhibiting α-ketoglutarate dehydrogenase. Depending on fermentation parameters, glutamate synthesis can be diverted toward L-glutamine or L-proline. Elevated levels of ammonium ions and biotin favor proline synthesis, whereas a slightly acidic medium, zinc ions, and excess ammonium promote glutamine formation.

Cell membrane permeability can be increased through mutations or by altering the culture medium composition. For example, biotin deficiency (1–5 µg/L) combined with penicillin (2–4 µg/L), detergents (e.g., Tween-40 or Tween-60), or fatty acid derivatives (e.g., stearic and palmitic acids) enhances amino acid excretion. Biotin affects phospholipid synthesis, while penicillin interferes with bacterial cell wall formation, facilitating amino acid leakage.

Mutant strains, such as auxotrophic and regulatory mutants, are developed to genetically alter biosynthetic control. Industrial strains usually carry multiple mutations affecting both the target amino acid and its metabolic precursors.

Auxotrophic mutants are unable to synthesize one or more amino acids due to the loss of specific enzymatic steps. They are particularly useful when the desired amino acid is the end product of a branched metabolic pathway, as in the production of L-lysine (*Corynebacterium glutamicum*) and L-threonine (*E. coli*). These pathways often feature dual regulation, such as the concerted inhibition of aspartate kinase by both lysine and threonine.

Individually, neither lysine nor threonine inhibits aspartate kinase, but together they do. Thus, hyperexpression of lysine is only achievable by limiting threonine or its precursor, homoserine. Auxotrophic mutants, lacking key enzymes, do not produce inhibitors and are therefore capable of overproducing precursors. Most lysine producers are auxotrophic for

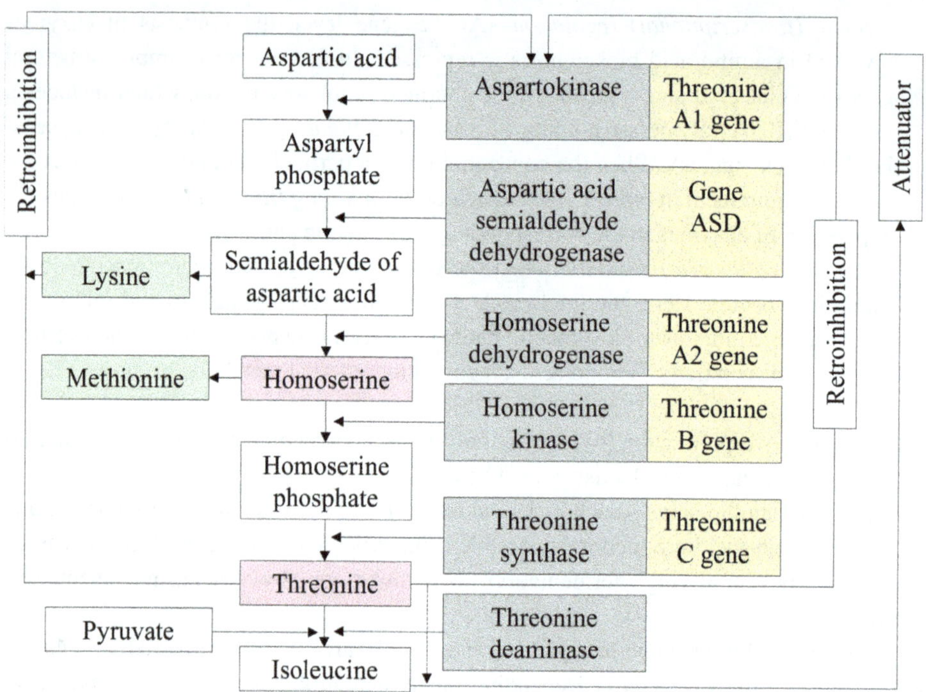

Fig. 6.15 Biosynthesis of amino acids of the aspartic acid family

threonine or homoserine and require supplementation with 0.5–1.5 g/L of these amino acids. To sustain long-term productivity, protein hydrolysates can be added as a complex amino acid source (Fig. 6.15).

In this case, there is an active growth of the biomass of the producer without the synthesis of lysine. Once the threonine disappears from the medium and biomass growth ceases, active lysine synthesis begins. Thus, this process has two stages of development: 1—growth of biomass; 2—synthesis of lysine. The duration of the synthesis is 2–3 days. The product's accumulation level is 50–100 grams per liter (Fig. 6.16).

The situation is different in the case of the regulation of threonine biosynthesis. Bacteria producers do not have a mechanism for coordinated inhibition of enzymatic activity; that is, if lysine inhibits the activity of its enzymes by the feedback principle, then threonine inhibits its enzymes. In addition, there is a "repression" of the entire complex of threonine enzymes with an excess of threonine or isoleucine, similar to "concerted repression." Independently (separately), neither threonine nor isoleucine represses the synthesis of enzymes.

To solve the problem of obtaining threonine in the required quantities, the following concerns must be addressed:

1. Changes to make threonine insensitive to the first enzyme of threonine biosynthesis.
2. Decrease the enzyme activity, synthesizing isoleucine from threonine.

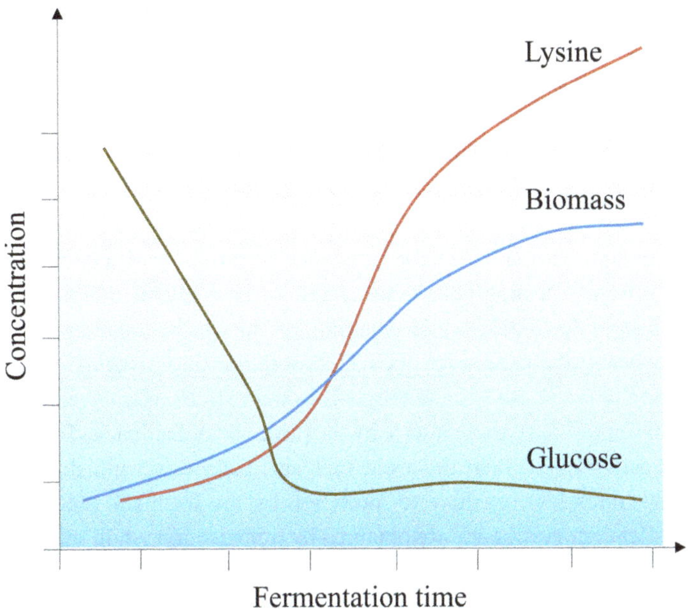

Fig. 6.16 Production of L-lysine in relation to biomass and substrate concentration

3. Remove the repression mechanism with an insufficient amount of isoleucine, despite the excess of threonine.
4. Applying genetic engineering (isolates threonine genes and multiplies them on plasmids in a microbial cell, dramatically increasing threonine synthesis by producer cells).

In this case, threonine synthesis differs from lysine's in that its synthesis co-occurs with biomass growth.

6.2.4   Features of Strain Cultivation Producing Amino Acids Lead to the Following Results

1. The highest possible rates of synthesis of amino acids by the producer cells are achieved.
2. The maximum duration of the producer's activity is achieved.
3. Minimally formed byproducts of the biosynthesis of amino acids.

The first task is addressed by cultivating highly active biomass under optimal conditions with sufficient nutrients in the medium, including carbon sources, ammonium nitrogen, mineral salts, and growth factors. This also involves optimizing parameters such as pH (acidity of the medium), temperature, and the controlled or fractional supply of substrates. To prevent acidification of the medium, automatic pH-stabilization can be conducted using ammonia water and carbon sources.

Auxotrophic mutants capable of accumulating the final products of unbranched chains of biosynthesis, for example, L-arginine, are not feasible. In such cases, it is necessary to obtain mutants with partially impaired biosynthesis regulation mechanisms, since this leads to increased yield of the target product. Such organisms are called *regulatory mutants*.

Regulatory mutants are selected for resistance to amino-acid analogs or among the auxotrophic revertants. Analogs of amino acids act as artificial inhibitors of enzymes, working principle of feedback, while ensuring the biosynthesis of the required amino acids and suppressing the process of their incorporation into proteins. Thus, the sulfur-containing analog of lysine S-(2-aminoethyl)-L-cysteine of *Brevibacterium flavum* is false and acts as an inhibitor of aspartate kinase by the principle of feedback. Therefore, mutants resistant to its activity, in which the yield of lysine reaches 33 g/l, that synthesize the enzyme are 100 times less sensitive to inhibition by the feedback mechanism than the original strain. Regulatory mutants are obtained by transduction while selecting individual mutations, causing a complete mismatch of the regulatory mechanisms and then combining these characteristics by co-transduction. As a result of this, one strain can consistently consolidate resistance to several analogs.

Recently, the latest biotechnology methods have been adopted to obtain new effective strain-producers of amino acids. Genetic engineering methods allow increasing the number of biosynthesis genes by cloning them on plasmids containing genes that control the biosynthesis of amino acids to the detriment of the synthesis of biomass and other cellular components. This leads to an increase in the number of enzymes responsible for synthesizing amino acids, increasing the yield of the desired product. Furthermore, with the introduction of hybrid plasmids, the by-product impurities usually disappear automatically, while the productivity of the biomass and the utilization rate of the substrate substantially increase.

Cloning the system's genes for the synthesis of amino acids in microbial cells with a different type of nutrition, in comparison with the donor organism, allows for expansion of the raw material base and replaces the expensive sugar-containing substrates with cheaper ones.

With the help of superproducer strains from the genera *Brevibacterium, Corynebacterium, Mycrococcus*, etc., large-scale production of glutamate and threonine, lysine, valine, histidine, and other amino acids is established. In super-production, the expression level of a cloned gene is expressed in the synthesis of a specific protein in an amount of at least 2% of all soluble host cell proteins. Currently, there are super producers, in which the amount of synthesized specific protein reaches 10–50% (here, the most crucial role is played by multicopy plasmids carrying built-in genes). A vivid example of the great possibilities of genetic engineering is the creation of the VNII of genomes and the selection of industrial microorganisms of the *E. coli* strain to produce threonine. As a result, the regulatory properties of the enzyme aspartate kinase were altered, as were the nutritional requirements of the strain. Furthermore, introducing a new gene into the bacterium's genome allowed the bacterium to use sucrose, the main disaccharide of traditional industrial raw materials, such as beet molasses, as a carbon source. Manipulations listed above, along with the amplification of plasmids containing the threonine operon, significantly increased the bacterial strain's productivity and obtained 100 h of fermentation of 100 g of L-threonine per liter of culture liquid.

Industrial biotechnological processes to produce amino acids are realized under periodic submerged fermentation conditions in aerobic conditions. The synthesis rate of amino acids does not coincide in time with the growth rate of the production culture. The maximum production of an amino acid usually occurs when biomass growth practically ceases. Therefore, the nutrient medium in the first fermentation stage should provide a balanced growth of cells. Second, the conditions for hyperexpression of the target amino acid.

As a source of carbon and energy, sugar-rich substrates—primarily molasses—are commonly used. In addition, more affordable substrates such as acetate, sulfite liquor, and hydrocarbons can be utilized. Depending on the taxonomy and physiological requirements of the microorganisms, nitrogen sources may include ammonium salts, nitrates, amino acids, or even molecular nitrogen.

The composition of the medium includes appropriate concentrations of carbon and nitrogen sources, phosphates, other essential salts, growth stimulants (e.g., vitamins, yeast extract), surface-active substances, and antibiotics. High levels of amino acid accumulation (typically 50–100 g/L) are achieved by fractional feeding of carbon and nitrogen substrates.

Although amino acids themselves are generally neutral, their biosynthesis often leads to significant acidification of the culture medium due to disruption of the ionic balance. This is because microbial producers convert ammonium ions (NH_4^+), supplied as salts (e.g., NH_4Cl, $(NH_4)_2SO_4$), into amino groups within amino acids. The resulting excess of counterions (Cl^- and SO_4^{2-}) remains in the medium, causing acidification.

To maintain optimal ammonium ion concentration and pH during fermentation, automatic pH stabilization is implemented using ammonia or ammonia water, often in combination with carbohydrates. With the appropriate ratio, both carbon and nitrogen levels can be kept constant throughout the process.

Amino acid biosynthesis is energy-intensive and must therefore be conducted under aerobic conditions with intensive aeration and agitation to ensure an oxygen transfer rate of 3–7 g/L. Once a sufficient amount of active biomass has accumulated in the fermenter, conditions must be adjusted to ensure that the cells continue to produce amino acids for as long as possible. As the biosynthesis process progresses, cells gradually lose viability, and strategies must be employed to prolong the active fermentation phase.

For auxotrophic amino acid producers (e.g., lysine-producing strains), productivity and biosynthesis duration can be increased by gradually adding carbon sources along with growth factors, such as protein hydrolysates, during fermentation.

In some cases, the synthesis of a target amino acid may be inhibited by toxic metabolites produced by the microorganism itself. For instance, during the biosynthesis of phenylalanine by *Bacillus subtilis*, by-products such as acetoin and butanediol can cause cell lysis, sporulation, and cessation of phenylalanine production. This issue can be avoided by carbon-source limitation, which ensures that all available sugar is directed towards phenylalanine synthesis. This approach may double both the yield and the purity of the final product.

The effectiveness of the substrate depends on how it is metabolized in relation to biomass growth:

- In processes where amino acid synthesis is decoupled from biomass growth (e.g., lysine production), efficiency is higher when the culture continues producing amino acids after cell growth stops.
- In contrast, when biosynthesis proceeds in parallel with biomass growth (e.g., threonine), efficiency can be improved by redirecting intracellular metabolic fluxes toward amino acid biosynthesis and reducing by-product formation.

Batch fermentation using complex media requires strict sterility during inoculation and fermentation. Therefore, all components of the process—including media, air, and equipment—must be sterilized. After fermentation, the culture broth undergoes treatment: cells are separated from the liquid, which is then purified from colored impurities and suspended particles using sorption techniques.

The isolation and purification of amino acids depend on their intended application:

- For pharmaceuticals and food, they are released as dried, pure crystalline products.
- For feed and technical purposes, stabilized and concentrated culture fluid or whole-cell biomass may be used.

The high and growing demand for amino acids continues to drive the development of new, more efficient biotechnological production methods, thereby expanding industrial capacities and increasing production volumes.

Brainstorming

1. In what ways can the yield of amino acids be increased in microbiological synthesis?
2. What kinds of microorganisms are super-producers of amino acids?
3. How is intracellular regulation of amino acid biosynthesis carried out?
4. What types of mutants are used in industry to produce amino acids and why?
5. Describe how to intensify the process of amino acid biosynthesis by optimizing the cultivation conditions.
6. What are the advantages of the microbiological method for producing amino acids?

Take-Home Messages

- Microbial fermentation is the most effective method for amino acid production. Unlike chemical synthesis, microbial methods yield optically pure L-amino acids, which are essential for biological activity and free from racemic mixtures.
- Hyperproduction of amino acids relies on disrupting natural feedback regulation. To achieve high yields, regulatory mechanisms such as gene repression and enzyme inhibition must be bypassed using wild strains under optimized conditions or by employing auxotrophic and regulatory mutants.
- Through targeted mutations, selection for analog resistance, and gene amplification via plasmids, modern biotechnology significantly boosts amino acid yields and minimizes by-product formation. By blocking specific biosynthetic branches or feedback loops, mutant strains can be fine-tuned to overproduce amino acids like lysine and threonine under controlled fermentation conditions.
- Amino acid synthesis is often decoupled from biomass growth, requiring specific timing, nutrient supply (e.g., carbon and nitrogen), aeration, pH control, and sterile conditions for optimal production.

6.3   Vitamins, Production, and Technological Design

Vitamins are low-molecular-weight organic compounds with high biological activity. In the natural environment, they are primarily derived from plants and microorganisms. Industrially, vitamins are produced via chemical synthesis or microbial biotechnology (see Table 6.7).

Microorganisms naturally contain many vitamins, often as components of enzyme systems. The vitamin content in microbial biomass depends on the biological characteristics of the culture and the cultivation conditions. Some vitamins are synthesized de novo by microorganisms, while others are assimilated in their complete form from the surrounding environment. From a biotechnological standpoint, vitamins such as A, D, B_2, B_{12}, and C are among those commonly produced.

Vitamin B_{12}

Also known as *cyanocobalamin*, vitamin B_{12} is produced exclusively through microbial synthesis. Commercially important producer strains include *Propionibacterium acidipropionici, P. freudenreichii, Pseudomonas denitrificans, Acetobacterium* sp., *Citrobacter freundii, Xanthomonas campestris*, and *Lentinula edodes*.

These organisms are typically cultivated under anaerobic (oxygen-free) batch fermentation conditions. The medium contains glucose, corn extract, ammonium and cobalt salts, and is maintained at pH ~7.0. The fermentation process lasts up to 7 days. Cyanocobalamin accumulates intracellularly, which makes downstream processing technically demanding (see Fig. 6.17).

Table 6.7 The manufacturing process and application of vitamins

Vitamin	Production type	Application
A β-carotene	Chemical synthesis	Colorant, animal feed
B$_1$ thiamine	Chemical synthesis	Health
B$_2$ riboflavin	Fermentation, chemical synthesis, chemo-enzymatic synthesis	Medicine, animal feed
B$_6$ thiamine	Chemical synthesis	Medicine, animal nutrition
B$_{12}$ cyanocobalamine	Fermentation	Medicine, animal nutrition
C ascorbic acid	Chemical-fermentative synthesis	Food industry, medicine, animal feed
D$_2$ calciferol	Photochemical synthesis	Health
E α-tocopherol	Plant oil extraction, algae	Health

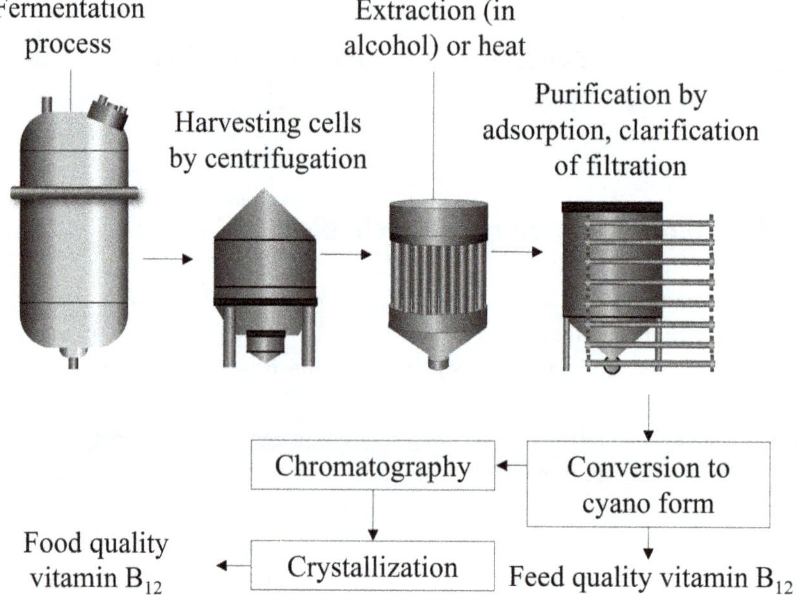

Fig. 6.17 Downstream processing of cyanocobalamin

For extraction, cells are separated and treated with hot water (85–90 °C, pH 4.5–5.0) in the presence of a stabilizer (e.g., 0.25% sodium nitrate). After one hour of extraction, the solution is cooled, neutralized with sodium hydroxide, and clarified with protein coagulants. The mixture is filtered, evaporated, and further purified using ion exchange and chromatography. Final crystallization of vitamin B$_{12}$ occurs at 3–4 °C from a water–acetone solution. Due to its high light sensitivity, all steps must be performed in darkness or under red light.

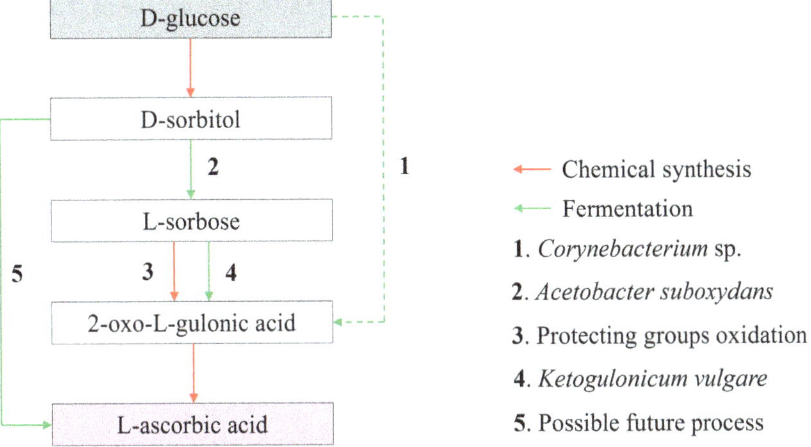

Fig. 6.18 Synthesis of vitamin C

Vitamin B_{12} is an essential anti-anemic factor for humans and animals. The annual global production of cyanocobalamin reaches approximately 20 tons, and it is primarily used for animal feed and pharmaceutical applications.

Vitamin C

Also known as *L-ascorbic acid*, vitamin C is an antiscorbutic vitamin naturally present in higher plants and animals. However, humans are unable to synthesize it endogenously. Industrial production of vitamin C involves both biological and chemical steps. One of the key intermediates, *L-sorbose*, is biosynthesized by specific microorganisms such as acetic acid bacteria.

The biotechnological process involves *membrane-bound polyol dehydrogenase* and follows a chemical–enzymatic route (see Fig. 6.18). *Gluconobacter oxydans* is fermented in media containing 20% sorbitol and enriched with corn or yeast extract under conditions of intense aeration. This yields up to 98% L-sorbose within 2 days.

Fermentation can be carried out in batch or continuous mode. L-sorbose production using immobilized cells has also been demonstrated. Other commercial strains used in ascorbic acid biosynthesis include *Ketogulonicigenium vulgare*, *Saccharomyces cerevisiae*, *Cryptococcus terreus*, and *Pichia fermentans*.

Ascorbic acid is widely used in the food and pharmaceutical industries, primarily for its antioxidant properties.

Vitamin B_2

Riboflavin, or vitamin B_2, is present in the cells of various microorganisms as a coenzyme in flavoproteins—especially oxidoreductases such as FMN and FAD. It is produced by bacteria, yeasts, and filamentous fungi. Through strain selection and genetic engineering, highly productive mutants have been developed, including *Ashbya gossypii*, *Pichia guilliermondii, Bacillus subtilis, Eremothecium ashbyii, Lactococcus lactis, Pichia pastoris, Bacillus megaterium, Enterobacter aerogenes, Candida tropicalis, C. famata,* and *C. guilliermondii.*

These strains can produce up to 20 g of riboflavin per liter of rich medium, which typically contains molasses, protein-vitamin concentrates, and hydrolysates. Common carbon sources include glucose, sucrose, yeast extract, corn extract, soy flour, and vegetable oils. An example of an inoculum medium includes sucrose, peptone, corn extract, potassium dihydrogen phosphate, magnesium sulfate, and sunflower oil. Cultivation usually lasts several days at ~30 °C. Fermentation is carried out for 5 days at pH 5.5–7.7, after which the biomass is harvested and dried. If the riboflavin yield is high, the vitamin can be isolated and purified for use in pharmaceuticals alongside synthetic riboflavin.

Carotenoids

Carotenoids are produced industrially by various microorganisms, including species of *Monascus, Penicillium, Ashbya, Nannochloropsis, Dietzia, Sporobolomyces, Paracoccus, Chlorella, Gordonia, Rhodotorula, Saccharomyces, Scenedesmus,* and *Rhodosporidium.* Over 700 microbial carotenoids are known, present either as esters and glycosides in membranes or as free pigments in cytoplasmic lipid granules. Their biosynthesis—primarily for protective functions—is stimulated by light. Some strains can produce up to 3–4 g of carotenoids per liter of culture.

The nutrient medium for carotenoid production is complex and includes carbon and nitrogen sources, vitamins, trace elements, and stimulants such as corn-soybean meal, vegetable oil, kerosene, and β-ionone. Production involves fermentation: first, strains are grown separately at 25–30 °C with aeration, then transferred to a bioreactor for 6–7 days. Carotenoids are extracted using polar solvents (e.g., acetone) and re-extracted into nonpolar solvents. Protein–carotenoid complexes are extracted using surfactants at 1–2% concentration. Vitamin A is formulated for pharmaceutical use in capsules or pellets.

Vitamin D

Vitamin D refers to a group of related compounds based on ergosterol, a sterol found in the cell membranes of eukaryotic microorganisms. Baker's and brewer's yeast are widely used sources, containing up to 11% ergosterol. Under UV irradiation, ergosterol is converted into vitamin D_2, which can be further transformed into vitamin D_3. Both forms exhibit similar physiological activity. Additional ergosterol-producing organisms include *Aspergillus* and *Penicillium* species, with ergosterol contents up to 2.2%.

Vitamin D deficiency can lead to rickets in children and osteomalacia in adults. Industrial vitamin D_2 production involves the following steps:

1. Cultivation and accumulation of producer biomass.
2. Cell separation and UV irradiation.
3. Drying and packaging of irradiated cells for use (e.g., in animal feed).

To produce crystalline vitamin D_2, the irradiated cells are hydrolyzed with hydrochloric acid at 110 °C. The mixture is cooled to 75 °C, ethanol is added, and the solution is filtered at 10–15 °C. The precipitate is washed, dried, and treated with ethanol again at 80 °C. The combined extracts are evaporated, and the resulting lipid concentrate is saponified with caustic soda. Ergosterol is then crystallized from the unsaponifiable fraction at 0 °C and purified through multiple recrystallizations. The purified crystals are dissolved in diethyl ether, irradiated with UV light, and then concentrated and crystallized to yield vitamin D_2. The residual acidic filtrate is evaporated to 50% solids and used as a vitamin concentrate.

Brainstorming

1. Which techniques dominate industrial vitamin production?
2. Describe the microbiological synthesis of vitamin B_{12}.
3. List and explain the main technological stages in the production of vitamin B_{12}. Can you sketch a basic flowchart?
4. What is known about the industrial production of vitamin B_2 (riboflavin)?
5. Which microorganisms are used for ergosterol production?
6. How is ergosterol produced on an industrial scale?
7. What are the key technological features of ergosterol production?
8. Which microorganisms are commonly used in industrial carotenoid production?
9. Which stage in ascorbic acid (vitamin C) production is biotechnological?
10. How is biotransformation implemented in the synthesis of vitamin C?

Take-Home Messages
- Vitamin B_{12} (Cyanocobalamin) is produced exclusively by microbial fermentation, using anaerobic bacteria such as *Propionibacterium freudenreichii* and *Pseudomonas denitrificans*. Its production requires strict anaerobic conditions, complex downstream processing, and light-protected handling due to high light sensitivity.
- Vitamin C (L-ascorbic acid) production combines chemical and biotechnological steps. The key microbial stage involves converting sorbitol to L-sorbose using *Gluconobacter oxydans*. Immobilized cells and membrane-bound enzymes improve process efficiency.
- Vitamin B_2 (Riboflavin) is synthesized by various bacteria, yeasts, and filamentous fungi, including genetically engineered strains such as *Ashbya gossypii* and *Bacillus subtilis*. Yields can reach up to 20 g/L in enriched media, and riboflavin is widely used in both food and medicine.
- Carotenoids (provitamin A) are produced by diverse microorganisms like *Monascus*, *Rhodosporidium*, and *Chlorella*. Their synthesis is often light-induced and carried out under aerobic fermentation. Products are extracted using polar/non-polar solvents depending on their form.

- Vitamin D production relies on ergosterol, a provitamin found in baker's yeast and fungi like *Penicillium*. Industrial production involves UV irradiation, acid hydrolysis, and solvent extraction, followed by purification and crystallization of vitamin D_2.
- Biotechnological vitamin production is guided by strain optimization and process integration, including fermentation control, aeration, pH regulation, and downstream purification. Vitamins produced microbially are key components in pharmaceuticals, food fortification, and animal feed.

6.4   Biotechnological Production of Hormones

6.4.1   Biotransformations of Steroids

Steroid hormones—such as corticosteroids, progestogens, estrogens, and androgens—are major representatives of biotechnological products. They exhibit a broad spectrum of biological activity and play key roles in nearly all vital physiological processes. In medical practice, steroid hormones are widely applied as anti-inflammatory, diuretic, anabolic, contraceptive, and anticancer agents. This extensive group of compounds includes over 10,000 natural and synthetic substances.

The synthesis of steroid hormones primarily relies on biotransformation techniques, in which metabolites are converted into structurally related compounds under the influence of microorganisms or their enzymatic systems. Microorganisms typically catalyze specific reactions that would be highly complex and inefficient using purely chemical synthesis. Common transformations include oxidation, reduction, decarboxylation, deamination, hydrolysis, methylation, condensation, and isomerization.

Among all areas of biotransformation, the modification of steroid compounds is especially promising. Due to the bulky and complex structures of steroid molecules, even minor modifications via chemical means are often impractical. Microorganisms, however, are capable of catalyzing unique and selective reactions that are essential for synthesizing steroid-based pharmaceuticals. The development of microbiological methods has made the industrial production of important steroid drugs—such as hydrocortisone, prednisolone, and dexamethasone—feasible.

One of the most commonly used techniques in microbial steroid synthesis is microbiological hydroxylation, since the presence and position of hydroxyl groups significantly influence the physiological activity of steroid hormones. The microbial synthesis of major corticosteroids, including cortisol and its synthetic analogs prednisolone and dexamethasone, is well-established on an industrial scale (see Fig. 6.19).

For the biosynthesis of hormones, natural compounds with a steroidal structure are used as starting materials and are chemically modified prior to introducing specific func-

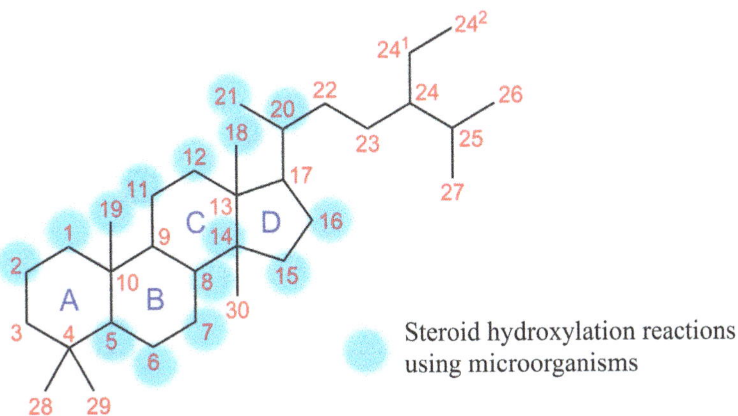

Fig. 6.19   Steroid ring system, indicating hydroxylation reactions

tional groups. Phytosterols, such as ergosterol, are commonly used raw materials and are found in various plants, yeasts, fungi, and bacteria.

Another important phytosterol is stigmasterol, which is abundantly present in soybean oil and sugarcane. The transformation pathway of stigmasterol into steroid hormones is outlined in Fig. 6.20.

Sitosterol, also widely used industrially, is found in large quantities in cottonseed oil, wheat germ, natural rubber, and sugarcane. Under manufacturing conditions, androsta-4-ene-3,17-dione (AD) is derived from sitosterol through microbial side-chain oxidation, using mutant strains of *Mycobacterium vaccae*. In this complex bioprocess, both chemical and microbiological steps are employed. Microbial transformation is specifically used to introduce hydroxyl groups at the 11β position and to form a double bond at the 1,2-position. The high regioselectivity of microbial side-chain cleavage ensures that the steroid backbone remains intact, resulting in high yields of the desired transformation product.

The ability to transform steroids is found in a wide variety of microorganisms. Typically, indigenous strains isolated from soil or other natural sources are used. The best-known 11α-hydroxylating cultures belong to the genera *Curvularia*, *Cunninghamella*, *Absidia*, and *Beauveria*. Dehydrogenation at the 1,2-position is primarily carried out by *Corynebacterium simplex*. For isomerization accompanied by oxidation of the 3-hydroxy group to the 3-keto form, strains of *Corynebacterium mediolanum* are widely used.

The bioconversion of steroids is an aerobic process performed under submerged fermentation conditions. Therefore, equipment that ensures a high degree of mass transfer is preferable. The industrial production of steroid products using biotechnological methods offers several advantages over traditional chemical synthesis. These include the ability to carry out reactions that are difficult or impossible to achieve chemically, the conversion of substrates into biologically active forms in a single step, greater convenience, and enhanced economic and environmental sustainability.

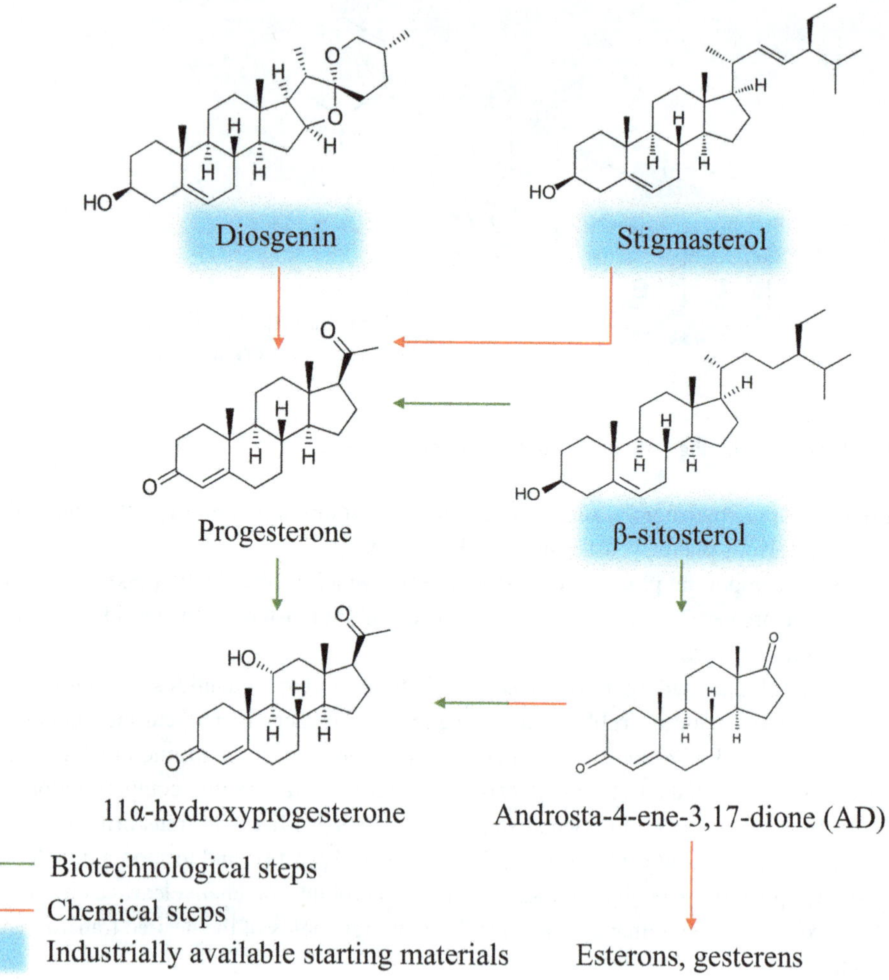

Fig. 6.20 Examples of steroid transformations

6.4.2   Hormones Produced by Genetic Engineering

Genetic engineering techniques make it possible to construct genetic systems that function in both prokaryotic and eukaryotic cells. These capabilities are essential for creating organisms with new and valuable properties, such as bacterial strains capable of synthesizing eukaryotic proteins. Among the most important targets for such work are protein and peptide hormones, which play indispensable roles in medicine and healthcare.

Attempts to produce relatively short peptide hormones through chemical synthesis have proven inefficient and uneconomical for molecules consisting of several dozen amino acid residues. The success of genetic engineering has raised expectations for the production of a wide range of hormones in microbial systems—expectations that have largely been fulfilled, particularly in the case of peptide hormones.

Through genetic engineering, microorganisms and super-producer strains have been developed that are capable of producing high yields of animal and viral proteins. In some engineered strains, up to 20% of the total cellular protein consists of recombinant gene products..

6.4.3   Recombinant Human Insulin

Insulin is a pancreatic hormone that regulates carbohydrate metabolism and maintains normal blood glucose levels. A deficiency of this hormone leads to the serious condition known as diabetes mellitus. Insulin is a small globular protein consisting of two polypeptide chains: the A-chain (20 amino acids) and the B-chain (30 amino acids), connected by disulfide bonds.

Research into the synthesis of both chains and their linkage by disulfide bonds began in the 1960s in the United States, China, and Germany. In 1980, the Danish company *Novo Industri* developed a method for converting pig insulin into human insulin by substituting the 30th residue of alanine in the B-chain with threonine. Both forms of insulin showed no difference in activity or duration of action.

In the same year, a team led by Walter Gilbert isolated insulin mRNA from a rat pancreatic β-cell tumor. A DNA copy of this mRNA was inserted into plasmid pBR322, specifically into the middle of the penicillinase gene, and then introduced into *E. coli*. It was later found that the resulting plasmid coded for *proinsulin*, rather than preproinsulin. Translation of this construct in *E. coli* led to the production of a fusion protein containing sequences from penicillinase and proinsulin. The active hormone was released from this fusion protein via trypsin cleavage. The resulting product demonstrated metabolic activity equivalent to pancreatic insulin.

In 1979, the genes for the A and B chains of insulin were synthetically constructed in the US within just three months. The production of proinsulin in *E. coli* (Fig. 6.21) offers advantages by simplifying extraction and purification processes.

Advancements in genetic engineering techniques have allowed for the development of high-yield producer strains, achieving up to 40% of cellular protein content as insulin (Fig. 6.22). For instance, a 40-liter bioreactor can yield approximately 100 grams of purified insulin, which covers around 1% of the annual global demand.

For example, a 40 L bioreactor provides ca. 100 g of pure insulin, which is about 1% of the annual world demand.

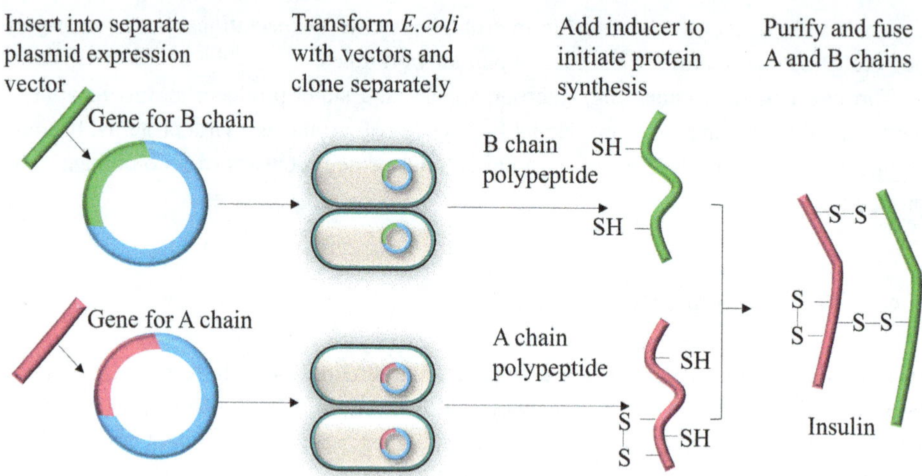

Fig. 6.21   Schematic representation of insulin production via recombinant gene transfer

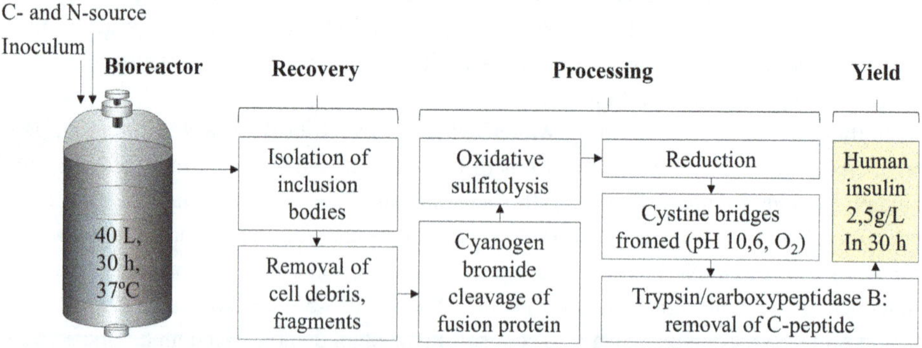

Fig. 6.22   Manufacture of insulin using E. coli

6.4.4   Synthesis of Somatotropin

Somatotropin, also known as human growth hormone (hGH), is secreted by the anterior pituitary gland. It was first isolated and purified in 1963. A deficiency of hGH leads to pituitary dwarfism. Historically, the hormone was obtained from the pituitary glands of cadavers, but this source proved inadequate and posed safety concerns.

Recombinant hGH offers several advantages: it is available in large quantities, is homogeneous, and is free from potentially hazardous contaminants such as viruses. In 1979, the genetic synthesis of hGH—comprising 191 amino acids—was successfully achieved in the United States. Using chemo-enzymatic synthesis, a gene encoding the hGH precursor was constructed and inserted into a suitable expression system (Fig. 6.23).

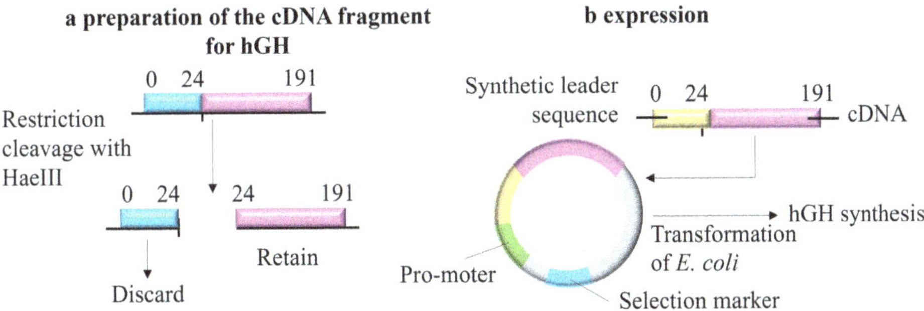

Fig. 6.23 Cloning of human somatotropin (hGH)

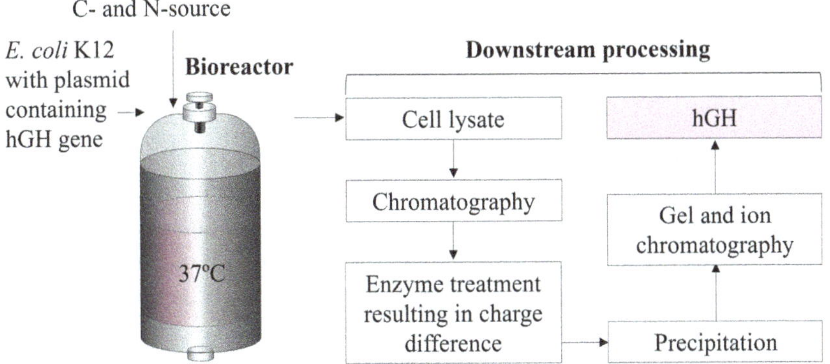

Fig. 6.24 Manufacture of hGH using E. coli

Industrial production of natural human insulin also uses recombinant *E. coli* strains, and the same system ensures the safe, continuous production of recombinant hGH. The hormone is subsequently purified through multiple chromatographic steps (Fig. 6.24).

Microorganisms capable of utilizing a wide variety of substrates allow the production of numerous biologically active compounds and enable valuable processes such as biotransformation, biodegradation, and bioconversion. Today, around 500 commercially valuable compounds are produced through various industrial microbiological processes.

Microbiological synthesis is a non-traditional yet fundamentally innovative approach to producing biologically active substances. Microorganisms grow significantly faster than plants or animals, enabling efficient use of material and energy resources. This method also requires less land, is independent of climate, and is environmentally sustainable.

Brainstorming

1. What is biotransformation, and how is it used in biotechnology?
2. What types of chemical modifications can be made to steroid molecules through biotransformation?

3. What are the criteria for selecting microorganisms capable of transforming steroids?
4. What is the main raw material for the production of steroid drugs, and why is it preferred?
5. What are the technological challenges in applying microbial transformation to steroid hormone production? What strategies can improve the efficiency of this process?
6. What are the advantages of biotechnological methods in steroid drug production compared to chemical synthesis?
7. What is insulin, and what is its biological significance in humans?
8. What methods are used for the industrial production of insulin?
9. How is insulin obtained through biotechnological (genetic engineering) methods? Describe the process.
10. What is the function of growth hormones (e.g., somatotropin) in the human body?
11. How is somatotropin produced for medical use?
12. Which biotechnological techniques are applied in the production of somatotropin? Describe the method.

Take-Home Messages
- Microorganisms can perform precise structural changes—such as hydroxylation, oxidation, and isomerization—on complex steroid molecules that are difficult to achieve through conventional chemical synthesis, enabling the industrial production of vital drugs like hydrocortisone and prednisolone.
- Plant-derived sterols such as stigmasterol, sitosterol, and ergosterol are widely used as starting materials in steroid hormone production due to their structural similarity to animal hormones and availability from agricultural by-products like soybean oil and sugarcane.
- Recombinant DNA technologies have enabled the large-scale, safe, and cost-effective production of protein hormones like human insulin and human growth hormone (hGH) in Escherichia coli and other microbial hosts, replacing limited and ethically challenging animal-derived sources.
- Compared to traditional methods, microbial and recombinant techniques offer faster growth rates, reduced dependency on animal tissues, smaller ecological footprints, and year-round production under controlled conditions.
- Engineered strains can produce hormones as high as 20–40% of total cell protein, allowing significant yields even in relatively small-scale fermenters—highlighting their role as efficient biofactories for pharmaceutical proteins.
- The fusion of fermentation, strain optimization, and enzyme-based biocatalysis enables tailored modification of bioactive molecules for medical applications, making this a cornerstone of pharmaceutical biotechnology.

6.5   Production of Antibiotics

6.5.1   General Characteristics of Antibiotics

The term *antibiotic* refers to any substance produced by microorganisms that suppresses the growth of other microorganisms. Unlike other microbial metabolic products such as alcohols or organic acids, which may also inhibit microbial growth, antibiotics are distinguished by their exceptionally high biological activity.

Over time, the definition of antibiotics has expanded to include chemically modified natural compounds, now referred to as *semisynthetic antibiotics*. Some antibiotics, especially those known as *antitumor antibiotics*, are also capable of inhibiting the growth of tumor cells. Therefore, antibiotics can be defined as low-molecular-weight bioactive compounds of natural origin that inhibit the growth of living cells.

Many antibiotics are used in chemotherapy as antimicrobial agents. They can be classified into:

- Broad-spectrum antibiotics, which are effective against a wide range of pathogens (e.g., cephalosporins, tetracyclines).
- Narrow-spectrum antibiotics, which are used for targeted treatments (e.g., rifampicin for tuberculosis, amphotericin B for fungal infections).

Antitumor antibiotics are also valuable as cytostatic agents, though their application is often limited by high toxicity—for instance, adriamycin.

Some antibiotics are employed in plant protection. These compounds are effective at much lower concentrations than herbicides and generally show low toxicity in mammals. Examples include blasticidin S and kasugamycin.

In food preservation, only a limited number of antibiotics are used. For example, pimaricin is occasionally applied as an antifungal agent in cheese production.

In animal husbandry, feed antibiotics are sometimes added to improve growth and feed efficiency—e.g., monensin in broiler chicken production. However, their use is often restricted to avoid promoting cross-resistance in clinical pathogens.

In molecular biology, antibiotics serve as research tools for the selective inhibition of cellular functions, enabling controlled experimental conditions.

Several systems exist for classifying antibiotics. One common approach is based on biological origin, which is particularly useful in microbiological and biotechnological studies.

Antibiotics represent the largest class of pharmaceutical compounds synthesized by microorganisms. Dozens of microbial producers have been identified and studied in laboratories worldwide. These producers are typically soil microorganisms, including mold fungi, actinomycetes, and spore-forming bacteria.

Fig. 6.25 The simplest
β-lactam

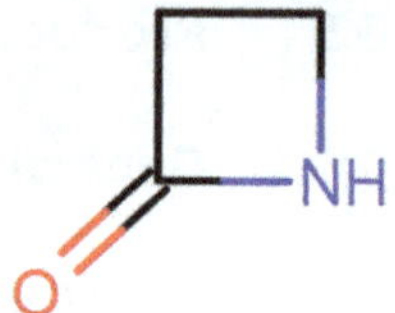

Mold fungi are multicellular organisms with a complex life cycle, producing various forms of mycelium, sporophores, and spores. Their developmental cycle typically lasts 6–7 days. While mold fungi are capable of synthesizing hundreds of different antibiotics, only a limited number have found practical application in medicine.

Among the most important antibiotics produced by fungi are penicillins and cephalosporins. These are collectively known as β-lactam antibiotics, as their antimicrobial activity is based on the presence of a reactive four-membered β-lactam ring—a cyclic amide (see Fig. 6.25).

β-lactam antibiotics are formed by two genera of mold fungi: *Penicillium* (penicillins) and *Cephalosporium* (cephalosporins). Currently, instead of *Cephalosporium*, the name *Acremonium* is preferred. Two principal producers of beta-lactams are widely known: *Penicillium chrysogenum* and *Acremonium chrysogenum*. The former produces benzylpenicillin, the later cephalosporin C. Structurally, penicillins with β-lactam structure exhibit a five-membered ring containing sulfur, whereas cephalosporins usually display a six-membered ring (Table 6.8).

β-Lactam antibiotics are among the most important in both therapeutic value and production volume due to their high efficacy, low toxicity, and the wide variety of chemical modifications available to create semi-synthetic derivatives. The two principal natural β-lactams are penicillin G and cephalosporin C, which serve as starting materials for the production of semi-synthetic penicillins and cephalosporins.

Another medically significant fungal producer is *Fusidium coccineum*, which synthesizes fusidic acid, an antibiotic with a steroid-like structure.

In the 1970s–1980s, a cyclopeptide was isolated from the fungus *Tolypocladium* species. Initially dismissed due to weak antimicrobial activity, it was later found to be highly effective as an immunosuppressant. This compound, now known as cyclosporine (specifically cyclosporin G), has become widely used in organ and tissue transplantation, as well as in the treatment of certain autoimmune diseases.

Actinomycetes (order *Actinomycetales*) produce a vast number—approximately 4000—of different antibiotics. Although they are prokaryotes and closely related to bacteria, actinomycetes differ from typical eubacteria in that they are multicellular and possess a complex life cycle, forming sporophores and spores.

Actinomycetes, especially species from the genera *Streptomyces* and *Micromonospora*, are the main source of antibiotics used in clinical medicine. For instance, *Streptomyces griseus* and *Micromonospora purpurea* produce aminoglycoside antibiotics, including gentamicin, neomycin, kanamycin, tobramycin, and others. These compounds exhibit a broad spectrum of antibacterial activity and are particularly effective against many

Table 6.8 Chemical structure of the most popular antibiotics

Name	Chemical structure
Benzylpenicillin	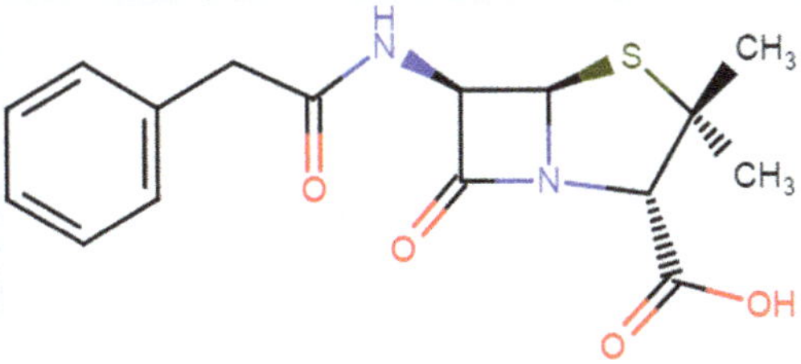
Cephalosporin C	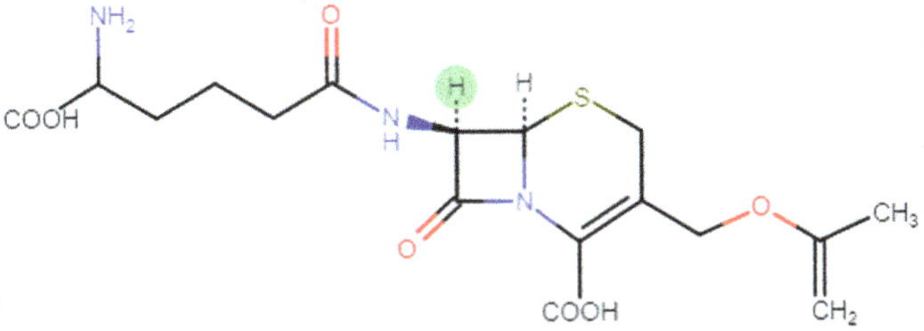
Streptomycin	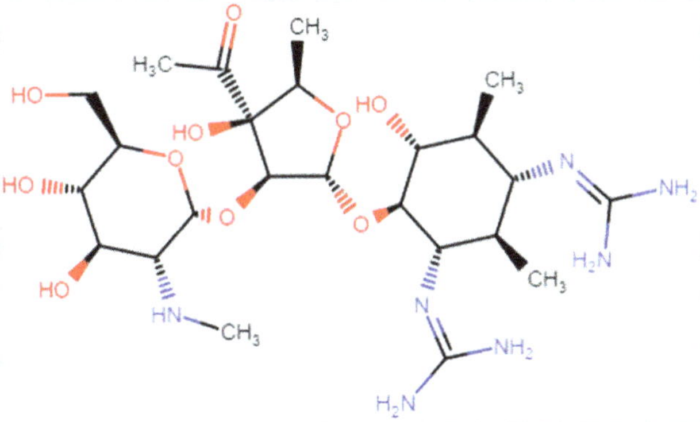
Gentamicin	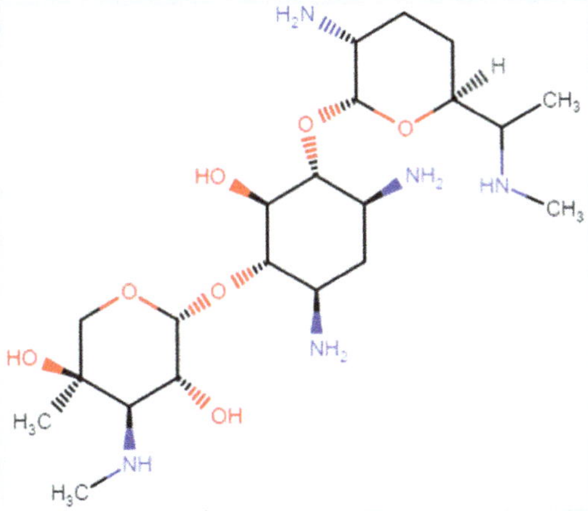

(continued)

Table 6.8 (continued)

Name	Chemical structure
Chlortetracyclin	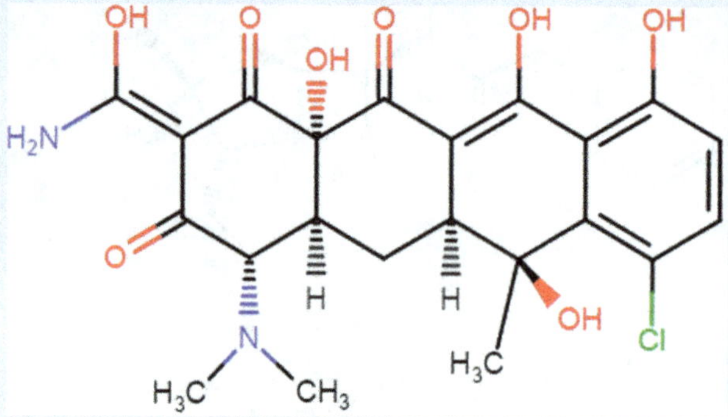
Oxytetracyclin	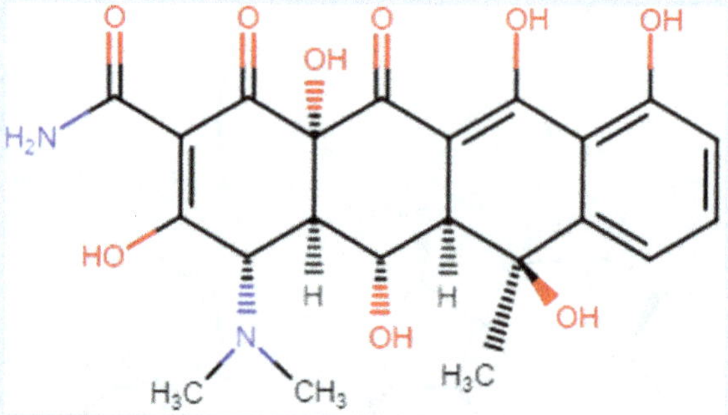
Oleandomycin	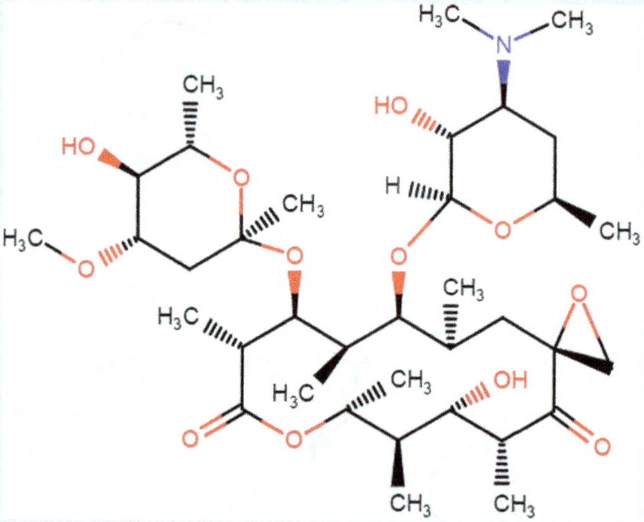

(continued)

Table 6.8 (continued)

Name	Chemical structure
Rifampicin	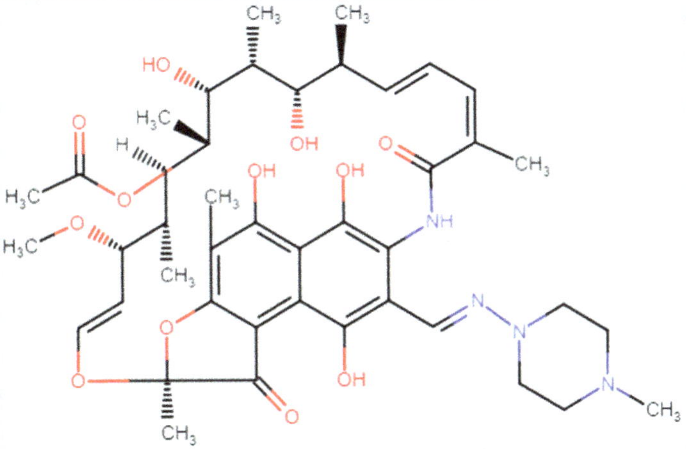
Amphotericin B	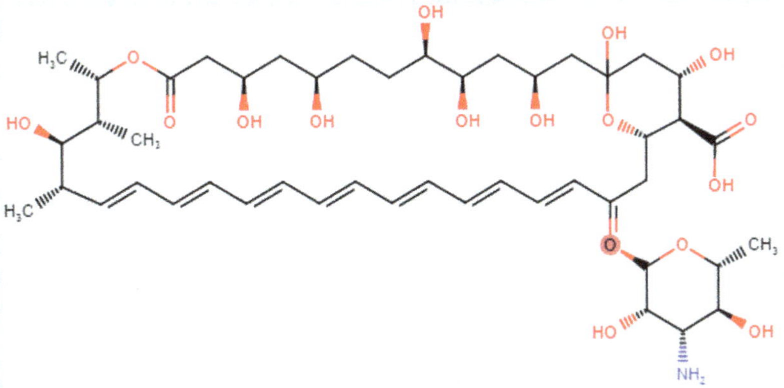
Nystatin	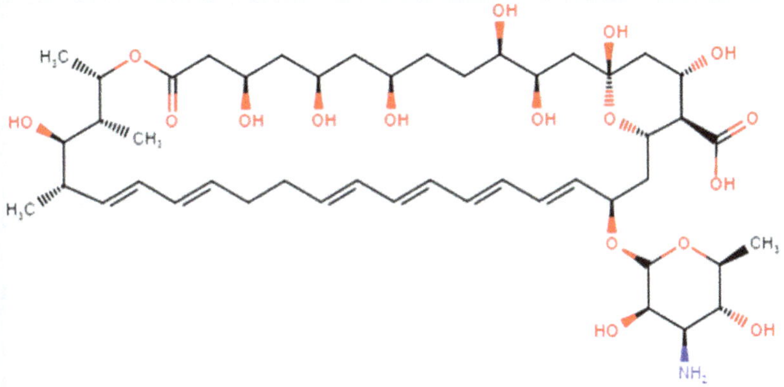

(continued)

Table 6.8 (continued)

Name	Chemical structure
D-phenylalanine	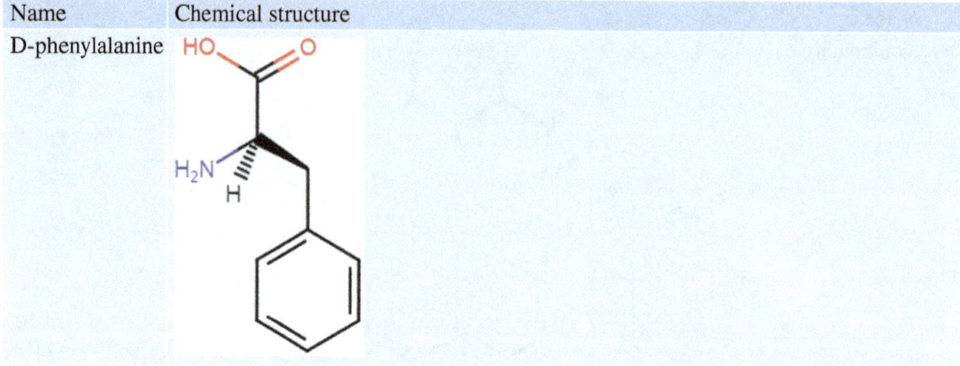

Gram-negative pathogens. Despite their high toxicity and the increasing prevalence of resistant strains, aminoglycosides remain vital for treating severe infections. For example, spectinomycin is effectively used against penicillin-resistant Neisseria gonorrhoeae.

Some aminoglycosides are also applied in agriculture and veterinary medicine:

- Kasugamycin is used to combat rice blight.
- Hygromycin is utilized in veterinary practice.

The aminoglycoside molecule typically contains a six-membered aminocyclitol ring and residues of sugars and/or amino sugars. In addition to natural aminoglycosides, semi-synthetic aminoglycosides are used clinically. These are chemical derivatives of natural compounds, often modified at the amino groups to enhance potency. Examples include sisomicin, amikacin, and tobramycin.

Attempts to develop semi-synthetic aminoglycosides via pathway engineering or combinatorial biosynthesis have so far achieved only limited success. Although most biosynthetic pathway genes have been cloned and suitable expression cassettes constructed, the individual regulation of key enzymes often limits successful pathway integration and manipulation.

Species of the genus *Streptomyces*, including *S. aurofaciens* and *S. rimosus*, also produce the well-known tetracycline antibiotics, such as tetracycline, chlortetracycline (C7-Cl), and oxytetracycline (C5-OH). These antibiotics have a broad spectrum of activity and are widely used in both human and veterinary medicine.

Unfortunately, the extensive use of tetracyclines as feed additives, particularly in poultry and swine farming, has contributed significantly to the emergence of resistant bacterial strains.

Macrolide antibiotics, produced by actinomycetes, are characterized by a macrocyclic lactone ring to which one or more sugars and/or amino sugars are attached. Notable examples include erythromycin A, synthesized by *Streptomyces erythraeus*, and oleandomycin, produced by *Streptomyces antibioticus*. These antibiotics are known for their low toxicity and are therefore commonly used in pediatric medicine. They are particularly effective against Gram-positive bacteria.

Another important antibiotic group is the ansamycins, originally isolated from *Streptomyces mediterranei*, later reclassified as *Nocardia mediterranea*. These antibiotics feature a naphthalene core connected to a long aliphatic chain through ester and amide linkages, forming a characteristic macrocyclic structure. The most prominent member of this group is rifampicin (or rifampin), a semisynthetic antibiotic with potent activity against Gram-positive bacteria and mycobacteria.

Actinomycetes also produce glycopeptide, polyether, and nucleoside antibiotics:

- Vancomycin, a glycopeptide antibiotic, is produced by *Amycolatopsis orientalis* and related strains.
- Avoparcin, produced by *Streptomyces candidus*, was used as a feed additive in higher quantities than vancomycin.
- Lincomycin, synthesized by *Streptomyces lincolnensis*, is active against Gram-positive pathogens and is used primarily in veterinary medicine.

Monensin, a polyether antibiotic produced by *Streptomyces cinnamoensis*, acts as an ionophore, incorporating into cell membranes and inducing osmolysis by facilitating sodium ion (Na^+) influx. Its antimicrobial activity extends to bacteria, fungi, and protozoa, including *Eimeria* and *Toxoplasma* species, which are important in poultry farming. Although toxic to humans and horses, monensin is well-tolerated in poultry and cattle when dosed properly, making it one of the most widely used broad-spectrum feed antibiotics.

Nucleoside antibiotics have limited commercial applications. For instance, blasticidin S, a cytosine analog produced by *Streptomyces griseochromogenes*, is used as a fungistatic agent to control rice blight. However, due to its toxicity to plants, fish, animals, and humans—particularly its effects on mucous membranes, skin, and lungs—its usage is highly restricted.

Some actinomycetes also synthesize antifungal antibiotics, effective against fungi and yeast, including human pathogens. *Streptomyces noursei* is a well-known producer of polyene macrolides, which are characterized by macrocyclic lactone rings with multiple conjugated double bonds. The most recognized representatives are:

- Amphotericin B (a heptaene macrolide)
- Nystatin (a tetraene-diene macrolide)

Furthermore, actinomycetes produce a range of antitumor antibiotics. These are used in the treatment of various cancers. For example, *Streptomyces verticillus* produces bleomycin, a complex glycopeptide antibiotic with antineoplastic activity.

The anthracycline group is especially significant in modern oncology and includes:

- Daunomycin ($R_1 = R_2 = CH_3$)
- Adriamycin or doxorubicin ($R_1 = CH_3$, $R_2 = CH_2OH$)
- Carminomycin ($R_1 = H$, $R_2 = CH_3$)

These compounds, along with their semi-synthetic derivatives, are synthesized by various *Streptomyces* species and remain vital in anticancer chemotherapy.

More than a thousand antibiotics of bacterial origin have been discovered, many of which are peptides or cyclopeptides. During the Second World War, a cyclic decapeptide, gramicidin C, was introduced into medical practice for wound treatment. This antibiotic contains both L-amino acids (valine, ornithine, leucine, proline) and D-phenylalanine. It is produced by *Bacillus brevis*.

Later, polymyxin B, another clinically used antibiotic, was isolated from the soil-dwelling spore-forming bacterium *Bacillus polymyxa*. Polymyxins are a group of more than 20 structurally related compounds differing in their amino acid residues and fatty acid components. The structure of polymyxin B includes three key segments: a cyclopeptide, a linear tripeptide, and a 6-methyloctanoic acid residue (a non-peptide moiety).

Among antibiotics originating from secondary amino acid metabolism, several have found application in medicine, wound care, and agriculture. These include cycloserine and phosphinothricin, cyclic peptide antibiotics like gramicidin and bacitracin, chelating peptides (e.g., bleomycin), chromopeptides (e.g., actinomycin), and depsipeptides (e.g., virginiamycin). While many of these were isolated from *Streptomyces* species, others originate from Gram-positive bacteria such as streptococci and bacilli. However, compared to actinomycetes, spore-forming bacteria produce fewer structurally diverse antibiotics and play a smaller role in clinical medicine.

More than 25,000 microbial metabolites are known, and approximately 50% exhibit antibiotic activity under suitable conditions. Additionally, around 4000 antibiotics have been isolated from higher organisms such as lichens, plants, and animals. Nonetheless, actinomycetes far surpass all other organisms in their capacity to synthesize antibiotics.

The second principle of antibiotic classification is by chemical structure (Table 6.9).

This classification is convenient for chemists studying the structure of antibiotics and the ways they are synthesised.

The third principle of classification is based on the type and mechanism of biological action (adopted in medical practice). The interactions of individual antibiotics are diverse (Fig. 6.26).

Antibiotics Can Affect

1. Biosynthesis of the cell wall (penicillins, cephalosporins, bacitracin, thienamycin).
2. Biosynthesis and function of the cytoplasmic membrane, or as in Gram-negative bacteria, the outer cell membrane (polymyxins, polyenes).
3. Biosynthesis of the components and functions of cell genetic machinery (rifampicins, actinomycins, mitomycin C, anthracyclines, nitrofurans, nalidixic acid).

Table 6.9 Classification of antibiotics by chemical structure

Chemical nature	Groups	Examples
Carbohydrate antibiotics	Aminoglycosides	Streptomycin (medicine), kasugamycin (rice fungicide)
Macrocyclic lactones	Macrolides	Erythromycin (medicine)
	Polyene antibiotics	Primaricin (cheese production)
	Ansamycines	Rifamycin (against tuberculosis)
Chinones and related antibiotics	Tetracyclines	Tetracycline, chlorotetracycline (medicine, feed antibiotics)
	Anthracyclines	Doxorubicin (cancer therapy)
Amino acid and peptide antibiotics	Amino acid derivatives	Cyclosporine (organ transplantation)
		Phosphinothricin (plant protection)
	β–lactam antibiotics	Penicillins, cephalosporins (medicine)
	Peptide antibiotics	Bacitracin (medicine), virginiamycin (medicine), actinomycin (cancer therapy), bleomycin (cancer therapy), vancomycin (medicine), avoparicine (cattle feed antibiotics)
	Chromopeptides	
	Glycopeptides	
N-heterocyclic compounds	Nucleoside antibiotics	Polyoxins, blasticidin S (fungicides for plant protection)
Q-heterocyclic compounds	Polyether antibiotics	Monensin (chicken feed)
Alicyclic compounds	Cycloalkane derivatives	Cycloheximide (leaf fungicide)
Aromatic antibiotics	Benzene derivatives	Chloramphenicol (medicine) Griseofulvin (fungicide)

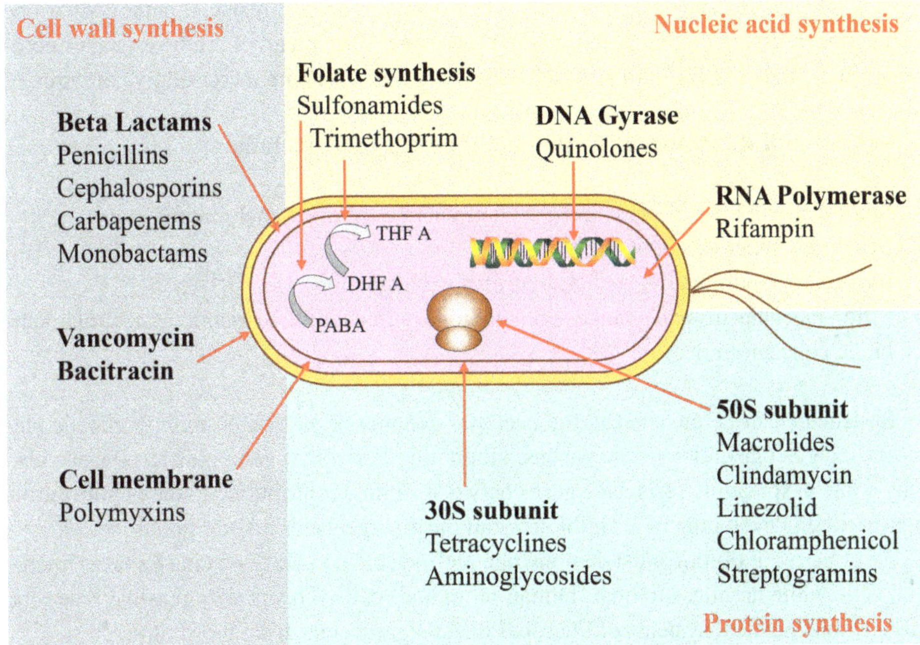

Fig. 6.26 Mechanisms of antibiotics action

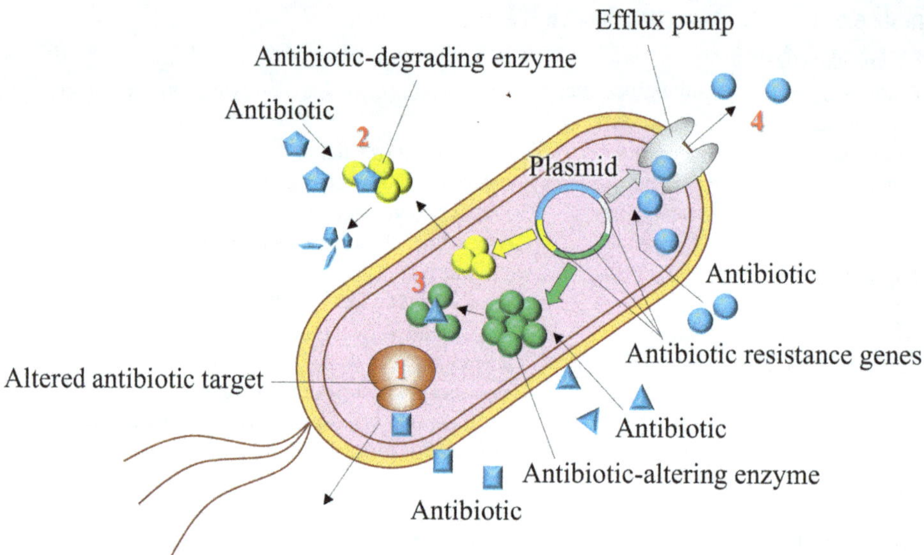

Fig. 6.27 Antibiotic resistance mechanisms

4. Biosynthesis and function of proteins (macrolides, lincomycin, aminoglycosides, tetracyclines, levomycetin).

Due to their rapid generation time and environmental adaptability, microorganisms can quickly develop resistance to antibiotics. The rise of antibiotic-resistant pathogens is one of the most pressing challenges in modern medicine. The spread of cross-resistant microorganisms—those resistant to multiple antibiotics—has become increasingly common.

Mechanisms of Resistance to Antibiotics (See Fig. 6.27) Include:

1. Modification of antibiotic targets, such as changes to ribosomal binding sites
2. Enzymatic inactivation of antibiotics
3. Enzymatic modification, such as hydrolysis or acylation
4. Efflux mechanisms and changes in membrane permeability, reducing intracellular antibiotic concentrations

Resistance can be encoded in the bacterial genome or carried on mobile genetic elements such as plasmids or transposons, facilitating horizontal gene transfer. Phages also play a role in spreading resistance genes between strains, contributing to the ongoing arms race between developing new antibiotics and the emergence of resistant microbes.

An effective antibiotic must demonstrate high selective toxicity—being lethal to microbial cells while harmless to host (human or animal) cells. This is only feasible when the targeted biochemical systems of microbes differ significantly from those of the host.

A critical limitation is the lack of effective antibiotics against pathogenic fungi, yeasts, and viruses that are also non-toxic to humans. This is because eukaryotic microbes share many metabolic pathways with human cells, and viruses utilize host enzymes for replication. Therefore, the search for novel antibiotics and new antimicrobial mechanisms remains a top priority in pharmaceutical biotechnology.

Traditionally, inhibition of microbial growth in the presence of a producer strain or its culture filtrate is used to identify antibiotic activity. If a promising biological effect is observed, the antibiotic is enriched and purified from the culture broth, and its chemical structure is determined. However, using this traditional approach today often leads to the rediscovery of known antibiotics.

To increase the number of novel "hits," various advanced screening procedures have been developed:

1. Precursor-directed biosynthesis—adding synthetic precursors to the fermentation medium to promote the production of modified antibiotics.
2. Screening in previously neglected taxa, such as *Myxobacteria*, rare *Actinomycetales*, lichens, and marine sponges.
3. Modified screening protocols based on novel detection assays.
4. Search for active intermediates in known antibiotic biosynthetic pathways.
5. Discovery of new intermediates using reverse genetics.
6. Recombination of related gene clusters through cell fusion techniques.
7. Combinatorial biosynthesis—in vitro shuffling of gene clusters to generate novel biosynthetic routes.
8. Targeted screening using genomic data and bioinformatics to identify pathogen-specific targets.

If a new antibiotic shows promise, its production yield must be optimized early, since antibiotics are secondary metabolites typically produced in low quantities. Yield improvement often follows empirical approaches, such as repeated rounds of mutation and selection, supported by cellular and genetic engineering techniques. Occasional backcrossing with wild-type strains can enhance the robustness of production strains. Using these methods, yields of important antibiotics have been increased by 1000- to 1000,000-fold compared to the original wild-type isolates.

Brainstorming

1. What are antibiotics, and how do they differ from other microbial metabolites?
2. Which fungi are known producers of antibiotics? Provide examples.
3. List examples of bacterial and actinomycete producers of antibiotics.
4. Why are actinomycetes considered prolific producers of antibiotics?
5. What criteria are used to classify antibiotics?
6. Describe the main classification systems of antibiotics.

7. What are the primary mechanisms of antimicrobial action of antibiotics?
8. What are the key mechanisms by which microorganisms develop resistance to antibiotics?
9. What are the risks associated with chromosomal and plasmid-based resistance?
10. Which screening procedures are used to identify novel antibiotics and their microbial producers?
11. On what basis are antibiotics commonly classified in medical and scientific contexts?

Take-Home Messages
- Antibiotics are biologically active compounds—primarily produced by microorganisms—that inhibit or kill other microorganisms with high specificity and potency, distinguishing them from general metabolic by-products like alcohols or organic acids. Modern antibiotics include both natural and semi-synthetic compounds, expanding the chemical diversity and therapeutic range of antimicrobial drugs. Some also possess antitumor, immunosuppressive, or antifungal activity.
- Microbial producers of antibiotics are mainly soil organisms, including mold fungi (*Penicillium*, *Cephalosporium*), actinomycetes (*Streptomyces*, *Micromonospora*), and some spore-forming bacteria (*Bacillus*, *Brevibacillus*). Actinomycetes alone account for over 70% of known antibiotics.
- Antibiotics are classified based on biological origin, chemical structure or mechanism of action.
- Mechanisms of antibiotic resistance include genetic mutations and horizontal gene transfer via plasmids, transposons, or phages, leading to enzymatic degradation, efflux pumping, or altered target sites.
- Modern antibiotic discovery has moved beyond traditional bioassays to include approaches like combinatorial biosynthesis, reverse genetics, and genomic target-based screening, aiming to overcome rediscovery of known compounds. Selective toxicity is critical for antibiotic effectiveness.

6.5.2   Production of Antibiotics: Methods and Technological Stages

There are three primary approaches to antibiotic production: biological synthesis, chemical synthesis, and a combined (semi-synthetic) method. The chemical structures of most antibiotics are complex and typically contain multiple stereocenters, making total chemical synthesis extremely challenging. While the chemical synthesis of some antibiotics has demonstrated remarkable achievements in organic chemistry, these processes are rarely feasible for industrial-scale production. As a result, biotechnological biosynthesis remains the most efficient and practical method.

Biological synthesis is conducted by direct fermentation using antibiotic-producing microorganisms. Developing the biotechnology of antibiotic production requires understanding the general properties of the microbial producers, as each antibiotic results from a long chain of specific enzymatic reactions. The general technological scheme of microbial antibiotic production is presented in Fig. 6.28.

Most antibiotics are produced by submerged aerobic fermentation, typically in batch mode under strict aseptic conditions. The fermentation process usually lasts 7–10 days. Submerged fermentation allows for modern large-scale biotechnological production, enabling high yields of the desired product. This stage aims to create optimal conditions for the development of the producer and the maximum possible biosynthesis of the required antibiotic. The peculiarity of antibiotic production is the producer development's two-phase nature. In the first phase of the development of the culture, which is called trophophase (phases of balanced growth of the microorganism), the producer biomass is intensively accumulated. The producer synthesizes proteins, nucleic acids, carbohydrates, enzymes, and other substances necessary for microbial growth. There is a rapid consumption of the main components of the substrate and intensive absorption of oxygen. In the culture medium, the pH can decrease due to the accumulation of organic acids.

Antibiotics, as a rule, are not formed in trophophase, or their amount is insignificant. In this phase, the synthesis of enzymes that take part in the formation of an antibiotic is likely suppressed. In the second phase, called the idiophase (the phase of unbalanced growth of microorganisms), biomass accumulation is slowed down. The culture medium is already depleted of components necessary for the development of the producer and enriched with the products of vital activity. Proteolytic processes predominate in culture, leading to its alkalization. The metabolic products of the microorganism are partially used for the construction of mycelial cells and partly for the synthesis of antibiotics. The maximum biosynthesis of the antibiotic in the culture medium occurs, as a rule, after the maximum accumulation of biomass; this maximum is different for various microorganisms and under different cultivation conditions.

In the growth phase (trophophase), all medium components must be contained in quantities that ensure a high growth rate, i.e., with a known excess. Therefore, this phase should be fast, and the nutrient medium should be cheap.

In the second phase, a start and active synthesis of the antibiotic is carried out. When the process goes into the idiophase, the concentration of the main components of the nutrient medium (carbohydrates, nitrogen, phosphorus) is much lower than in the growth phase. However, an essential condition is the presence of a minimum of substances from which the antibiotic molecule is built and provides this process with energy. Thus, the decisive parameter for the transition of the process from the first to the second phase and the creation of conditions for the maximum rate of biosynthesis of the antibiotic is the ratio of the components in the initial medium.

To create a controlled fermentation method, it is necessary to determine the range of the producer's growth limitation and the biosynthesis of the antibiotic by various substrates, to observe the parameters of the biosynthetic process regulation, to determine the optimal regulatory concentrations of substrates, etc. For this purpose, experimental fermentations and a comparative analysis of the dependence of culture productivity are carried out under

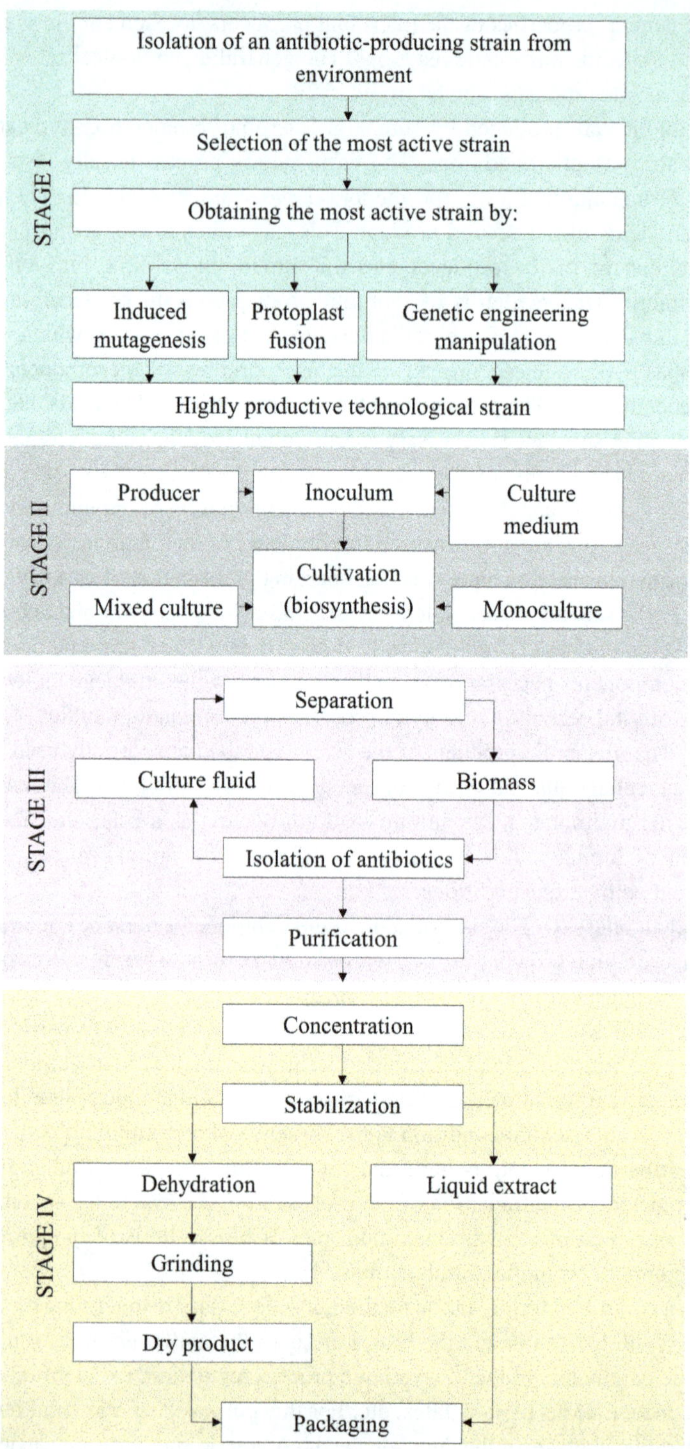

Fig. 6.28 Production scheme of antibiotics in the process of microbial synthesis (according to Yegorov)

various parameters of the process introduction. During the fermentation process, the culture liquid is sampled periodically, and the following points are determined:

- Amount of producer biomass.
- The content of amine and ammonium nitrogen in the medium by biochemical analysis or by measuring sensors.
- The amount of carbohydrates (total and glucose) by biochemical analysis.
- The pH of the medium by the potentiometric method.
- Amount of dissolved oxygen with a pO_2 sensor or by biochemical analysis.
- The amount of biosynthetic product formed.

The fermentation process is conducted until the intensive synthesis of the desired product ceases, and the nutrients are completely exhausted in the medium. While determining the end of fermentation, it is also necessary to consider the data of the microscopic control of the culture state.

Controlled Processes of Fermentation

The first type of process refers to batch (periodic) fermentations, in which no additional nutrients are supplied during cultivation, but physicochemical parameters such as temperature, pH, and aeration mode (including mixing and airflow) are actively regulated. For example, in erythromycin biosynthesis, both temperature and pH are adjusted according to profiles developed through mathematical modeling. Maintaining these optimized parameter profiles significantly increases process productivity, reducing the time required to accumulate the same amount of erythromycin from 300 to 200 h, compared to control conditions with constant temperature and pH.

The second type involves fed-batch (semi-periodic) processes, where nutrients are added during cultivation. This strategy is especially useful when high initial substrate concentrations inhibit either culture growth or the biosynthesis of the target metabolite. In such cases, it is essential to:

- Determine the optimal substrate concentration in the initial medium.
- Define the optimal timing and mode of substrate addition.
- Understand how biomass growth and product accumulation depend on substrate concentration in the culture liquid—or at least identify the concentrations that maximize both.

A typical example is tetracycline biosynthesis, where glucose, ammonium sulfate, and corn extract are added during the process. Preliminary experiments determine the optimal concentrations of these components for microbial growth and antibiotic production. The selected initial medium composition ensures rapid biomass accumulation in the first phase and an early transition to the intensive production phase. In the second phase, a nutrient dosing regimen is applied to maintain:

- Glucose concentration at 10–20 mg/ml
- Ammonium nitrogen at 0.4–0.8 mg/ml
- Inorganic phosphorus at 25–30 µg/ml

Laboratory-scale experiments demonstrated that, under these optimized conditions, the final antibiotic yield increased 5–7-fold compared to an unregulated process, while the fermentation duration decreased by 2–2.5 times.

Tasks related to determining the optimal profiles of control variables and nutrient dosing strategies are typically addressed when simpler means of process intensification are exhausted.

The composition of the medium and cultivation conditions are critical for both biomass growth and target product accumulation. These factors must be tailored to the physiological characteristics of the producing organism. Notably, intensive antibiotic biosynthesis is often facilitated by a marked reduction of easily metabolized carbon and nitrogen sources in the medium. This reduction triggers derepression of antibiotic biosynthetic enzymes.

However, initiating fermentation in such minimal media is impractical. Low biomass formation during the trophophase (growth phase) ultimately leads to insufficient antibiotic accumulation, since too few producer cells are available to sustain high biosynthetic activity.

To ensure highly productive fermentation, certain conditions must be met. Antibiotic producers are cultivated on various media, ranging from relatively simple compositions to complex (mixed) media. The latter often include natural multicomponent nutrient sources such as soy or cottonseed flour, corn extract, and others. In addition, defined organic compounds and mineral salts are added to optimize growth and biosynthesis.

For each producer strain, the optimal medium composition must be determined individually, even for strains of the same species producing the same antibiotic. However, some general principles apply across most producers.

Carbon source regulation is one key mechanism influencing the biosynthesis of secondary metabolites. While glucose is a highly efficient source of carbon and energy, its rapid catabolism often strongly inhibits antibiotic biosynthesis. Studies have shown that glucose suppresses the production of β-lactams, aminoglycosides, and many other antibiotics. This suppression occurs because glucose catabolites inhibit enzymes involved in antibiotic biosynthetic pathways.

By contrast, slowly metabolized polysaccharides (e.g., starch) are more favorable for antibiotic production. Lactose, also a slowly utilized sugar, does not repress biosynthesis. During lactose hydrolysis, glucose is gradually released, which in turn represses β-galactosidase. As a result, lactose hydrolysis slows down, supporting a more controlled nutrient release and better biosynthetic outcomes.

High levels of inorganic phosphate in the medium negatively affect the biosynthesis of most antibiotics. The main reason is that phosphate enriches the cell with high-energy phosphate compounds (primarily ATP), which stimulate rapid mycelial growth at the expense of antibiotic accumulation. For example, high-yielding strains of tetracycline producers often contain lower ATP levels and exhibit slower growth compared to low-yielding strains.

For β-lactam antibiotic producers, phosphate inhibition occurs through a well-defined mechanism: the key precursor LLD-tripeptide, from which penicillins and cephalosporins are synthesized, is inhibited by glucose-6-phosphate. Additionally, the interaction between easily oxidized sugars and phosphates further reduces biosynthetic efficiency. Nevertheless, phosphorus cannot be excluded entirely, as it is still essential for cellular function. The optimal phosphate concentration must therefore be determined for each producer strain.

Ammonium and other easily metabolized nitrogen sources can also enhance cell growth but negatively affect antibiotic biosynthesis, especially for β-lactam and polyene antibiotics such as erythromycin and rifamycins. Complex nitrogen sources such as soy or cotton flour and protein-vitamin concentrates degrade slowly during fermentation. This leads to a gradual release of amino acids and ammonium ions, supporting sustained biosynthesis without overly stimulating biomass accumulation.

Thus, when designing media for antibiotic production, the negative effects of readily available nitrogen sources must be carefully considered, and their concentrations must be tightly controlled.

Some primary metabolites serve as direct precursors of antibiotics. For example, valine is incorporated into the tripeptide that forms the core of β-lactam antibiotics. However, when valine accumulates in excess within the mycelium, it inhibits its own biosynthesis through feedback regulation. Specifically, excess valine suppresses the activity of acetohydroxyacid synthase, the first enzyme in its biosynthetic pathway. As a result, the synthesis of the tripeptide—and consequently the β-lactam antibiotic—is reduced.

Similarly, some primary metabolites are final products of branched biosynthetic pathways, where one branch ends in a primary metabolite, and the other in an antibiotic. For example, α-aminoadipic acid is a precursor for both lysine and β-lactam antibiotics, as it is part of the tripeptide used in their synthesis. When lysine is in excess, it suppresses α-aminoadipic acid formation via feedback inhibition, thereby reducing the biosynthesis of both lysine and the β-lactam antibiotic. These examples illustrate that genetically engineered high-yield producer strains must have disrupted feedback regulation mechanisms for the primary metabolites required in antibiotic biosynthesis. For instance, in low-activity penicillin producers, lysine inhibits penicillin synthesis. In contrast, highly active isogenic strains are no longer sensitive to lysine excess and maintain high antibiotic production.

Aeration is another crucial factor during fermentation, as most antibiotic producers are strict aerobes. Oxygen is required not only for growth but also for the biosynthesis of many antibiotics. In the case of β-lactam antibiotics, molecular oxygen is essential for the formation of isopenicillin N from the LLD-tripeptide. The process requires a stoichiometric 1:1 ratio of oxygen, with a typical oxygen saturation in the culture liquid of about 30%. During efficient fermentation, oxygen consumption may reach 1 mmol/L/min, depending on the biomass concentration and metabolic activity. Optimization of oxygen supply is achieved by improving oxygen transfer rates through agitation and aeration.

In addition to oxygen, several other factors influence antibiotic biosynthesis:

- pH
- Redox potential
- Temperature
- Fermenter design (e.g., diameter, height, stirrer type, impeller configuration)

Each biosynthetic process may also require specific parameters or additions. For example, precursors or inducers can be introduced to stimulate antibiotic production. One such case is phenylacetic acid (PAA), a precursor of penicillin G (benzylpenicillin). PAA is typically added at 100–500 µg/mL after 24 hours of *Penicillium chrysogenum* cultivation. Under these conditions, benzylpenicillin yields after 72 h may reach 500–1000 µg/mL. In contrast, cultivation without PAA results in a mixture of penicillin G (~45%) and penicillin K (~53%). Adding PAA shifts the composition strongly in favor of penicillin G (up to 99%). Moreover, in the absence of PAA, sulfur-containing non-β-lactam compounds (related to cysteine and methionine) accumulate. PAA addition redirects sulfur metabolism toward β-lactam biosynthesis.

This precursor feeding strategy also applies to other antibiotics. For example:

- Propionic acid and propyl alcohol stimulate macrolide biosynthesis via methylmalonyl-CoA.
- L-phenylalanine induces the synthesis of gramicidin S.
- Suppressing chlorination in *Streptomyces aureofaciens* leads to tetracycline production instead of chlortetracycline.
- Inhibiting methylation results in demethylated derivatives of chlorotetracycline.

Thus, direct fermentation with appropriate precursors activates secondary metabolism enzymes during idiophase and enables targeted antibiotic biosynthesis.

Another method of directed antibiotic biosynthesis is using mutants that do not contain a certain link (or are blocked) in the chain of reactions leading to antibiotic synthesis. Blocked mutants cannot form the desired antibiotic. Instead, using low substrate specificity of secondary metabolism enzymes and introducing analogs of antibiotic precursors, they are converted into analogs of the antibiotic itself during the process known as "Mutational biosynthesis" or "mutasynthesis":

1. The expected sequence of reactions leading to the synthesis of the antibiotic:

$$A \rightarrow B \xrightarrow{\text{enzyme}} C \rightarrow D \rightarrow E \rightarrow \text{Antibiotic}$$

2. Absence of antibiotic synthesis on "blocked mutant":

$$A \rightarrow B \xrightarrow{\text{enzyme}} C \rightarrow D$$

3. Synthesis of the modified antibiotic after the introduction of the precursor analogue:

$$A \rightarrow B \xrightarrow{\text{enzyme}} \ldots D^* \rightarrow E^* \rightarrow \text{Modified antibiotic}$$

This form of mutational biosynthesis led to the discovery of new natural antibiotics, among them an antibiotic belonging to the aminoacetic group.

An important condition for successful antibiotic biosynthesis is selecting an appropriate defoamer—one that does not inhibit the growth or metabolism of the producer organism and may even stimulate biosynthesis. During fermentation on complex media containing pea flour, soy flour, or other organic components, excessive foam formation often occurs. This can hinder aeration and risk contamination. To control foaming, natural defoamers (e.g., cachalot fat) or synthetic ones (e.g., laprol, propenol, polydimethylsiloxane) are added in controlled doses.

Throughout the process, strict aseptic conditions must be maintained. Contamination with foreign microorganisms drastically reduces antibiotic yield.

The isolation and primary purification stages depend on the chemical nature of the antibiotic, the type of production, and the intended application of the final product. After the fermentation step, the culture liquid contains dissolved antibiotic, the producer mycelium, the products of its lysis, several unused nutrient medium components, high and low molecular weight organic substances, and inorganic salts. Sometimes an antibiotic is contained not only in the culture liquid but also in the mycelium. In addition, the culture liquid often has a high viscosity. Therefore, isolating antibiotics from such a complex heterogeneous system is not easy.

Current methods and sequence of isolation and purification operations are developed concerning a specific antibiotic and are determined by its physicochemical properties: localization, the composition of the culture liquid, and its rheological and other characteristics.

In the preliminary stage of culture liquid treatment, the dissolved antibiotic is separated from the mycelia suspension and culture liquid components in the colloidal state. If a part of the antibiotic is in the mycelium, it is transferred to the aqueous phase, for example, by changing the pH of the culture liquid (in the case of tetracyclines). Sometimes, on the contrary, the dissolved and mycelia-associated antibiotic is combined in a typical deposit, from which the antibiotic is then extracted. Next, separate the native solution from the mycelium and colloidal particles by filtration or centrifugation using drum vacuum filters, filter presses, separators of various designs, etc.

In the final stage, the task is to obtain antibiotics as an individual substance. When carrying out secondary purification, concentration, and preparation of the medical form, it is necessary to consider the relatively high lability of many antibiotics. During purification, impurities are usually separated, as well as further concentration of the product. Traditionally, extraction and adsorption are used for this purpose.

The extraction method is applied to isolate antibiotics of hydrophobic nature. The extraction is mainly conducted using organic solvents from the aqueous phase when isolating antibiotics. They are isolated from the native aqueous solution by extraction with an organic solvent that is immiscible with water (chloroform, ethyl acetate, butane), using separator funnels. In particular, penicillin (at acidic pH) is extracted from the acidified native solution into chloroform and then re-extracted from chloroform to alkaline water.

Extraction in two-phase aqueous systems is also carried out by two aqueous phases, which are formed by dissolving two incompatible polymers, for example, polyethylene glycol and dextran. The resulting phases contain more than 75% of water. Polymer-containing phases can be separated by decantation or short-term centrifugation. Multistage extraction provides both concentration and purification of released substances; however, large amounts of organic solvents are required. Then, the desired product is isolated (for example, by ultrafiltration) while simultaneously regenerating the polymer. The principle of extraction with organic solvent is used in the purification of such important antibiotics as penicillin, erythromycin, and others. The corresponding antibiotics are released immediately from many impurities when passing into the organic solvent. By varying the pH and thus changing the solubility of the antibiotic in water or buffer solution, it is possible to repeatedly transfer the antibiotic from one phase to another, releasing each time from a certain amount of impurities. However, not all antibiotics are isolated and purified by extraction. Thus, streptomycin is isolated from the culture liquid by ion exchange.

Sorption is the distribution of a solute between liquid and solid phases (usually a porous or large surface material). The sorption method based on surfactants is more cost-effective since the sorbent requires about 10% of the volume of the native solution as sorbents, activated carbon, polysorbates, and ion exchange resins (cation exchangers, and anion exchangers). These techniques played a unique role in solving the problem of obtaining highly purified aminoglycoside antibiotics—streptomycin and others.

Streptomycin is sorbed from the native solution to activated charcoal, followed by filtration, desorbing the antibiotic from the coal with carbon dioxide. Then, the antibiotic from the eluate is precipitated with a solution of sulfuric ester to give the hydrochloride salt of streptomycin. Finally, antibiotics with pronounced basic properties (aminoglycosides) are readily sorbed onto ion exchange resins (cation exchangers) and replaced by an aqueous mineral acid solution.

In addition to the traditional extraction and sorption methods during isolation and purification of antibiotics, a growing number of techniques, combined under the name of membrane technology, have become increasingly important.

After this procedure (or several procedures), the product is ready to be mixed with other constituent parts of the composition or directly sent to the consumer. At this stage, centrifugation is used, followed by drying the crystalline substance, spray drying, or in a drum, drying by lyophilization (freezing) or distilling off the organic solvent. Antibiotic stability depends on the degree of purity of the preparation, humidity, and pH of the solvent.

Since the biosynthesis of antibiotics is carried out under aseptic conditions, the maximum possible precautions against contamination are also observed in the stages of isolation, purification, and preparation of medical forms. Nevertheless, the problem of the sterility of injection preparations and the contamination of products for external use remains one of the most difficult to produce antibiotics and medicines in general. Therefore, radiation sterilization is sometimes used when the prepackaged, non-sterile series of preparations are detected. Certain types of ionizing radiation are permissible for the sterilization of medicinal products. Relevant guidance is available in official pharmacological documents. With this sterilization (in minimum doses), microorganisms contaminating the product lose their ability to reproduce and die due to DNA damage (crosslinks between nucleotides occur, as well as DNA breaks). With thermal sterilization, unlike radiation, denaturation of many cell proteins occurs, because of which the damages become more numerous; with membrane filtration sterilization, the microbial cells do not perish but are removed from the medicine.

Brainstorming

1. What production methods are used in the industrial manufacture of antibiotics?
2. Describe the main technological stages involved in the microbial synthesis of antibiotics.
3. Outline the production scheme of antibiotics via microbial fermentation.
4. What are the biotechnological characteristics of antibiotic production related to the biphasic growth of producer strains?
5. What is controlled (or regulated) cultivation? Describe the main strategies used in the controlled fermentation of antibiotics.
6. What are the key features of antibiotic biosynthesis during the fermentation stage?
7. Describe the composition of the nutrient medium and the optimal fermentation conditions for antibiotic production.
8. What biological and physicochemical factors influence the biosynthesis of antibiotics?
9. What is mutational biosynthesis, and how is it applied in antibiotic production?
10. Describe the main methods used for the isolation and purification of antibiotics from fermentation broth.

Take-Home Messages

- Microbial fermentation is the dominant industrial method for antibiotic production due to its efficiency, sustainability, and ability to produce structurally complex compounds that are difficult to synthesize chemically. Antibiotic biosynthesis occurs in two distinct growth phases: the trophophase, where biomass accumulates, and the idiophase, where secondary metabolites like antibiotics are synthesized. The transition between these phases must be carefully managed.
- Controlled fermentation processes, including batch, fed-batch, and semi-continuous modes, allow for precise regulation of nutrients, pH, temperature, and aeration to maximize antibiotic yields and reduce fermentation time.
- The composition of the nutrient medium significantly influences biosynthesis. Slowly metabolized carbon and nitrogen sources (e.g., starch, soybean flour) promote antibiotic production, while rapidly metabolized substrates like glucose and ammonium can suppress it. Oxygen supply, pH, redox potential, and precursor addition (e.g., phenylacetic acid for penicillin) are critical for optimizing biosynthesis and must be tailored to the specific metabolic needs of the producing microorganism.
- Antibiotic isolation and purification are complex multi-stage processes involving cell separation, extraction, adsorption, and, increasingly, membrane-based technologies. These methods are adapted to the antibiotic's chemical properties and final use.
- Sterility is essential throughout fermentation and downstream processing, as contamination can significantly reduce product quality. In some cases, radiation sterilization is used to decontaminate the final product.

6.5.3 Semisynthetic Antibiotics

Semisynthetic antibiotics are natural antibiotics modified through enzymatic or chemical methods. Due to increasing challenges in isolating new, effective antibiotics and the widespread emergence of resistance among pathogenic bacteria, modifications of existing antibiotic structures have become a central strategy. Most developments focus on penicillins and cephalosporins, both of which contain a four-membered β-lactam ring as their core structure.

Semisynthesis typically involves replacing the side chain of the β-lactam ring in the fermentation-derived molecule with an alternative one. The development of new, more effective penicillin analogs is based on modifying the acyl side group, while maintaining the unchanged penicillin core, 6-aminopenicillanic acid (6-APA). Although 6-APA has only weak antibiotic activity, it serves as a valuable precursor for attaching side groups that enhance antimicrobial potency.

In industrial settings, 6-APA is produced by enzymatic hydrolysis of natural penicillins using penicillin acylase, an enzyme from specific microbial strains. These acylases differ in substrate specificity, and some can catalyze reverse reactions—i.e., acylation of the 6-APA amino group to form new penicillins. In many cases, 6-APA is not isolated as an intermediate. For example, in the conversion of benzylpenicillin to ampicillin, the process proceeds directly:

$$+H_2O,\ \text{Penicillinacylase}$$

Benzylpenicillin   6-aminopenicillanic acid

Benzylpenicillin is hydrolyzed by the acylase of the mutant *Kluyvera citrophila* at pH 7,8-8,0, and at 40–50 °C. Then mutant *Pseudomonas melanogenum* and phenyl glycine are added to the fermenter. The fermentation conditions are thus changed (pH 5,0-5,5) so that the acylase of the second mutant organism synthesizes ampicillin:

Phenylglycine   6-aminopenicillanic   Ampicillin
                              acid

Replacement of the acyl residue leads to the synthesis of other semisynthetic antibiotics (Fig. 6.29).

In the case of cephalosporin C, a semisynthetic approach consists of attaching the side chains to 7-α-aminocephalosporinic acid (7-ACA), which is obtained by deleting the amide side chain of semisynthetic cephalosporins (Fig. 6.30).

The addition of the meta-acid (CH_3O-) group to the ß-lactam ring led to the appearance of cephamine, close to cephalosporins and effective against both gram-negative and penicillin-resistant microbes. Similarly, one natural antibiotic can create 100 semi-synthetic medicines with different properties.

Antibiotics produced via biotechnological methods constitute a significant portion of modern pharmaceuticals. However, the risk of counterfeit or substandard products highlights the importance of stringent quality control. Biotechnologists play a key role in ensuring the purity, efficacy, and safety of antibiotic products through analytical and regulatory oversight.

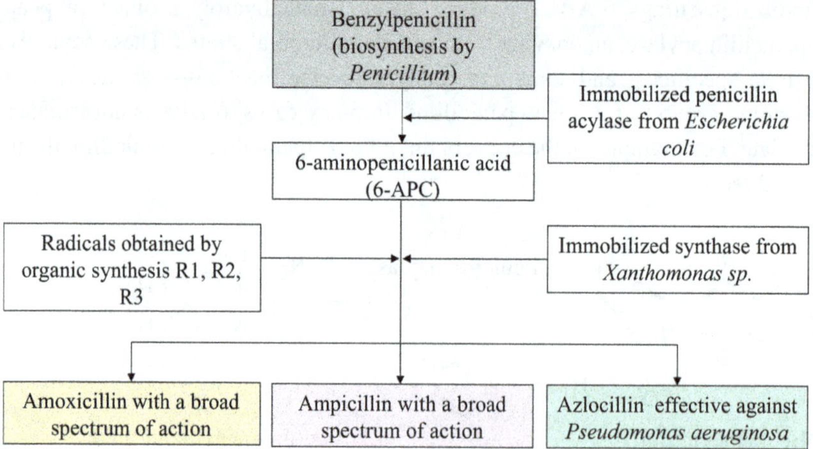

Fig. 6.29 Schematic representation of using benzylpenicillin as a precursor

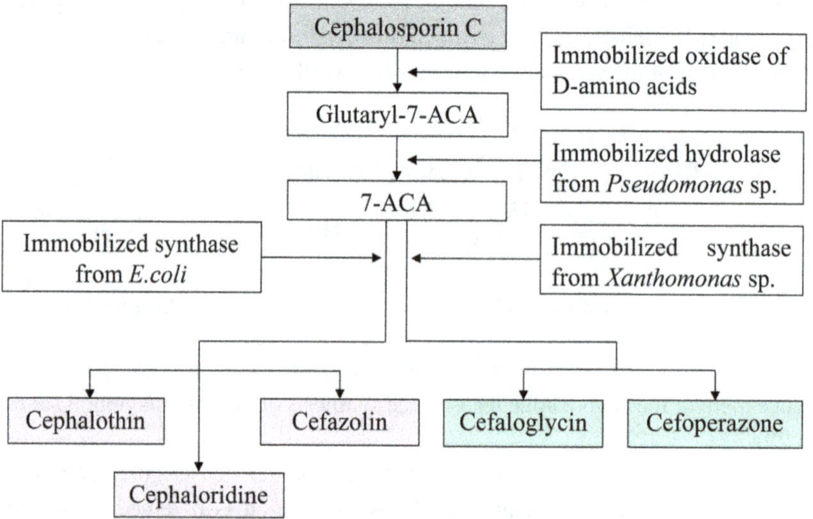

Fig. 6.30 Schematic representation of using cephalosporin as a precursor

All antibiotics are subject to biological and pharmacological quality control. Biological control ensures the sterility of the final product. Pharmacological control involves determining the spectrum and potency of antimicrobial activity using test microorganisms. This is typically done via biological titration, such as the three-dose method (comparing test and standard solutions) or the single-dose method using a calibration curve.

Pharmacological testing also includes assessing toxicity, both acute and chronic. This involves determining:

- The maximum tolerated dose (MTD)

- The lethal dose for 50% of test animals (LD_{50})
- And the lethal dose for 100% (LD_{100})

Products must meet the requirements of relevant regulatory frameworks (e.g., GMP, ISO 9000) before they can be recommended for clinical use.

Brainstorming

1. What are semi-synthetic antibiotics?
2. What is the fundamental difference between semi-synthetic and natural antibiotics?
3. Justify the expediency of obtaining semi-synthetic antibiotics.
4. Is it possible to use biosynthesis and mutasynthesis when creating semi-synthetic antibiotics? How?
5. What is the basis for obtaining new, more effective analogs of penicillin?
6. What enzymes and microorganisms are used to produce semi-synthetic penicillin derivatives?
7. Which semi-synthetic approach is used to create cephalosporin C derivatives?
8. How and why is the biological control of antibiotic drugs carried out?
9. How and why is the pharmacological control of antibiotic drugs conducted?
10. What control functions does the biotechnologist perform to determine the quality of the target product?

Take-Home Messages
- Semisynthetic antibiotics are created by chemically or enzymatically modifying natural antibiotics, particularly β-lactams like penicillins and cephalosporins, to improve efficacy, spectrum of activity, and resistance to bacterial enzymes.
- Rigorous biological and pharmacological quality control—including sterility checks, potency assays, and toxicity testing—is essential to ensure the safety and effectiveness of semisynthetic antibiotics before their approval for medical use.

6.6   Biotechnology of Lipids and Proteins

Lipids, in conjunction with proteins and carbohydrates, constitute the major components of living cells. All cells, excluding specific obligate parasites reliant on their hosts for lipids, possess the capability to synthesize lipids. Consequently, lipid synthesis is regarded as a fundamental biochemical requirement for life. From a chemical standpoint, lipids are defined broadly based on their properties rather than their structure. A chemical definition of lipids is "compounds insoluble in water but soluble in organic solvents like chloroform, ethers, and alcohols." This definition encompasses a wide range of chemicals, including sterols, carotenoids, polyhydroxylalkanoates, triacylglycerols (TAGs), and phospholipids. Apart from their solubility characteristics, these chemicals share little similarity in terms of chemical structure or biosynthesis.

Fatty acids serve as the fundamental building blocks of fats or lipids, similar to how amino acids function as the structural units of proteins. However, a notable distinction exists: while amino acids in proteins are covalently linked by amide bonds, fatty acids in lipids form covalent bonds with a glycerol molecule through ester linkages. The chemical structure of fatty acids varies widely, leading to diverse chemical characteristics among them.

The composition of fatty acids, commonly referred to as the fatty acid profile, significantly influences the physicochemical properties of the lipid. Consequently, fatty acids play crucial roles in the biological functions of lipids within cells, and specific fatty acids are essential for human and animal development and health. Fatty acids, integral components of lipids, hold considerable commercial importance in both the food industry (e.g., cooking oils/fats, cocoa butter) and the chemical industry (e.g., detergents, surfactants, and lubricants).

Although the overwhelming majority of living cells have the ability to produce a range of fatty acids, not all accumulate significant quantities of storage lipid (in the form of TAG/neutral lipid). The production of fatty acids and their incorporation into phospholipids are physiologically necessary for biological membrane synthesis/function and required for cell growth and, therefore, is part of cellular primary metabolism. In contrast, the accumulation of storage lipid is not a physiological necessity, is not universal among biological systems, and is an example of secondary, or overflow, metabolism.

While significant storage lipid is relatively restricted (although not unknown) among prokaryotes, it is more commonly a feature of eukaryotic cells. Even among eukaryotes, however, the ability to accumulate substantial amounts of cell lipid is not universal. Some species do not elaborate lipid as a storage material. These species are designated nonoleaginous and, irrespective of growth conditions, contain less than 25% (dry weight) as lipid—the majority of which is phospholipid. In contrast, oleaginous cells contain greater than 25% (dry weight) cell lipid under suitable conditions, the majority being in the form of storage TAG.

Certain microbial species have been recorded with cell lipid contents exceeding 60% of their dry weight. In oleaginous cells, specific biochemical processes have evolved to redirect excess carbon toward fatty acid synthesis, facilitating extensive storage lipid accumulation.

Describing storage lipid as a secondary metabolite, significant lipid accumulation typically commences after cell growth ceases due to the depletion of a nutrient other than carbon, usually nitrogen, while a continued supply of carbon is available. Under these conditions, the citric acid cycle (tricarboxylic acid cycle or TCA) is downregulated. However, in many organisms, glucose uptake and glycolysis persist, necessitating a suitable fate for the metabolized sugar.

Nonoleaginous organisms channel this carbon into various compounds, including intracellular carbon/energy storage molecules like starch or glycogen, or they secrete it as organic acids, alcohols, or other secondary metabolites, including antimicrobial compounds. In contrast, oleaginous organisms convert the majority of the metabolized carbon source into intracellular lipid. This lipid serves as an internal source of both carbon and energy if the environmental carbon source becomes limited. In oleaginous yeast and fungi, as well as in animal cells, lipid synthesis occurs in the cytosol, where the fatty acid synthase (FAS) complex is located.

Research on lipid accumulation in yeasts and fungi has been extensively conducted and well-documented. The motivation behind studying lipid accumulation in these microbial systems stems from two main factors. Firstly, yeast and fungi are conducive to large-scale cultivation and are known for producing a diverse range of fatty acids in significant quantities, making them promising candidates for the commercial production of specialty fats and oils. Secondly, the similarity in the organization of lipid biochemistry between yeasts/fungi and higher animals positions them as excellent model systems for understanding the processes underlying lipid accumulation in animal cells.

The synthesis of fatty acids and lipids initiates with the generation of the primary substrate acetyl-CoA, derived from pyruvate through the action of pyruvate dehydrogenase as part of central metabolism. However, this reaction occurs within the mitochondria, and since acetyl-CoA is too large to traverse the inner mitochondrial membrane unassisted, active processes are necessary to transport acetyl-CoA into the cytosol for the synthesis of lipids and other metabolites.

Acetyl-CoA does not directly cross the mitochondrial membrane; instead, it undergoes metabolism to citrate through aldol condensation with oxaloacetate as part of the citric acid cycle. This reaction, catalyzed by citrate synthase, initiates a series of transformations. In conditions of unrestricted growth, where all necessary nutrients are available for cell growth, the formed citrate is further metabolized. It undergoes conversion to isocitrate via aconitate, a reaction facilitated by aconitase, and subsequently to α-ketoglutarate through the activity of isocitrate dehydrogenase, an enzyme that generates NADH and involves decarboxylation.

This metabolic pathway continues through the tricarboxylic acid (TCA) cycle, ultimately leading to the regeneration of oxaloacetate for subsequent condensation with another acetyl-CoA molecule. However, in the absence of a suitable nitrogen source, the activity of NAD-dependent isocitrate dehydrogenase (NAD-ICDH) is significantly reduced. This downregulation essentially halts carbon flux through the citric acid cycle. Consequently, the restricted conversion of isocitrate results in the accumulation of intra-mitochondrial citrate, as the isomerization of citrate to isocitrate is a reversible reaction with an equilibrium strongly favoring citrate formation.

The accumulation of mitochondrial citrate prompts its export into the cytosol through a citrate malate antiport mechanism, where mitochondrial citrate is exchanged for cytosolic malate. In the cytosol, ATP citrate lyase plays a crucial role by cleaving citrate. This enzyme, particularly vital for lipid accumulation in oleaginous eukaryotic microorganisms, breaks down citrate, regenerating oxaloacetate and acetyl-CoA, but in this case, the process occurs in the cytosol. The resulting acetyl-CoA becomes available for fatty acid biosynthesis, and there is potential for malate production (via cytosolic malate dehydrogenase) for exchange back into the mitochondria to acquire more citrate.

Another pathway for the cytosolic malate produced, through the combined activity of ATP citrate lyase and cytosolic malate dehydrogenase, involves its decarboxylation by cytosolic malic enzyme. This decarboxylation yields pyruvate and NADPH, providing the necessary reducing power for fatty acid synthase, contributing to both keto reduction and enoyl reduction. Although the decarboxylation of cytosolic malate by malic enzyme appears essential in supplying reducing power for fatty acid synthesis, it's worth noting that the stoichiometry of this process alone cannot account for all the NADPH required. The remaining NADPH is likely provided by the cytosolic hexose monophosphate pathway.

Brainstorming

1. What are the major components of living cells, and why is lipid synthesis considered essential for life?
2. How are lipids chemically defined, and in what types of solvents are they soluble?
3. How do fatty acids differ from amino acids in terms of their linkage and roles in lipids and proteins?
4. Why are specific fatty acids essential for human and animal health, and how do they affect lipid function in cells?
5. How does the fatty acid profile influence the physical and chemical properties of lipids?
6. What distinguishes fatty acid and phospholipid synthesis as part of primary metabolism from the accumulation of storage lipids in secondary metabolism?
7. Why is significant lipid accumulation more common in eukaryotes than in prokaryotes?
8. Under what conditions do oleaginous organisms accumulate large amounts of lipid, and how do they redirect excess carbon toward lipid synthesis?
9. Why are yeasts and fungi widely used as model systems for studying lipid accumulation?
10. What is the role of acetyl-CoA in fatty acid synthesis, and how is it made available in the cytosol?
11. Describe the pathway by which acetyl-CoA is metabolized through the TCA cycle and how this contributes to lipid biosynthesis.
12. How does reduced activity of NAD-dependent isocitrate dehydrogenase affect the carbon flow through the citric acid cycle?
13. What is the citrate–malate antiport system, and how does it support fatty acid synthesis?
14. What role does ATP citrate lyase play in lipid accumulation?

15. How does cytosolic malic enzyme support fatty acid synthesis, and what is its connection to NADPH generation?

Take-Home Messages
- Fatty acids are the primary building blocks of lipids and determine the physical, chemical, and biological properties of lipids. Their composition (fatty acid profile) plays a vital role in human and animal health as well as in industrial applications.
- While all cells synthesize lipids, only certain eukaryotic organisms—known as oleaginous species—accumulate lipids as storage compounds. Yeasts and fungi are model organisms for studying lipid metabolism due to their similarity to higher eukaryotes and their ability to produce valuable specialty lipids in industrial quantities.
- Storage lipid accumulation is an example of secondary metabolism, triggered when nitrogen becomes limiting but carbon remains available. In this state, excess carbon is redirected toward lipid synthesis rather than cell growth.
- The metabolic routing of carbon toward lipid biosynthesis in oleaginous organisms involves citrate export from mitochondria to the cytosol, where ATP citrate lyase generates cytosolic acetyl-CoA for fatty acid synthesis.
- NADPH is crucial for fatty acid synthesis and is supplied by both the cytosolic malic enzyme and the hexose monophosphate pathway, supporting reductive biosynthetic reactions.

6.7 Natural Products as Anticancer Drugs

Natural products, particularly secondary metabolites, have long served as a valuable reservoir for discovering medicinal compounds. Today, over 50% of available drugs are either directly derived from natural sources or are synthetic derivatives closely mimicking their chemical structures. Among these sources, plants are especially prolific—having evolved complex chemical defenses against pathogens and predators—and are a primary source of anticancer agents.

Plant-derived anticancer compounds can be grouped into distinct classes, including vinca alkaloids, taxane diterpenoids, podophyllotoxin (PTOX) lignans, and camptothecin (CPT) quinoline alkaloids. This chapter highlights recent advancements in enhancing the production of these four compound classes. These developments span both homologous and heterologous culture systems, leveraging biotechnological innovations such as in vitro culture, metabolic engineering, and biosynthetic pathway elucidation. The focus lies in optimizing production using modern biotechnological strategies.

There are five principal approaches to obtaining plant-derived natural products:

1. Direct extraction from plants
2. Total chemical synthesis
3. Plant cell and tissue culture
4. Endophytic fungal culture
5. Heterologous biosynthesis

As secondary metabolites, natural products are typically present in very low concentrations in native plants, making direct extraction labor-intensive and costly. Factors such as slow growth rates, limited plant availability, and seasonal variation further hinder efficient production. Despite these challenges, plant extraction remains the main commercial method for most plant-derived anticancer agents, owing to the lack of scalable alternatives. Among the four major drug classes mentioned, only taxane diterpenoids (e.g., paclitaxel) are currently produced through plant cell culture; the other three are still commercially obtained via conventional plant extraction.

Chemical synthesis has also been investigated for these compounds. However, from a commercial perspective, total synthesis is economically and environmentally unviable due to its multi-step complexity, use of harsh solvents, and low yields with frequent formation of structurally similar byproducts. As a result, semi-synthetic approaches—which involve chemically modifying extracted natural products or metabolic intermediates—are preferred for generating derivatives with improved pharmaceutical properties.

Biotechnological advances in plant in vitro culture systems offer promising alternatives for producing valuable secondary metabolites. Systems such as suspension cultures and hairy root cultures have been studied extensively and show potential for industrial-scale applications. Suspension cultures, in particular, demonstrate biosynthetic totipotency—the capacity to produce the full range of compounds found in the parent plant. In some cases, these cultures have been shown to yield higher levels of secondary metabolites than whole plants. A prominent example is paclitaxel, which is commercially produced via plant cell culture.

Despite the promise of in vitro cultures, several challenges persist, including low product yields, genetic instability, and metabolite variability. Hairy root cultures, which exhibit growth rates comparable to cell suspensions but with improved biochemical and genetic stability, represent a viable alternative. Their organized differentiation allows for the compartmentalization of cellular pathways, which can facilitate the production of toxic intermediates and specialized metabolites. Moreover, hairy roots retain many of the physiological properties of natural roots and synthesize root-specific secondary metabolites. This is a crucial advantage, as they are not involved in the transport of metabolites to and from other plant tissues, unlike whole-plant roots.

Endophytic fungi also play key roles in the biosynthesis of plant secondary metabolites, with several species capable of producing the same compounds as their host plants. For the four major classes of anticancer drugs, many endophytic fungal strains have been isolated and studied for their biosynthetic potential. Although commercial-scale fermentation based on endophytes has not yet been realized due to low and often unstable yields, this approach remains promising. Fungal fermentation is well established in the food, pharmaceutical, and agricultural industries. To enable the commercial production of anticancer compounds by endophytic fungi, a deeper understanding of their secondary metabolic pathways and the dynamic interactions with host plants is required.

Another promising avenue is the heterologous production of plant metabolites in microbes—a concept under investigation for decades. Recent advances in synthetic biology and improved knowledge of the biosynthetic pathways in medicinal plants are bringing this strategy closer to practical application. Microbial hosts such as *Escherichia coli* and *Saccharomyces cerevisiae* are favored due to their rapid growth, well-characterized genetics, and industrial scalability. However, several obstacles remain. These include incomplete knowledge of all pathway enzymes and issues with protein folding and functionality of plant-derived proteins in microbial systems. To date, the highest reported microbial titer for an anticancer precursor is that of taxadiene, a key intermediate in paclitaxel biosynthesis.

Elicitation is another strategy commonly employed to enhance the production of secondary metabolites in whole plants, tissues, and cell cultures. Jasmonates, which include jasmonic acid, its precursors, and conjugates, are well-known elicitors that activate defense responses and stimulate the biosynthesis of a broad range of secondary metabolites. In *Taxus* cell suspension cultures, treatment with methyl jasmonate (MeJA) significantly boosted the production of paclitaxel and baccatin III, a biosynthetic precursor. Similarly, jasmonic acid has been identified as the most effective elicitor for enhancing camptothecin (CPT) levels in *Camptotheca acuminata* cell cultures, outperforming other biotic and abiotic elicitors.

Precursor feeding offers another potential strategy for increasing product titers. However, its success is limited by the tight genetic regulation of the biosynthetic pathways; increasing precursor availability does not always result in proportionally higher metabolite accumulation. Lastly, cell immobilization—which promotes enhanced cell-to-cell contact—can stimulate the biosynthesis of certain secondary metabolites by inducing changes in cellular physiology.

In microbial production systems, metabolic engineering strategies are widely employed to enhance metabolite yields. In contrast, plant metabolic engineering remains in its early stages and still requires the development of comprehensive genetic toolkits. Particle bombardment and Agrobacterium-mediated transformation are the most commonly used methods for introducing foreign DNA into plant cells. Well-established protocols exist for transforming *Catharanthus roseus* (producer of vinca alkaloids) and *Taxus* species (producer of taxane diterpenoids). More recently, *Agrobacterium*-mediated transformation has been successfully applied to *Podophyllum hexandrum* (producer of podophyllotoxin lignans), marking a significant step toward expressing targeted genes to improve podophyllotoxin titers and optimize additional traits in *P. hexandrum*.

Beyond production in native organisms, heterologous synthesis—particularly in microbial hosts—offers both promising opportunities and technical challenges. The advantages of heterologous expression include:

1. The ease of genetic manipulation and culture handling in well-characterized hosts.
2. The absence of complex endogenous regulation seen in native producers.
3. Simplified product purification due to fewer structurally similar metabolites.

However, several challenges remain. Engineering a microbial host for the production of plant-derived anticancer drugs typically requires the coordinated expression of multiple genes, and plant-derived enzymes may not fold or function properly in microbial systems. Moreover, imbalanced expression can lead to the accumulation of toxic intermediates, and some biosynthetic reactions require plant-specific organelles, necessitating the construction of alternative biosynthetic compartments within the host.

Despite these difficulties, substantial progress has been made in developing microbial platforms such as *E. coli* and *S. cerevisiae* for the heterologous production of anticancer compounds. Advances in synthetic biology continue to open new avenues for overcoming technical hurdles—for example, the engineering of *E. coli* strains as chassis for terpenoid biosynthesis.

A key prerequisite for successful metabolic engineering is comprehensive knowledge of the biosynthetic pathways. However, the biosynthetic routes for most anticancer drugs remain incompletely elucidated. Characterizing a pathway requires the identification of intermediate metabolites and the corresponding genes and enzymes. While the full genome sequences of many medicinal plants are still unavailable, a wealth of expressed sequence tags (ESTs) and cDNA sequences can be accessed in public databases. These resources have facilitated the prediction and experimental validation of gene function via homology searches. One notable example includes the identification of members of the cytochrome P450 monooxygenase family, a group of enzymes known for their well-defined sequence similarities and subfamily classifications.

Although the genomes of key medicinal plants remain unsequenced and many genes are poorly annotated, a considerable body of omics data—including transcriptomic, proteomic, and metabolomic profiles—has been accumulated. Comparative analysis of these datasets, either separately or in combination, from different plant tissues or under specific elicitation conditions, represents a powerful strategy for narrowing down candidate genes involved in secondary metabolism. A prime example is the successful identification of five enzymes involved in the podophyllotoxin biosynthetic pathway. Similar integrative "omics"-based strategies are being applied to elucidate biosynthetic routes for other plant-derived anticancer agents, as will be discussed in subsequent sections.

Brainstorming

1. How have natural products, particularly secondary metabolites, contributed to the discovery of medicinal compounds?
2. What are the main categories of anticancer natural products derived from plants, and how do they contribute to medicinal drug development?
3. What recent advancements have been made in enhancing the production of anticancer drugs from vinca alkaloids, taxane diterpenoids, podophyllotoxin lignans, and camptothecin quinoline alkaloids?
4. What are the five potential avenues for obtaining valuable natural products from plants, and what challenges are associated with direct extraction?
5. How does chemical synthesis compare to direct extraction in the production of plant-derived anticancer drugs, and what challenges does total chemical synthesis face from a commercial standpoint?
6. What biotechnological advancements in plant in vitro cell and tissue culture are being explored for the production of secondary metabolites, and what challenges are associated with these approaches?
7. How do suspension cells and hairy roots offer viable alternatives for the production of valuable secondary metabolites in plant in vitro cultures?
8. What roles do plant endophytic fungi play in the biosynthesis of plant secondary metabolites, and why is fungal fermentation considered a promising avenue for drug production?
9. How does heterologous production in microbes contribute to the synthesis of anticancer drugs, and what challenges are associated with this approach?
10. What strategies, such as elicitation and precursor feeding, are commonly used to boost the levels of secondary metabolites in intact plants and plant cell cultures?
11. What are the advantages and challenges of heterologous synthesis, particularly in microbes, for the production of anticancer drugs?
12. How do metabolic engineering strategies play a role in enhancing the production of anticancer compounds in both plant and microbial systems?
13. What is the significance of "omics" data, including transcriptomics, proteomics, and metabolomics, in elucidating biosynthetic pathways for anticancer drugs?
14. How can comparative analysis of "omics" data contribute to the discovery and characterization of genes involved in the biosynthesis of plant-derived anticancer drugs?
15. What are the key challenges and opportunities in the field of plant-derived anticancer drug production, considering the various avenues and biotechnological approaches discussed in the review?

Take-Home Messages
- Plant-derived secondary metabolites are a major source of anticancer agents, with over half of current drugs being natural products or their derivatives. Vinca alkaloids, taxanes, podophyllotoxin lignans, and camptothecin alkaloids are among the most prominent classes used in therapy.
- Traditional extraction from plants remains the main commercial method for producing these compounds, but it is often limited by low yields, plant availability, and ecological constraints. Alternative strategies include chemical synthesis, in vitro cultures, endophytic fungi, and heterologous production in microbes.
- Biotechnological approaches such as metabolic engineering, elicitation, and cell immobilization are increasingly applied to improve yields in plant tissue cultures and microbial hosts. Suspension cultures and hairy roots show great potential for scalable production.
- Advances in "omics" technologies and synthetic biology support the identification and manipulation of biosynthetic genes and pathways, opening promising avenues for the heterologous production of complex anticancer molecules in engineered organisms like E. coli and S. cerevisiae.

6.7.1 Important Anticancer Natural Products and Prospects of Anticancer Products Biotechnology

Vinca Alkaloids

Vinca alkaloids are a class of natural compounds derived from the Madagascar periwinkle (*Catharanthus roseus*) and are well known for their potent anticancer properties. The principal members of this class include vincristine, vinblastine, and the semisynthetic derivative vinorelbine, all of which are employed in cancer treatment.

Vincristine acts by inhibiting the formation of microtubules, which are essential components of the cellular cytoskeleton. This disruption interferes with mitosis and impedes the proliferation of cancer cells. It is widely used to treat various cancers, including leukemia, lymphomas, and solid tumors. Similarly, vinblastine, a structurally related compound, also disrupts microtubule formation and is used in the treatment of Hodgkin's and non-Hodgkin's lymphoma, testicular cancer, and other malignancies. Vinorelbine, another derivative, is applied in the treatment of non-small cell lung cancer and breast cancer.

Vinblastine and vincristine are classified as terpenoid indole alkaloids (TIAs) and are synthesized by *C. roseus*. This plant has a long history of use in traditional medicine for treating ailments such as diabetes and eye infections. In the late 1950s, researchers discovered that vincristine and vinblastine could significantly reduce white blood cell counts, leading to their adoption in chemotherapy protocols. To reduce toxicity and expand therapeutic applications, several semisynthetic derivatives have since been developed. Notably, vindesine and vinorelbine have been introduced to clinical practice. These four compounds—vincristine, vinblastine, vinorelbine, and vindesine—exhibit unique pharmaceutical profiles, making them valuable in the treatment of different cancer types.

Recently, proteomic profiling of *C. roseus* suspension cells identified 1663 proteins, including 63 enzymes potentially involved in secondary metabolism. Of these, 16 were predicted to function as transporters, and four were confirmed participants in the TIA biosynthetic pathway. Based on proteomic data and comparisons with related species, gene candidates for an additional seven reactions—primarily in the secologanin biosynthetic pathway (a TIA precursor)—were proposed.

Modern bioinformatics tools, particularly those utilizing RNA-Seq data, have enhanced the identification of novel genes in secondary metabolic pathways. For instance, the final four enzymes missing from the secologanin (iridoid) biosynthetic pathway were identified using extensive transcriptomic datasets from various tissues and elicitation conditions. By generating hierarchical clustering maps and matching gene expression profiles to known enzymes, researchers successfully identified and validated candidate genes. These results are compiled in CathaCyc, an online pathway database developed from *C. roseus* RNA-Seq data. CathaCyc offers tools for visualizing and analyzing "omics" data, aiding in the discovery of missing enzymes and insights into pathway evolution.

To elucidate gene function in the TIA pathway, two primary strategies are applied: forward genetics and reverse genetics. Forward genetics identifies genes associated with specific phenotypes, while reverse genetics starts from known gene sequences to determine their phenotypic effects. A notable success in forward genetics is the identification of the transcription factor ORCA3 using T-DNA activation tagging in *C. roseus* suspension cultures.

As more gene sequences become available, reverse genetics techniques have gained importance. Among these, RNA interference (RNAi) has proven to be a valuable tool for gene silencing and functional validation. RNAi has been successfully applied in *C. roseus* hairy root and suspension cultures. Furthermore, the adaptation of virus-induced gene silencing (VIGS) in *C. roseus* plants enabled the identification of the final two enzymes in the vindoline biosynthetic pathway, underscoring the utility of gene-silencing technologies in pathway elucidation.

Identification of Intermediates in the Biosynthetic Pathways of Anticancer Drugs

Metabolomics, or the profiling of small-molecule metabolites, plays a pivotal role in drug discovery, particularly in identifying intermediate compounds within biosynthetic pathways. In plants, many biosynthetic pathways responsible for anticancer compounds remain poorly characterized, largely due to the low concentrations of secondary metabolites. This challenge is especially evident in the camptothecin (CPT) biosynthetic pathway, which remains only partially understood.

Analytical techniques such as mass spectrometry (MS) coupled with high-performance liquid chromatography (HPLC) or gas chromatography (GC) are widely used in plant metabolomics, particularly when paired with high-resolution MS or tandem MS (MS/

MS). However, even advanced MS/MS techniques may not always provide sufficient structural information, especially for isomeric compounds. For example, Fourier-transform ion cyclotron resonance mass spectrometry (FTICR-MS) can determine elemental composition but may fall short in resolving structural details of unknown metabolites.

Nuclear magnetic resonance (NMR) serves as a valuable complement to MS-based metabolomics. In recent years, HPLC-NMR has proven effective for profiling natural products, particularly those of marine origin, and it holds promise for broader application in identifying plant metabolites.

Additionally, studying differential metabolite accumulation across various plant tissues, in vitro cultures, or environmental conditions aids in identifying target compounds. Going forward, the development of plant-specific metabolite databases will be essential. While gene expression databases such as ArrayExpress and GEO are well-established, standardized repositories for metabolite profiles remain limited. Establishing such databases in unified formats will significantly facilitate data sharing and integration across studies.

Discovery of Unknown Genes in Biosynthetic Pathways

In the context of plant-derived anticancer drugs, most genes in the biosynthetic pathways of vinca alkaloids, taxane diterpenoids, and PTOX lignans have been identified using a variety of characterization methods. However, several unresolved reaction steps still hinder the full reconstruction of these pathways, particularly for transfer into heterologous microbial hosts.

To identify unknown genes, comprehensive "omics" approaches, especially transcriptomics and proteomics, are indispensable. Differential expression analysis across diverse explant types, developmental stages, and induction conditions can help uncover novel gene candidates. For example, the work by Alex et al. led to the creation of the CathaCyc database and incorporated transcriptomic data from 23 different tissues and treatments to cluster genes involved in secondary metabolism.

Expanding such datasets with additional RNA-Seq data from *Catharanthus roseus* could enable more refined clustering and enhance the identification of missing genes. This strategy can also be applied to proteomics, where differential protein expression supports gene function assignment. Even without extensive omics data, cDNA library screening remains a practical technique, especially for identifying stress-responsive genes related to secondary metabolism.

Production of Anticancer Drugs in Microbial Hosts

Beyond plant cell and tissue cultures, producing plant-derived secondary metabolites in microbial hosts presents a highly promising—but technically demanding—approach. Among the various anticancer drug classes, taxane diterpenoids are among the most likely candidates for microbial production, due to the relatively well-characterized nature of their biosynthetic pathways.

One of the primary obstacles is the expression of plant cytochrome P450 enzymes in microbial systems such as *E. coli* and *Saccharomyces cerevisiae*. These enzymes often require specific post-translational modifications and subcellular localization, which are difficult to replicate in microbial hosts.

Researchers, such as Ajikumar et al., have addressed these issues through transmembrane engineering, creating chimeric enzymes composed of P450 and cytochrome P450 reductase (CPR). This enabled the conversion of taxadiene to taxadien-5α-ol. Additionally, co-expression of P450 enzymes with CPR and cytochrome b5 has been shown to enhance enzymatic activity.

The central challenge in heterologous production lies in maximizing metabolic flux toward the desired product while minimizing byproduct formation. To achieve this, researchers increasingly rely on systems biology tools, including computational modeling, flux balance analysis, and genome-scale metabolic models, which together streamline strain design and reduce development time.

Brainstorming

1. What are vinca alkaloids, and how do they exert anticancer effects?
2. What are the primary vinca alkaloids, and how do vincristine and vinblastine inhibit cancer cell proliferation?
3. How has the medical use of vinca alkaloids, derived from *Catharanthus roseus* (Madagascar periwinkle), evolved over time?
4. What is the pharmaceutical significance of semisynthetic derivatives like vindesine and vinorelbine?
5. What recent insights have been gained from proteome mining of *C. roseus* suspension cells, particularly regarding enzymes in the terpenoid indole alkaloid (TIA) biosynthetic pathway?
6. How do bioinformatics approaches, such as RNA-Seq analysis, contribute to identifying genes in the secologanin biosynthetic pathway?
7. What are the key differences between forward genetics and reverse genetics in functional gene analysis within the TIA pathway?
8. How is RNA interference (RNAi) used in reverse genetics to study gene function in *C. roseus* hairy root and cell cultures?
9. How does metabolomics contribute to drug discovery, especially in identifying intermediate compounds in anticancer biosynthetic pathways?
10. What analytical techniques are commonly used in plant metabolomics, and how do they complement each other in identifying and characterizing metabolites?
11. What are the main challenges in identifying unknown genes in anticancer biosynthetic pathways, and how can omics techniques help overcome them?
12. What potential does microbial heterologous expression hold for producing plant-derived secondary metabolites, and what barriers must be addressed?

13. How can transmembrane engineering, such as creating chimeric enzymes, improve the functional expression of plant cytochrome P450 enzymes in microbial hosts?
14. What role do computational models, metabolic flux analysis, and genome-scale simulations play in optimizing microbial production of anticancer compounds?
15. What are the key prospects and ongoing challenges in the future development of anticancer drugs using metabolomics, gene discovery, and microbial production systems?

Take-Home Messages
- Vinca alkaloids, such as vincristine and vinblastine from *Catharanthus roseus*, remain key anticancer agents due to their ability to disrupt microtubule formation and inhibit cell division. Semisynthetic derivatives like vinorelbine and vindesine expand therapeutic options with improved properties.
- Modern "omics" tools, especially RNA-Seq, proteomics, and metabolomics, are essential for identifying unknown genes and metabolic intermediates in complex biosynthetic pathways of plant secondary metabolites.
- Heterologous production in microbial hosts (e.g., *E. coli, S. cerevisiae*) offers a promising alternative for sustainable synthesis of anticancer compounds but faces challenges such as correct protein folding and pathway balancing.
- Metabolic and genetic engineering, including RNA interference, transmembrane enzyme engineering, and genome-scale modeling, are accelerating the development of high-yield production platforms for plant-derived drugs.

7.1   Sources of Enzymes

Enzymes can be extracted from a wide variety of organisms, and today a broad spectrum of biological sources is used commercially. Over 50% of enzymes are derived from fungi and yeast, around 35% from bacteria, and the remaining portion is split between animals (8–9%) and plants (4–5%) (see Table 7.1).

The use of plant and animal sources for enzyme production has notable limitations. Plants generally contain low enzyme levels, and their availability is often seasonal. For animal-derived enzymes, extraction is most economically feasible when performed directly at meat processing facilities, tying production to the location of meat and poultry plants. Additionally, the storage and preservation of animal by-products pose challenges (Fig. 7.1).

In contrast, microbial synthesis overcomes these drawbacks and offers significant advantages. Microorganisms are considered preferable to plants and animals as enzyme sources for several reasons:

- Lower production costs
- Predictable enzyme yield and synthesis rates
- A wide variety of available raw materials as substrates
- Fewer potentially undesirable compounds (e.g., phenolics or endogenous enzyme inhibitors)

Furthermore, genetic engineering enables targeted improvement of microbial strains to enhance enzyme yields.

In practice, the strategy has demonstrated a high potential for rational immobilization by combining genetic engineering and immobilization of enzymes (Fig. 7.2).

I. Digel et al., *Introduction to Industrial Biotechnology*, Learning Materials in Biosciences,
https://doi.org/10.1007/978-3-032-07918-3_7

Table 7.1 Some industrially important enzymes and their sources

Enzyme	Source	Application
Animal enzymes		
Catalase	Liver of bovine, pigs	Dairy products, eggs, beverages, salads
Lipase	Pancreas	Cheese, fats, oil
Rennet	Abomasum of dairy calves and lambs	Cheese
Lactate dehydrogenase	The heart of cattle (bovine)	Medicine
Alkaline phosphatase	Intestines of cattle	Diagnostics, biomarker
Fumarase, transaminase	Pig's heart	Clinical studies
Trypsin, chymotrypsin, carboxypeptidase, elastase	Pig's pancreas	Food, biological research, lab studies
Pepsin	Pig's stomach	Food manufacturing, pharmaceuticals
Luciferase	Muscle tissues of fireflies	Biological research, lab studies
Acetylcholinesterase	Muscle tissues of electric eels	Biological research, lab studies
Plant enzymes		
Actinidin	Kiwi fruit	Food
Alpha and beta amylases	Malted barley	Brewing
Bromelain	Pineapple latex	Brewing
Beta-Glucanase	Malted barley	Brewing
Ficin	Fig latex	Food
Lipoxygenase	Soybeans	Food
Papain	Pawpaw latex	Meat
Acid phosphatase	Potatoes	Biological research, lab studies
Peroxidase	Horseradish	Manufacturing processes
Urease	Canavalia ensiformis	Biological research, lab studies
Bacterial enzymes		
Alpha and beta amylases	Bacillus	Starch, fats and oils, cheese, beverages, bakery,
Endo-beta-glucanase	Bacillus	Beverages
Glucose isomerase	Bacillus	Fructose syrup, cereals, fruits, vegetables, beverages
Hemicellulase	Bacillus	Cocoa, chocolate, coffee and tea
Protease	Bacillus	Meat, poultry, fish
Pullulanase	Klebsiella	Starch, sugar, honey, beverages
Fungal enzymes		
Alpha amylase	Aspergillus	Bakery, fruits, vegetables; beverages
Tannase	Aspergillus	Beverages
Glucoamylase	Aspergillus	Starch, beverages, confectionery, bakery, dietary foods

(continued)

Table 7.1 (continued)

Enzyme	Source	Application
Catalase	Aspergillus	Dairy products, egg, salads, beverages
Cellulase	Trichoderma	Waste processing, fruits, vegetables, beverages, dietary foods
Dextranase	Penicillium	Food
Glucose oxidase	Aspergillus	Food, beverages, salads
Lactase	Aspergillus	Dairy products, dietary foods
Lipase	Rhizopus	Food, cheese, fats and oils
Rennet	Mucor miehei	Cheese
Pectinase	Aspergillus	Drinks, fats and oils, fruits, vegetables, fish
Pectin lyase	Aspergillus	Drinks
Protease	Aspergillus	Baking
Raffinase	Mortierella	Food, beverages
Yeast enzymes		
Invertase	Saccharomyces	Confectionery
Lactase	Kluyveromyces	Dairy products, dietary foods
Lipase	Candida	Food, fats, and oils
Raffinase	Saccharomyces	Food, sugar, honey

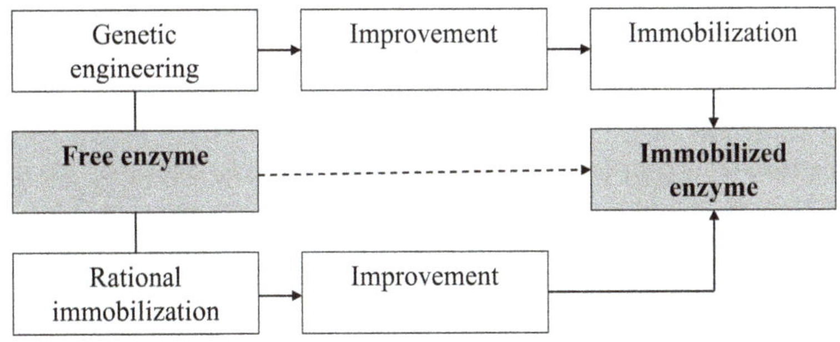

Fig. 7.1 Sources of enzymes and their processing

Fig. 7.2 Improvement of the biocatalysts by genetic engineering and immobilization

Table 7.2 Some major products manufactured applying immobilized enzymes

Enzyme	Product
Glucose isomerase	High-fructose corn syrup
Amino acid acylase	L-amino acids
Penicillin acylase	Semi-synthetic penicillins
Nitrile hydratase	Acrylamide
β-Galactosidase	Whey, lactose
Pectinases	Pectin
Laccase	Wine, fruit juice
Trypsin	Lactoglobulin
Lipase	Oil, grease

Industrial processes based on enzyme and cell immobilization are widely used for producing glucose-fructose syrups, L-amino acids, and L-malic acid, as well as for producing lactose-free milk, sugars from whey, 6-aminopenicillanic acid, etc. (Table 7.2).

7.2   Engineering Enzymology and Enzyme Immobilization

I Immobilized enzymes tend to be more stable and resilient to environmental fluctuations than their free counterparts in solution. Most importantly, their heterogeneous structure allows for the recovery and reuse of both the enzyme and the product, the continuous operation of biocatalytic processes, and flexibility in bioreactor design.

Enzyme immobilization refers to the binding of enzymes to a water-insoluble carrier, which restricts their movement in space and stabilizes their activity.

7.2.1   Fundamentals of Enzyme Immobilization

In the mid-20th century (see Table 7.3), a new interdisciplinary scientific and technical field emerged—engineering enzymology. This field draws from enzyme technology and engineering disciplines to develop methods for producing, purifying, stabilizing, and applying enzymes in industrial and environmental processes.

The main goal of engineering enzymology is to design bioorganic catalysts with tailored properties and to develop efficient and eco-friendly biotechnological processes based on these catalysts. Thanks to the high substrate specificity and catalytic efficiency of enzymes, these processes can achieve high product yields and minimal environmental impact.

Enzyme-based biotechnological processes are increasingly applied across industries, including the food industry, energy, medicine, microelectronics, etc. (Fig. 7.3).

Any immobilized enzyme, necessarily, must meet two vital functions, mainly the non-catalytic functions (NCF) and the catalytic functions (CF) (Fig. 7.4). NCF is related to the physical and chemical nature of the immobilized enzymes (e.g., the composition, shape,

Table 7.3 The chronological development of enzyme-immobilization techniques

Phases	Years	Methods developed and valuable observations
1916–1940s	1916	Nelson and Griffin showed that the invertase, which was adsorbed on activated carbon, retained its enzymatic activity
	Succeeding 40 years	Using bio-immobilization techniques to make adsorbents for the isolation of proteins on simple inorganic matrices (glass, alumina)
1950s	Beginning	The adsorption technique was gradually transformed from simple physical adsorption to specific ionic adsorption
	1953	The covalent binding technique of enzymes with a carrier was first applied by N. Grubhorfer and D. Schleit
	1955	For the first time, Dickey observed the technique of AMP deaminase entrapment in sol-gel glass
1960s	1960	Adsorption and entrapment techniques were further optimized/designed
	1964	Encapsulation of enzymes in semi-permeable spherical membranes (artificial cells) was first demonstrated by Chang
	1966	Adsorptive cross-linking of enzymes on films, membranes, and beads for the creation of enzyme envelopes was established
	By the end	First industrial application of an immobilized enzyme (L-amino acid acylase) for the manufacturing of L-amino acids by a Japanese company
1970s	Beginning	Affinity and coordination binding and many modern alternatives of enzyme immobilization were developed
	1971	The term "immobilized enzyme" was used at the first conference on engineering enzymology, which was held in Henniker (USA)
	Middle	Covalent coupling of an enzyme to a solid matrix or enzyme entrapment in a gel matrix can be conducted in organic solvents
	By the end	Various novel synthetic or natural functionalized polymers with chemical and physical nature were specifically developed or modified for covalent enzyme immobilization Bio-immobilization methods had matured to such an extent that every single enzyme could be immobilized by selecting a suitable technique of immobilization or a suitable carrier and immobilization conditions
1980s	The entire period	Many methods designed in the 1970s were further implemented to improve enzyme functions The idea of the water activity of the reaction medium was developed, and a reliable comparison of various catalytic processes in low-water media became attainable
1990s–date	The entire period	Development of plastic-immobilized enzymes in organic solvents and chemical post-immobilization methods Construction of the microenvironment and engineering the multipoint attachment, site-directed bio-immobilization. Initiation of tentacle carriers Imprinting-immobilization and enzyme deposition strategies

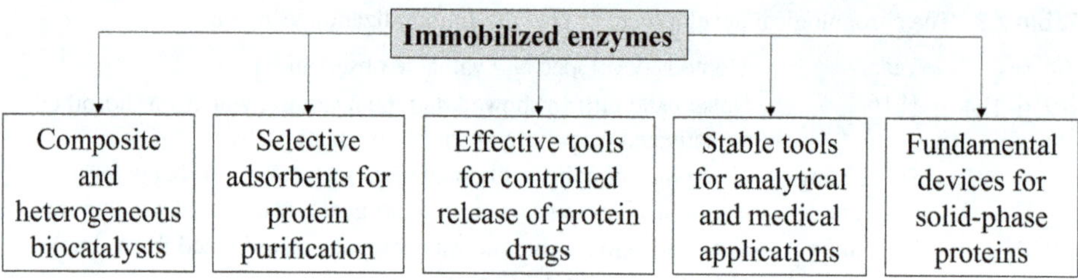

Fig. 7.3 Application of immobilized enzymes

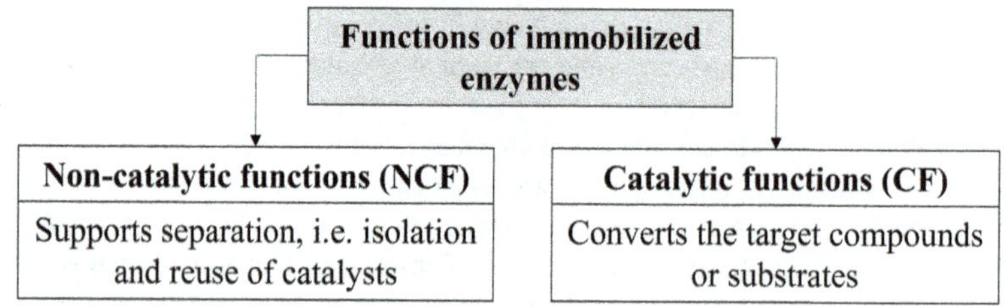

Fig. 7.4 Main functions of immobilized enzymes

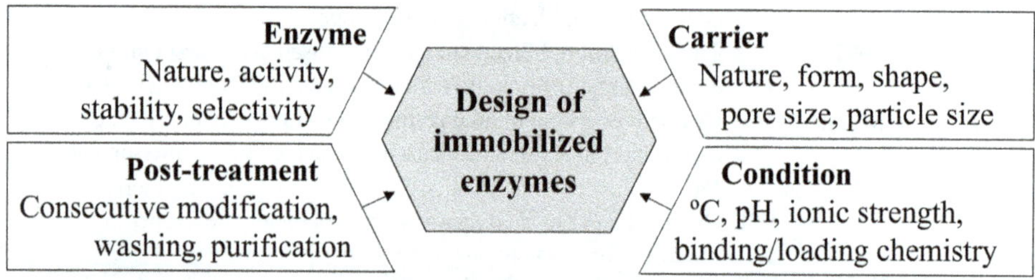

Fig. 7.5 General procedures for enzyme immobilization

size, thickness, and length of a carrier). In contrast, the CF relates to the catalytic properties (e.g., activity, selectivity, stability, pH, and temperature parameters).

Immobilization is attaching or loading enzymes to the surface of natural or synthetic materials, such as polymeric materials, fibers, membrane capsules, etc. Immobilization can also be characterized as a physical separation of catalyst and solvent, during which the substrate and product molecules are easily exchanged between phases. In addition to stability, immobilized enzymes acquire certain properties that are not characteristic of their free state, e.g., the ability to function in a non-aqueous medium and a broader range of optimum temperature and pH. The duration of maintaining catalytic activity and the number of enzyme properties are determined by the proper choice of carrier, the method, and conditions for immobilization, etc. (Fig. 7.5).

Table 7.4 Standard criteria for the choice of potential immobilized enzymes as catalysts

Requirements
1 Broad applicability and reproducibility
2 Recyclability
3 Easy and swift design
4 Suitable chemical, physical and mechanical properties
5 Broad substrate specificity
6 Stability in various (organic and inorganic) solvents
7 Thermostability, operational and conformational stabilities
8 Safety for use and easy disposal
Advantages
1 Great separation of enzymes from substrates and final products
2 Ensuring the continuity of enzymatic processes, regulation of enzymatic activity
3 Ensuring the stabilization of enzymes
4 The possibility of changing the properties of a carrier and the regulation of the immobilized
   enzyme activity
5 Tolerance through both intrinsic and extrinsic factors
6 High productivity and space-time yield, cost-effective
7 Flexibility of reactor design, easier down-stream processing, separation and purification
8 Less pollutions and meeting safety regulations

The main criterion for selecting a perfect immobilized enzyme is its ability to remain active for as long as possible. However, immobilized enzymes also have other vital requirements (Table 7.4).

Brainstorming

1. List the areas of enzyme application.
2. What are the producers of microbial enzymes known? Give examples.
3. Sources of enzymes and their processing.
4. Explain why microbes are privileged to plants and animals as sources of enzymes.
5. Which enzymes of microbial origin are used in medicine?
6. What types of compounds serve as the foundation for enzyme engineering?
7. What are the key disadvantages of using free (non-immobilized) enzymes?
8. What defines an immobilized enzyme?
9. Who first discovered that enzyme activity can be preserved outside the cell?
10. What are the main advantages of immobilized enzymes over free enzyme preparations?
11. What are the catalytic and non-catalytic functions of immobilized enzymes?
12. What criteria are used to select enzymes suitable for immobilization?

Take-Home Messages
- Microorganisms are the preferred source of industrial enzymes due to their cost-efficiency, consistent yield, scalability, and suitability for genetic modification—unlike plant or animal sources, which face seasonal and logistical constraints.
- Enzyme immobilization enhances stability, reusability, and process control, making it a cornerstone of modern biotechnological applications in medicine, food production, energy, and other industries.
- Engineering enzymology is an interdisciplinary field focused on designing biocatalysts with desired properties and developing eco-friendly, high-yield processes that minimize waste.
- The success of immobilized enzyme systems depends on both catalytic (activity, stability, specificity) and non-catalytic factors (carrier material, shape, binding method)—underscoring the need for careful optimization of immobilization strategies.

7.3   Carriers for Immobilization of Enzymes and Cells

Various materials and compounds are currently used as carriers or matrices for enzyme immobilization. Ideally, the matrix must satisfy several requirements, as shown in Fig. 7.6.

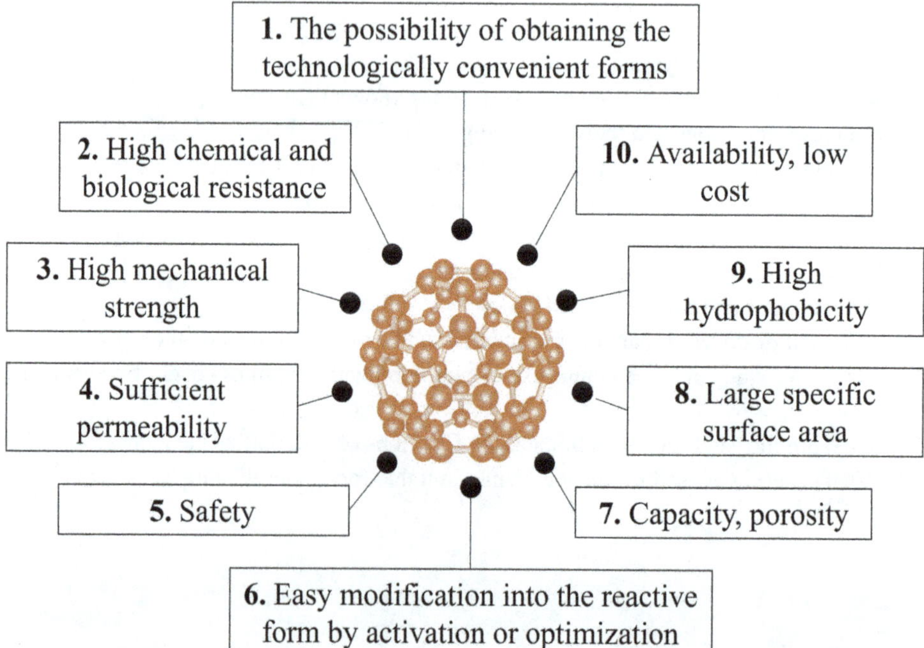

Fig. 7.6 The main requirements for the carriers used in enzyme immobilization

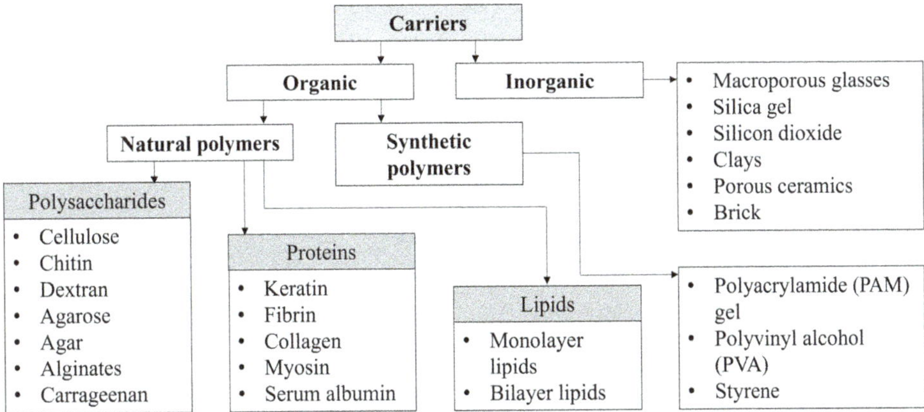

Fig. 7.7 Classification of carriers used in biotechnological purposes

However, currently, no matrix fully meets all the requirements for an "ideal carrier." This concern led to the development of various carriers, the choice of which is made depending on specific goals. All carriers can be categorized as natural and synthetic polymers (Fig. 7.7).

Natural polymers used as carriers have sufficient reaction-functional groups, can easily be modified, and undergo a reaction with biomolecules. However, the main disadvantage of natural polymers is their low stability to microbial action and a relatively high cost.

The various inorganic and organic carriers and the immobilized enzymes are listed in Table 7.5.

7.3.1   Polysaccharides

Cellulose, dextran, agarose, carrageenan, and their derivatives are the most often-used carriers for enzyme immobilization (Table 7.6).

Cellulose

Cellulose consists of poly-1,4-β-D-glucopyranosyl-D-glucopyranose. It has a high hydrophilicity. Many OH groups make it possible to modify it easily by introducing various substituents. Cellulose preparations are cross-linked with epichlorohydrin to give them chemical strength. The porous spherical granules are formed to increase cellulose's mechanical strength. There are two methods of granulation: precipitation in an aqueous-organic medium and hydrolysis with the destruction of amorphous regions. In this form, cellulose is easily converted into various ion-exchange derivatives: DEAE-cellulose, CM cellulose, etc. Granular cellulose has a high porosity, excellent mechanical strength, and apparent hydrophilicity and is widely adopted in engineering enzymology and chromatography. The main disadvantage is its instability in the action of strong acids and alkalis.

Table 7.5 Support carrier systems and immobilized enzymes

Material	Immobilized enzyme
Inorganic materials	
Zeolite	Glucose oxidase, α-chymotrypsin, lysozyme
Ceramics	Candida Antarctica lipase
Celite	Lipase, polyphenol oxidase, β-galactosidase
Silica	Lignin peroxidase, horseradish peroxidase, α-amylase
Glass	α-Amylase, nitrite reductase
Activated carbon	Acid protease, acidic lipase
Charcoal	Papain, amyloglucosidase
Natural Polymers	
Collagen	Tannase, catalase
Chitosan and chitin	D-Hydantoinase
Carrageenan	Lipase, α-galactosidase
Cellulose	Fungi laccase, penicillin G acylase, glucoamylase, α amylase, tyrosinase, lipase, β-galactosidase
Starch	Peroxidase
Pectin	Papain
Sepharose	Amylase, glucoamylase

Chitin

Chitin-based carriers are natural amino polysaccharides. In structure, they are similar to cellulose, in which an acetate residue replaces the CH_2OH group. Chitin forms the outer skeleton of insects, crustaceans, and the cell walls of some fungi. It is porous and insoluble in water, dilute acids, alkalis, and organic solvents. It is easily modified by glutaraldehyde. When chitin is treated with concentrated solutions of alkalis, it converts into chitosan. Having free amino groups, chitosan easily binds to proteins. Unlike chitin, chitosan dissolves readily in mineral and organic acids and can immobilize the enzyme in solutions at pH 3–7.

Dextran

Dextran is a polysaccharide consisting of α-1,6-linked glucose, a branched polysaccharide isolated from the cells of *Leuconostos mesenteroides*. Based on dextran with cross-linked epichlorohydrin, "Sephadex" is produced (Sweden, Pharmacia). Sephadex is easily compressed when dried, while in aqueous solutions, it strongly swells. The smaller the number of cross-links, the greater the ability to swell. Different brands of Sephadex contain a small amount of carboxyl groups, which ensures their affinity for cations.

Agarose

Because of its high cost, the search for modification methods has been conducted to provide an easy reaction of the carrier. If the 2–6% aqueous agarose solution is cooled to 45 ° C, large-porous gels are formed that cannot be autoclaved and dried. Therefore, various companies produce it known as "Sepharosa," "BioGel," "Ultragel," and "Superosa."

Table 7.6 Common polysaccharides used as carriers

Polysaccharides	Monosaccharide 1	Monosaccharide 2	Linkage type	Linkage type at branch points	Source	Function
Bacteria						
Dextran	D-Glc	–	PP$^\alpha$PP$(1 \rightarrow 6)$	PP$^\alpha$PP$(1 \rightarrow 3)$	Slime	Water-soluble polysaccharide
Peptidoglycane	D-GlcNAc	D-MurNAc	PP$^\beta$PP$(1 \rightarrow 4)$	–	Cell wall	Structural polysaccharide
Plant						
Agarose	D-Gal	L-aGal	PP$^\beta$PP$(1 \rightarrow 4)$	PP$^\beta$PP$(1 \rightarrow 3)$	Red algae	Water-soluble polysaccharide
Carrageenan	D-Gal	–	PP$^\beta$PP$(1 \rightarrow 3)$	PP$^\alpha$PP$(1 \rightarrow 4)$	Red algae	Water-soluble polysaccharide
Cellulose	D-Glc	–	PP$^\beta$PP$(1 \rightarrow 4)$	–	Cell wall	Structural polysaccharide
Amylose	D-Glc	–	PP$^\alpha$PP$(1 \rightarrow 4)$	–	Amyloplast	Reserve polysaccharide
Inulin	D-Fru	–	PP$^\beta$PP$(2 \rightarrow 1)$	–	Reserve cell	Reserve polysaccharide
Animal						
Chitin	D-GlcNAc	–	PP$^\beta$PP$(1 \rightarrow 4)$	–	Insects, crustacean	Structural polysaccharide

Agar

Agar is a component of some red algae's cell membranes, containing two polysaccharides: agarose and agaropectin. Gels are obtained by cooling a 2 to 5% hot aqueous solution.

Carrageenan

It is a heterogeneous polysaccharide. For immobilization, k-carrageenan is used, an insoluble fraction obtained by adding calcium ions to the aqueous carrageenan extract in physiological saline in the presence of K + ions.

Protein-Based Carriers

Natural proteins such as collagen, keratin, fibrin, and myosin can serve as matrices, especially for membrane-bound enzymes. They form fibrillar layers (~80 μm thick) and may be used with or without crosslinkers. However, proteins are prone to biodegradation and may be immunogenic.

Collagen is highly hydrophilic and reactive, allowing modification and derivation into gelatin for gel-based immobilization.

Keratin, from wool or horn, exists as α-keratin (rich in cysteine) or β-keratin (rich in glycine and alanine). α-Keratin can bind enzymes with free SH-groups. Collagen can be modified and obtained on the basis of matrices with specific properties, e.g., gelatin, which can also be used as a matrix for immobilizing enzymes. In addition, due to the ability to form a gel structure, gelatin can be a convenient carrier for enzymes.

Lipid-Based Carriers

In the composition of cell membranes, a large group of enzymes performs crucial functions (so-called membrane-bound enzymes). Lipid components of membranes play an essential role in forming the conformation of enzymes. Using lipids as carriers for immobilizing enzymes allows approaching the modeling of natural enzyme systems.

Lipids are often used as a monolayer (on different surfaces) or bilayers (liposomes). Polar lipids can form monomolecular films at the interfaces (water-air, water-non-polar solvent). In this state, the lipid molecules are located in such a way that their polar heads (the charged part) are immersed in the aqueous phase, and the hydrocarbon (uncharged "tails") are directed into the ambient air.

Such a lipid film can adsorb proteins (electrostatic and hydrophobic interactions) and ensure their functional activity. Silica gel, aerosol, or other solid carriers can be used as a surface for producing a lipid monolayer.

Liposomes

The categorization of liposomes based on the number of bilayers and size is the most common. Liposomes are classified as unilamellar liposomes (ULVs, 25 nm to 1 μm), multilamellar vesicles (MLVs, 0.1–15 μm), and multi-vesicular vesicles (MVVs, 1.6–10.5 μm). Furthermore, ULVs are categorized as large unilamellar vesicles (LUVs, 100 nm to 1 μm) and small unilamellar vesicles (SUVs, 25–50 nm) (Fig. 7.8).

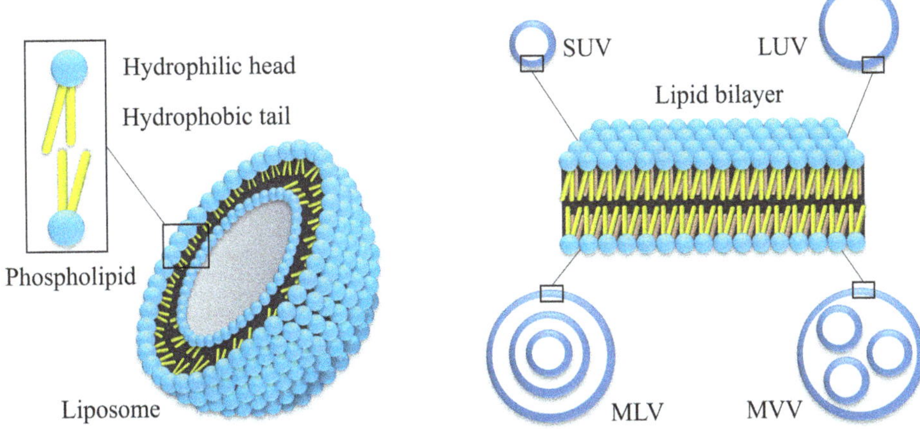

Fig. 7.8 Liposome structure and its variety

Synthetic Polymers

Most synthetic polymers used to immobilize enzymes and cells are polymers based on acrylic acid. First of all, this is acrylamide, a hydrophilic carrier. The enzymes and cells are typically incorporated into a polyacrylamide (PAM) gel, which is obtained by copolymerizing acrylamide with a cross-linking agent N, N-methylene bisacrylamide (MBA). Acrylamide strands cross-linked by MBA form a rigid and stable spatial gel network. Active chemical groups are introduced into the final polymer by chemical modification, or the corresponding functional monomer derivatives are activated to obtain reactive polymers for covalent immobilization. Carriers of synthetic and natural origin, such as PAM and agarose, are crosslinked for preparing gels with a more rigid and inert structure. Among other synthetic polymers, it is worth mentioning polyamide carriers. These are heterogeneous polymers containing repeating amide groups (C(O)-NH-), such as Nylon 6. In practice, the polyamides are first activated, partially hydrolyzed, followed by glutaraldehyde treatment. The advantage of these carriers is that they can be prepared in various forms—granules, powders, fibers, membranes, etc. Hydrophilic gels based on polyvinyl alcohol (PVA) have been widely used for immobilization. They are obtained using cross-linking agents (xylene dichloride, chlorohydrin in an alkaline medium, glutaraldehyde in an acid medium, or UV light).

Synthetic analogs of lipids (surfactants)—tripton, theine, and bridge—can also be used as potential carriers.

Inorganic Carriers

With the ease of regeneration and the possibility of giving various configurations, inorganic carriers are increasingly used in industrial processes. They are available in the form of powders, pellets, or monoliths. They can be porous or non-porous. The most distributed types are macroporous glasses, silica gel, and silicas. They have high mechanical strength, are chemically inert, and are resistant to the action of microorganisms. In addition, the pores of these materials have sufficient rigidity. *Silica gel* is an amorphous substance from orthosilicic acid (OSA) by polycondensation. Other inorganic carriers include clays, cyanites, and porous ceramics, which can also be modified.

7.4   Methods of Enzyme Immobilization

There are many techniques to immobilize enzymes. The activity of the resulting biocatalysts highly depends on the immobilization method. This must be considered when choosing carriers and methods of immobilization. All the variety of enzyme immobilization methods can be divided into two groups: **chemical methods**, which are based on the covalent binding of the enzyme with the carrier, and **physical methods,** which are based on the adsorption of enzymes on the matrix, on the incorporation of the enzyme into the forming matrix or the envelope (microcapsule, liposome).

7.4.1   Physical Immobilization of Enzymes

The immobilization method, in which an enzyme or a cell binds to a matrix without forming covalent bonds, is called physical. They can be divided into four groups:
- Adsorption on insoluble carriers
- Immobilization by incorporation into gels
- Immobilization by microencapsulation
- Double-phase type systems

Enzyme Immobilization by Adsorption

In this case, the enzyme or cell is retained on the carrier surface by electrostatic, hydrophobic, and hydrogen bonds. For this purpose, inorganic (alumina and titanium oxides, ceramics, coal) and organic (polysaccharides, ion exchange resins, proteins) insoluble carriers in the form of powders, pellets, and granules are used. The main characteristics of the carriers are mechanical strength, chemical inertness, pore size, and specific surface area (Fig. 7.9). Because of simplicity, availability, low cost of sorbents, and preservation of high catalytic activity of biocatalysts, adsorption immobilization is one of the most frequently used methods. However, its main drawback is the possibility of a weak binding of the enzyme and carrier.

In practice, the following methods are perhaps most employed:
1. The static carrier and the aqueous enzyme solution are mixed and left for a while. Immobilization is achieved due to the spontaneous diffusion of the enzyme to the carrier, followed by its adsorption. The process usually takes a day or more.
2. Agitation of the enzyme solution and the carrier by a magnetic stirrer.

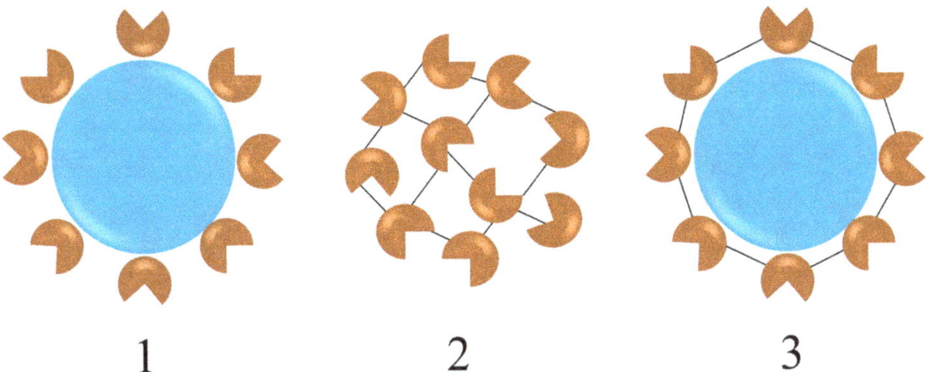

Fig. 7.9 Immobilization via adsorption: (1) physical adsorption; (2) cross linking; (3) adsorption and cross-linking

3. Method of electrodeposition—two electrodes are immersed in the enzyme solution, one of which is supported, and the electric current is passed.
4. Column method—the solution of the enzyme is pumped through a column filled with the carrier.

Retention of the enzyme on the carrier occurs due to van der Waals interactions, electrostatic forces, hydrogen bonds, and hydrophobic interactions between the carrier and the surface groups of the protein. The main factors affecting the adsorption of the enzyme are the specific surface and porosity of the carrier, the pH and the ionic strength of the enzyme solution, its concentration, and the temperature of adsorption.

To increase the adsorption efficiency, the carrier and enzyme are pre-modified: the carrier is maintained in a buffer solution, treated with an excess of a solution containing the ions of the corresponding metal or substances whose molecules contain many functional groups, for example, albumin, hydrophobic treatment, etc. To modify the enzyme, it is treated with compounds containing ionogenic groups—polyacids, carboxymethyl cellulose, succinic acid residues, etc. Electroretention can be used to hold the enzyme on the carrier further.

In industrial conditions, applying a column with carriers is more reliable. In this case, the enzyme solution is pumped through the column from below upwards or above downwards. The flow rate is selected so the carrier particles remain suspended, forming a fluidized bed. Using this technique, it is possible to carry out enzyme immobilization, washing, and fermentation in the same column (Fig. 7.10).

Enzyme Immobilization by Entrapment Within a Gel

The enzyme molecules are included in a three-dimensional network of closely intertwined chains forming a gel. The distance between adjacent chains is less than the enzyme molecule. Therefore, the enzyme is not able to pass out the surrounding solution. The two main techniques are described for this immobilization (Fig. 7.11):

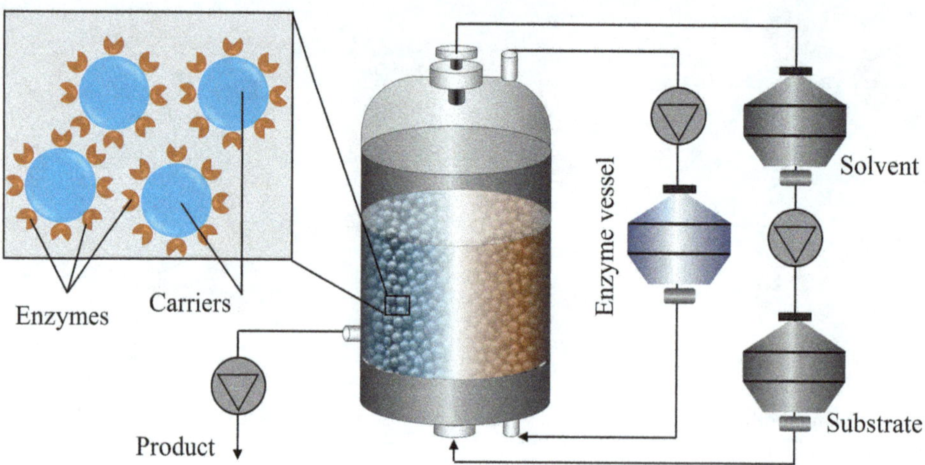

Fig. 7.10 Immobilization using a column

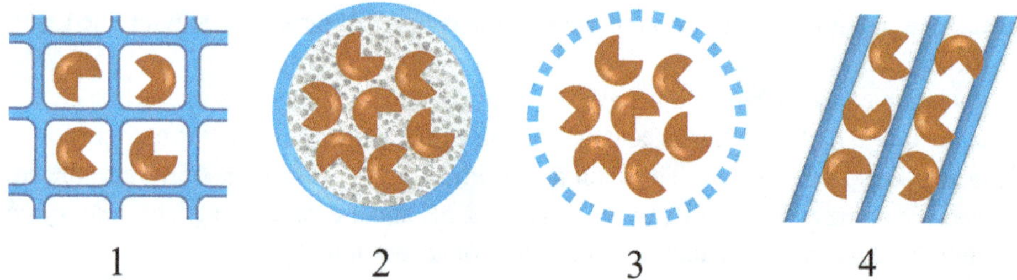

Fig. 7.11 Enzyme immobilization based on entrapment. (1) lattice entrapment type; (2) entrapment within a gel; (3) microencapsulation; (4) fiber entrapment type

1. The enzyme is entrapped in the aqueous monomer solution and then polymerized. Cross-linking agents are added to the reaction mixture, which provides a three-dimensional network structure. Trypsin, ribonuclease, and β-amylase are immobilized in this way.
2. The enzyme is placed in a prepared polymer solution, which is then converted into a gel-like state.

Gels are constructed from agar-agar, carrageenan, agarose, or starch. Sometimes collagen membranes are treated with glutaraldehyde, and cross-linked membranes are formed. However, cross-linked gels can be obtained by exposure to aqueous solutions of polyvinyl alcohol or polyvinylpyrrolidone with gamma radiation or electron current. The incorporation of enzymes into the matrix of photopolymerizable resins has also been used. They contain photosensitive functional groups, and covalent bonds and a three-dimensional network are formed under ultraviolet irradiation.

Poly silicic acid or calcium phosphate is used to produce inorganic gels. The activity of the enzymes included in the gel is affected by the content of the enzyme, the size of the gel particles, and the nature of the polymer matrix.

The advantages of this method of immobilization are the simplicity of the technique, the ability to create immobilized preparations of any geometric configuration (spherical particles, films), high chemical, mechanical and thermal resistance, and repeated application. The multi-purpose technique applies to the immobilization of practically any enzyme and poly-enzyme systems, cell fragments, and whole cells. In addition, enzymes are reliably protected from bacterial contamination. At the same time, the method also has a disadvantage; the polymer matrix creates barriers to the diffusion of the substrate to the enzyme, which reduces its catalytic activity.

Enzyme Immobilization Using Membranes and Fibers, Microencapsulation

With microencapsulation, the main goal is to hold the solution and not to create the physical or chemical forces necessary for immobilization, i.e., this process undergoes the whole initial solution containing the biocatalyst, but not individual molecules or cells. The essence of the method is that the aqueous solution of the enzyme is enclosed within microcapsules, which are closed spherical vesicles with a thin polymeric network.

There are several ways to achieve this immobilization:
1. **Microencapsulation**. The method was developed in 1964 by T. Chang. Microcapsules are produced, ranging in size from several tens to several hundreds of micrometers, with the thickness of the membranes ranging in tenths-hundredths of microns, with a pore diameter of several nm. Inside these microcapsules are various enzymes, such as asparaginase.
2. **Double emulsion system**. An emulsion of an aqueous solution with an enzyme is prepared in an organic polymer solution. Then again dispersed, but in water. As a result, an aqueous emulsion is obtained from drops of the organic polymer solution containing even smaller droplets. After some time, the organic solution solidifies, forming polymer spherical particles with the trapped enzyme.
3. **Inclusion in fibers.** The aqueous solution of the enzyme in the organic solution of the fiber-forming polymer (cellulose, polyvinyl chloride, etc.) is forced through the filters into a liquid (toluene), causing polymer coagulation. In this way, a tissue with enzymatic activity is produced (surgical wipes with immobilized trypsin).
4. **Inclusion into the liposomes**. Enzymes immobilized by this method are used mainly for medical purposes, as well as for basic research, as liposomes are naturally similar to natural biomembranes.
5. The advantages of this method with the use of membranes are simplicity and universality, high enzyme content included, and the catalytic activity maintained.

Double-Phase Type Systems

The substrate, under the action of the enzyme, is processed in the aqueous phase, and the product is extracted into the organic solvent phase. Similarly, starch can be hydrolyzed by a mixture of α-amylase and amyloglucosidase in a two-phase system from aqueous solutions of starch and polyethylene glycol. However, this method is rarely used since it is characterized by a low process speed, inactivation of the enzyme during its adsorption at the interface, contamination of the product, etc.

Chemical Techniques of Immobilization

The methods of enzyme immobilization using a covalent compound are based on forming chemical bonds between the enzyme molecules and the carrier. Therefore, the amino acids responsible for the catalytic activity of enzymes must not participate in covalent binding to the carrier.

The basic principles of constructing the final product with chemical immobilization are compiled by three means. This is because, regardless of the nature of the process, three components are included in this process: the proper biocatalyst (enzyme) molecule (E), the carrier (C), and the cross-linking reagent (R). In other words, such immobilization implies the creation of structures of chemically related three elements: C-R-E, C-E, and R-E.

If there are functional groups on the surface of the carrier capable of undergoing chemical reactions with the functional groups of the enzyme, with the formation of covalent bonds, the immobilization process comes down to irreversible C-E.

Due to steric and diffusion limitations, a tight bond between the enzyme and the carrier can be highly undesirable. This special cross-linking reagent of different lengths is used. This is a very convenient method, as it is possible to change the catalytic characteristics of the biocatalyst, the nature of the bond in the agent, etc.

In some cases, only an enzyme and a cross-linking agent are used for covalent immobilization. Here the enzyme itself serves as a carrier due to its multi-functionality; besides the active center, many reactive groups are presented. Usually, individual molecules cross-link with each other and form aggregates of a network structure, where the enzymes themselves serve as supports.

Often, to increase stability and strength, the immobilized enzymes are additionally treated with cross-linking agents. Further, it is possible to obtain a cross-linked enzyme film by removing a solid carrier. Finally, the enzyme's binding to the carrier can be carried out by forming various chemical bonds (Fig. 7.12).

Currently, the following main methods of chemical immobilization of enzymes are used:

1. The reaction of amide bond formation, the acylation of amino groups of the enzyme, is often due to the action of anhydrides:

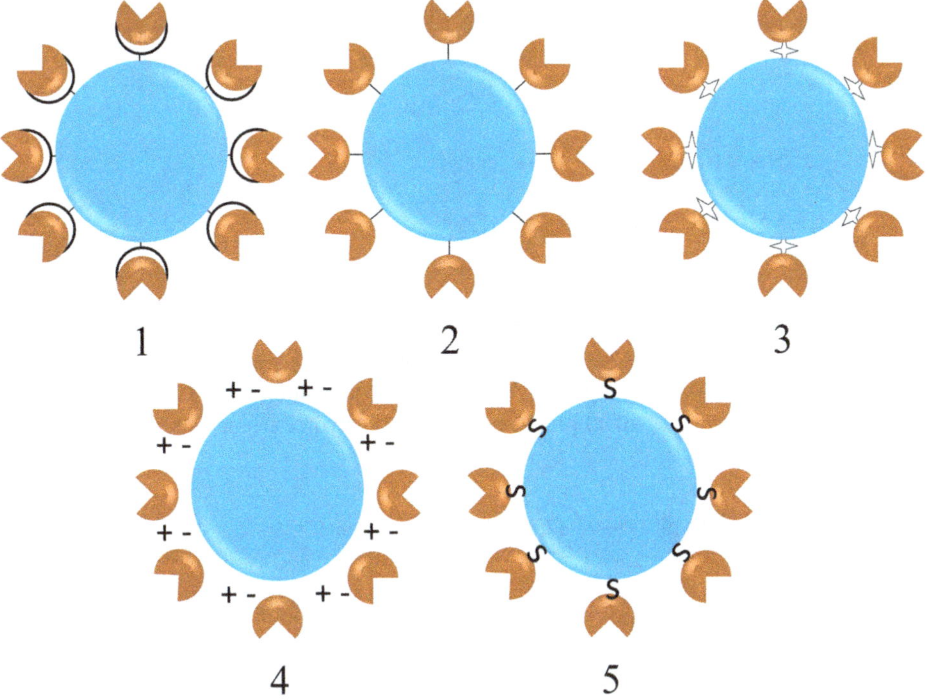

Fig. 7.12 Chemical immobilization: (1) affinity bonding; (2) covalent bonding; (3) chelation or metal bonding; (4) ionic bonding; (5) disulfide bonding

And carboxylic acid chlorides:

2. In the reaction of urea bond formation, the bromine method is common here. When natural carriers or synthetic polyols are treated with bromide, very reactive cyanate – $O – C \equiv N$ and imido-carbonate groups are formed. The protein interacting with such an activated carrier, the isoureas (1) and urethanes (2) are generated:

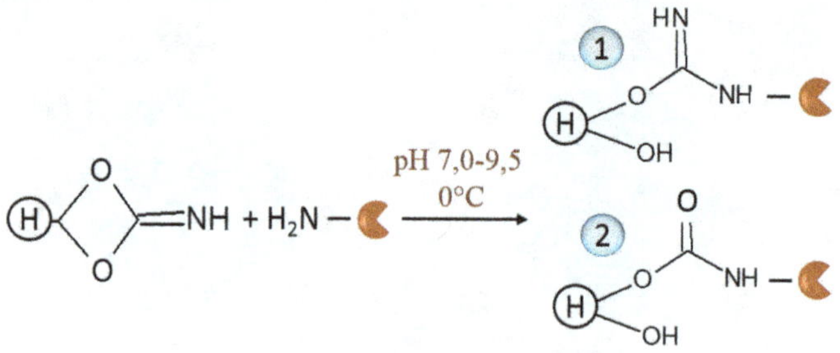

3. The reaction of secondary amine formation; α- and ε- Amino groups of proteins can be transformed into secondary ones during alkylation and arylation reactions, as well as by reduction of azomethine bonds (Schiff bases). Alkylating and arylating agents are halogenated derivatives of aliphatic and aromatic hydrocarbons, as well as methyl alkanes:

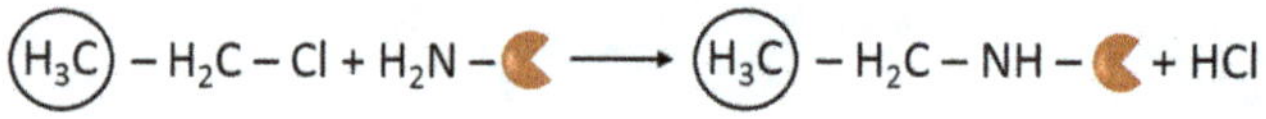

Under such conditions, thiol, imidazole, and hydroxyl groups are alkylated. Oxidative arylation is carried out with benzoquinone:

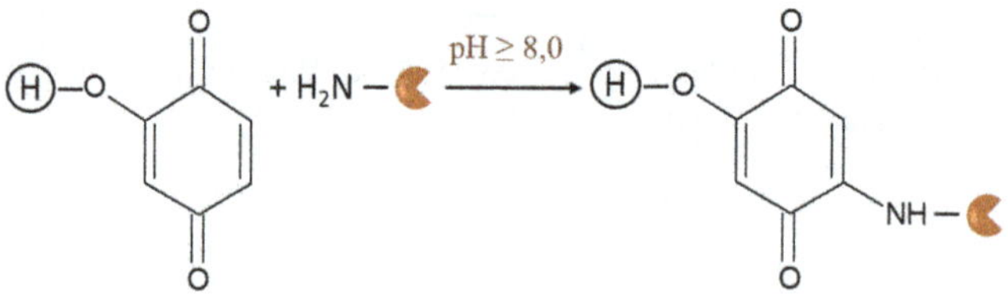

4. The reaction of thiol-disulfide exchange. The principle of immobilization is based on the formation of intermolecular disulfide bonds:

To increase the number of thiol groups, disulfide bonds are reduced by sodium borohydride, mercaptoethanol, cysteine, etc., and endogenous thiol groups (thiolation) are introduced, e.g., converting amino groups.

5. Modification of amino groups with the formation of azomethine bonds, $- CH \equiv N$—in the reaction of proteins with aldehydes:

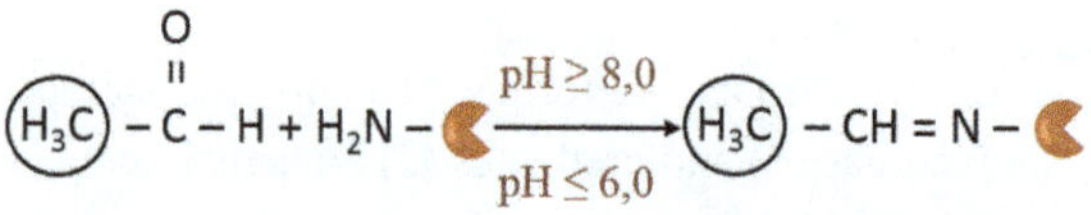

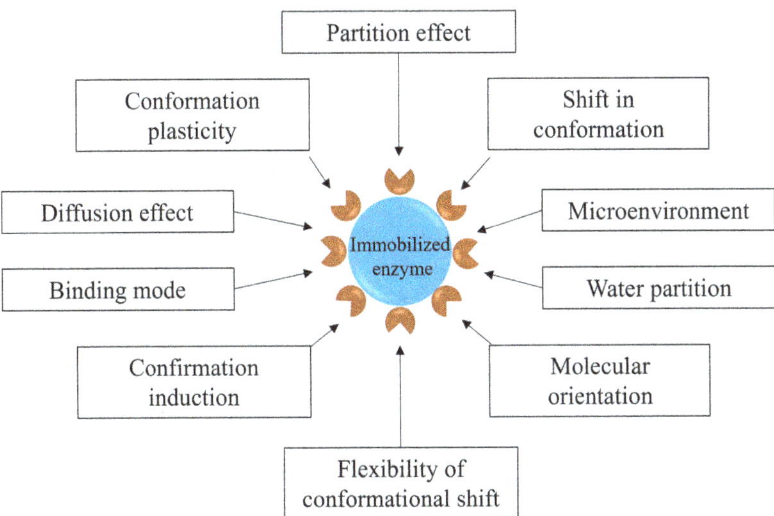

Fig. 7.13 Factors effecting the overall performance and activity of immobilized enzymes

The kinetics of the enzymatic reaction depends significantly on the substrate concentration, temperature, the concentration of the immobilized enzymes, the stirring rate, pH, and the polymer carrier may have an inhibitory or activating effect, etc. (Fig. 7.13).

7.5   Immobilized Enzymes in the Industry

7.5.1   Hydrolytic Enzymes

Hydrolases belong to the third class of enzymes and are characterized by their ability to catalyze hydrolysis reactions, i.e., the breakdown of more complex compounds into simpler ones through the addition of water. Many hydrolases are compartmentalized in various structural elements of the cell, separated from the cytoplasm by membranes. Gram-positive bacteria release many hydrolases extracellularly, while Gram-negative bacteria store them in the periplasmic space. In eukaryotic cells, hydrolases may be localized in specialized organelles—such as lysosomes or the periplasm—or excreted into the surrounding medium.

Most hydrolytic enzymes used in industry are extracellular waste products of microorganisms. However, some are found in the cytoplasm, where they participate in metabolic cycles.

Amylases (EC 3.2.1)

Enzymes that catalyze the hydrolysis of starch include α-amylase, β-amylase, glucoamylase, and others. These enzymes are widely distributed in nature and synthesized by various microorganisms, animals, and plants.

Among industrial enzyme applications, the enzymatic hydrolysis of starch to produce syrups with defined compositions has gained particular significance. The hydrolysates are classified according to their reducing sugar content, expressed as glucose, and are characterized by the dextrose equivalent (DE). Pure glucose has a DE of 100 (glucose = dextrose). By selecting appropriate enzyme combinations and process conditions, it is possible to manufacture sugar-based products with precisely defined physical and chemical properties (Fig. 7.14).

In the typical process, the raw material (e.g., corn) is ground, and the starch is dispersed or gelatinized in an aqueous solution. It is then fluidized with thermostable bacterial α- and/or β-amylases at around 100 °C for 2–4 h. This step usually continues until the DE value reaches 10–20. The process is then followed by saccharification: the addition of glucoamylase (amyloglucosidase) converts the partially hydrolyzed starch into a high-glucose syrup.

A necessary industrial process is glucose production to isomerize it into fructose. Saccharified starch should contain at least 95% glucose, as residual di- and oligosaccharides often taste unpleasant. After correcting the pH and ionic strength, the product is treated with isomerase immobilized by package into a column reactor. The glucose syrup

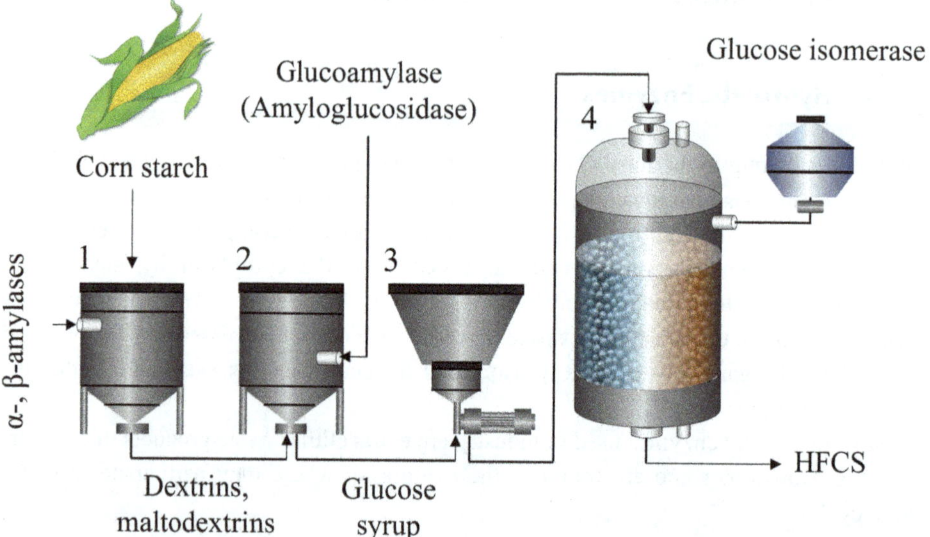

Fig. 7.14 Technological flowchart of high-fructose corn syrup (HFCS) production: (1) primary treatment tank; (2) secondary treatment tank for treatment with α-amylase; (3) filter; (4) bioreactor with an immobilized enzyme

is passed through the columns in a liquid chromatography stage that isomerizes glucose to a mixture of high fructose corn syrup (HFCS-90: 90% fructose and 10% glucose). Immobilized enzymes are obtained by covalent binding to porous silica gel through glutaraldehyde, incorporation of cellulose triacetate into hollow fibers, adsorption on DEAE-cellulose, etc.

Lactase (EC 3.2.1.23)

Also known as β-galactosidase and Exo-(1,4)-beta-D-galactanase, belongs to glucosidases. As a result of hydrolysis by the enzyme, the milk sugar (lactose) turns into a sweeter and well-soluble mixture of monosaccharides, glucose, and galactose, and therefore, the compounds based on β-galactosidase are widely used in the dairy industry and in those areas where waste from the dairy industry contains lactose.

Industrial production of lactose-free milk using immobilized lactase is well established (Fig. 7.15).

The skimmed milk from the reservoir is sterilized at 140 °C for 3 secs, quickly cooled down to 5 °C and pumped through the reaction column, then into an aseptic storage tank. A 20-Lbioreactor contains immobilized enzyme entrapped in cellulose triacetate fibers. Processing milk can be passed through many times at a speed of 7–9 L/min.

Proteases (EC 3.4)

Also called peptidases or proteinases, which catalyze the hydrolysis of peptides and proteins. The main reaction, carried out with the participation of enzymes, is the hydrolysis of the peptide bond. Proteases (trypsin, pepsin, renin, papain, protease, etc.) have found wide application in industry and in medicine.

Immobilized proteases can be used to coagulate milk. The process should be conducted in two stages: product exposure to the enzyme and coagulation. Only in this case is the steadiness of the process achieved.

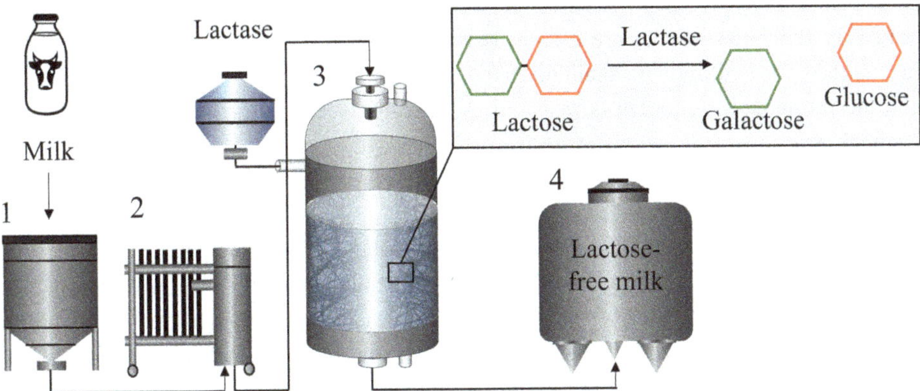

Fig. 7.15 Technological operations for obtaining lactose-free milk using immobilized lactase: (1) milk reservoir; (2) sterilizer; (3) bioreactor with immobilized enzyme; (4) aseptic storage tank

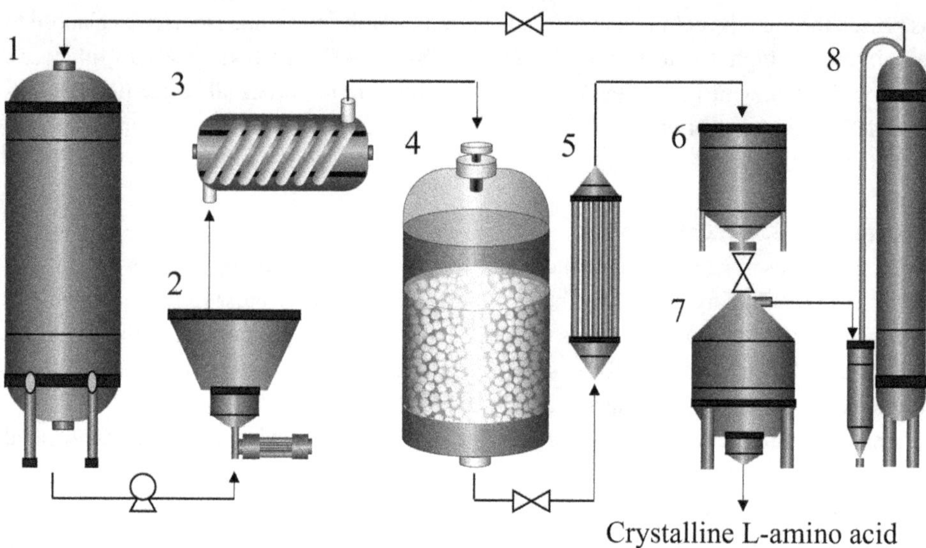

Fig. 7.16 Technological scheme to produce L-amino acids from a racemic mixture using immobilized aminoacylate: (1) Acetyl-D, L-amino acids reservoir; (2) filter; (3) heat exchanger; (4) immobilized enzyme reactor; (5) continuous evaporator; (6) crystallizer; (7) separator; (8) racemization tank

Aminoacylates (EC 3.5.1.14)

The ability of aminoacylates to distinguish L- and D-forms is used in industry to produce amino acids, including essential amino acids. The technological process of L-amino acids production is shown in Fig. 7.16.

As the starting material, acetyl D- and L-amino acids (N-acyl-D, L-amino acids) obtained by conventional chemical synthesis are used. They are exposed to aminoacylase (immobilized by ionic bonds on DEAE-Sephadex) that hydrolyses only one isomer, forming an unsubstituted L-amino acid and maintaining the uncleaved acyl-D-amino acid. Cleavage of the acyl group increases the solubility of the L-amino acid; the amino acids are easily separated, and a pure L-form is isolated. The remaining acyl-D-amino acid is racemized upon heating, i.e., again turns into a mixture of acylated D- and L-amino acids, and the fermentation process is repeated. As a result, the only product produced is the L-amino acid.

Penicillin Acylase (EC 3.5.1.11)

Alternatively, penicillin amidase catalyzes the hydrolysis of the side chains of penicillins and their derivatives, affecting the C-N bond. Enzymes isolated from different sources have different characteristics of substrates. If the enzyme hydrolyses a specific penicillin V, it is designated type I. It can be found in fungi and bacteria. Enzymes of type II are specific to penicillin G and are contained only in bacteria. Type III refers to all penicillin amidases that catalyze the hydrolysis of cephalosporins. In the enzymatic hydrolysis of penicillins and cephalosporins, 6-aminopenicillanic and 7-aminocephalosporanic acid are formed, which are then used to prepare semisynthetic antibiotics. Various companies produce different preparations based on immobilized penicillin amidase on a polyacrylamide gel, cellulose triacetate fibers, etc.

7.6   Immobilized Enzymes as Biosensors

The biosensor is a device in which a sensitive layer contains biological material (enzymes, tissues, microbes, antigens/antibodies, liposomes, receptors, DNA) directly responsive to the presence of the detectable component, generating a signal functionally associated with the concentration of this component. Structurally it consists of two transducers—biochemical and physical. The schematic outline of a device is shown in Fig. 7.17.

A biotransducer (biochemical converter) performs the function of the biological element of detection, transforming the component, i.e., information about chemical bonds, into a physical or chemical property or signal. The physical converter records this property with the help of special equipment. The biomaterial with unique properties in the device makes it possible to determine the necessary compounds in a complex mixture with high selectivity without additional operations and reagents. There is a wide variety of physical transducers, such as electrochemical, spectroscopic, thermal, piezoelectric, transducers on surface acoustic waves, etc. Figure 7.18 demonstrates conventional signal transducers used in biosensors. At present, the most widely used are electrochemical types.

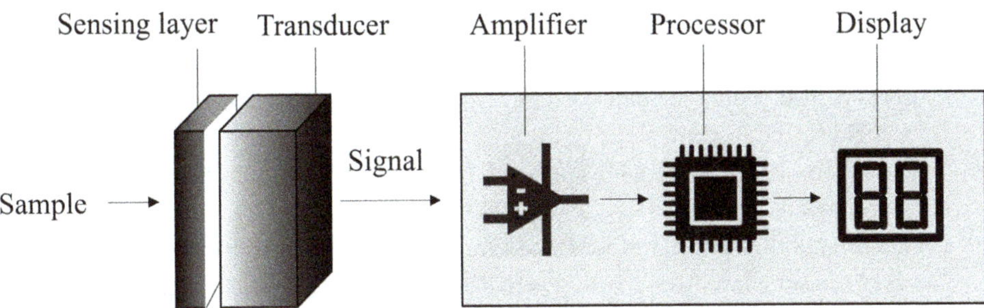

Fig. 7.17   Schematic diagram of biosensor

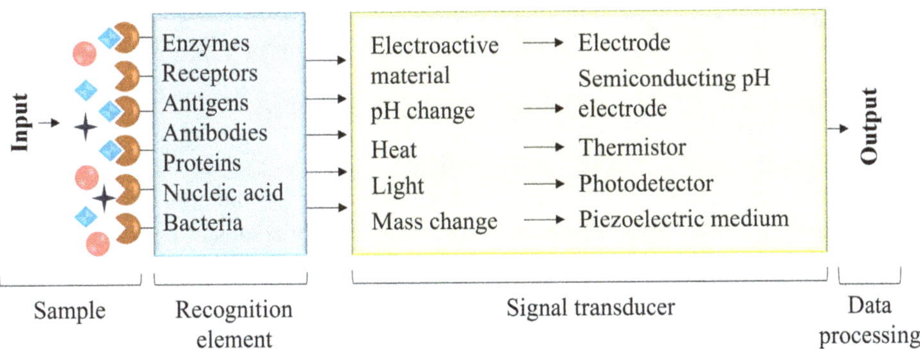

Fig. 7.18   Components of biosensors

Table 7.7 Application of enzymes for the detection of diagnostic analytes

Enzyme used for detection	Diagnostic use
Glucose oxidase	Glucose
Penicillinase	Penicillin
Urease	Urea
Aldolase	Muscle disorders
Cholesterol oxidase	Cholesterol
Alcohol oxidase	Alcohols

Table 7.8 Immobilized enzymes as biosensors for the detection of xenobiotics from food and environment samples

Enzymes	Substrate	Immobilization support	Application
Acetylcholinesterase	Organic liquids, water samples	Graphite, multiwall carbon nanotubes	Pesticide detection
Urease	Water samples	Sol-gel matrix and film	Heavy metals detection
Glucose oxidase	Water and soil samples	Glutaraldehyde and BSA	Heavy metals detection
Butyrylcholinesterase	Fruits and vegetables	Glutaraldehyde and BSA	Detection of other chemical components

Biosensors have potential applications in medical diagnostics (Table 7.7), food and water quality control, pharmaceuticals, environmental assessment, and industry based on food and agriculture. The emergence of biosensors with immobilized enzymes has solved various concerns, including enzyme stability and quick enzymatic response, and provides a disposable tool that can be simply used.

Biosensors also enable the detection of xenobiotics, such as heavy metals, glycoalkaloids, pesticides, and insecticides. For instance:

- Glucose oxidase, immobilized by electropolymerization, is used to detect heavy metals in environmental samples.
- Urease, immobilized on cellulose triacetate, helps monitor mercury levels.

Novel biosensors have been developed to detect organochlorine and organophosphorus pesticides (Table 7.8).

7.6.1　Brainstorming

1. What properties should the carriers used for enzyme immobilization have?
2. What are natural polymer-carriers?
3. What are synthetic polymeric natural carriers?
4. Explain why and how liposomes can be used as carriers for immobilizing enzymes.
5. Which inorganic compounds can be used for the immobilization of enzymes?
6. What methods are used to immobilize enzymes?
7. What is the essence of physical immobilization methods?
8. What physical methods of immobilization are used in practice?

9. How can the adsorption efficiency of enzymes on the carrier be increased?
10. How is enzyme immobilization by entrapment within a gel carried out?
11. Describe such methods as enzyme immobilization using membranes and fibers and microencapsulation.
12. Describe the chemical methods of immobilization.
13. What types of chemical immobilization methods are used in practice?
14. What are the application areas of immobilized enzymes?
15. In what industries are immobilized enzymes used?
16. How are immobilized amylases obtained and used?
17. In what production are immobilized lactases used?
18. How are immobilized enzymes used in the production of antibiotics?
19. What are biosensors?
20. Which biomaterials are used for obtaining biosensors?
21. Talk about the principles of designing biosensors.
22. What are the components of the biosensor?
23. What types of microorganisms are used to create biosensors?
24. Where are the biosensors used?

Take-Home Messages
- Immobilized enzymes are widely used in industrial processes due to their reusability, stability, and ease of separation from reaction mixtures, improving process efficiency and cost-effectiveness.
- Hydrolytic enzymes, such as amylases, proteases, lactases, and aminoacylases, are essential in food processing, pharmaceuticals, and fine chemical industries, often enabling stereoselective or high-yield transformations.
- Immobilized lactase is used in the dairy industry to produce lactose-free milk, while aminoacylases are employed to selectively synthesize optically pure L-amino acids. Penicillin acylase plays a crucial role in the production of semisynthetic antibiotics, allowing for selective hydrolysis of penicillin side chains under mild and controlled conditions.
- Biosensors combine biological recognition (e.g., enzymes) with physical signal transduction, enabling rapid, specific detection of analytes in complex samples.

  Immobilized enzymes enhance the stability, reusability, and operational robustness of biosensors, making them suitable for point-of-care diagnostics and environmental monitoring.
- Electrochemical biosensors, particularly those employing glucose oxidase or urease, are widely used for detecting heavy metals, toxins, and pesticide residues.

  Biosensors enable real-time, selective, and reagent-free detection of compounds in medical, food, environmental, and industrial applications.

7.7   Cell Immobilization

Alongside the successful immobilization and industrial application of many enzymes, researchers have encountered certain challenges—particularly those related to enzyme dependence on cofactors. In some cases, transformations require multiple enzymes. As a solution, attempts have been made to exploit the activity of complex enzymatic systems by immobilizing whole cells. This approach enables the use of not only individual enzymes but also entire enzymatic pathways, which is often difficult when working with isolated, immobilized enzymes.

Cell immobilization refers to the physical confinement or localization of free cells to a defined area, while preserving their biological activity.

The metabolism of immobilized cells differs from that of free cells, which can be exploited to control biochemical transformations. In some cases, immobilized cells are more effective than either free cells or immobilized enzymes. Most enzyme immobilization methods can also be applied successfully to whole cells. The effectiveness of any immobilized cell system depends heavily on the intended application and the nature of the carrier material. Key selection criteria for immobilization matrices are shown in Fig. 7.19.

There are four types of immobilization of whole cells: attachment, entrapment, inclusion, and aggregation (Fig. 7.20).

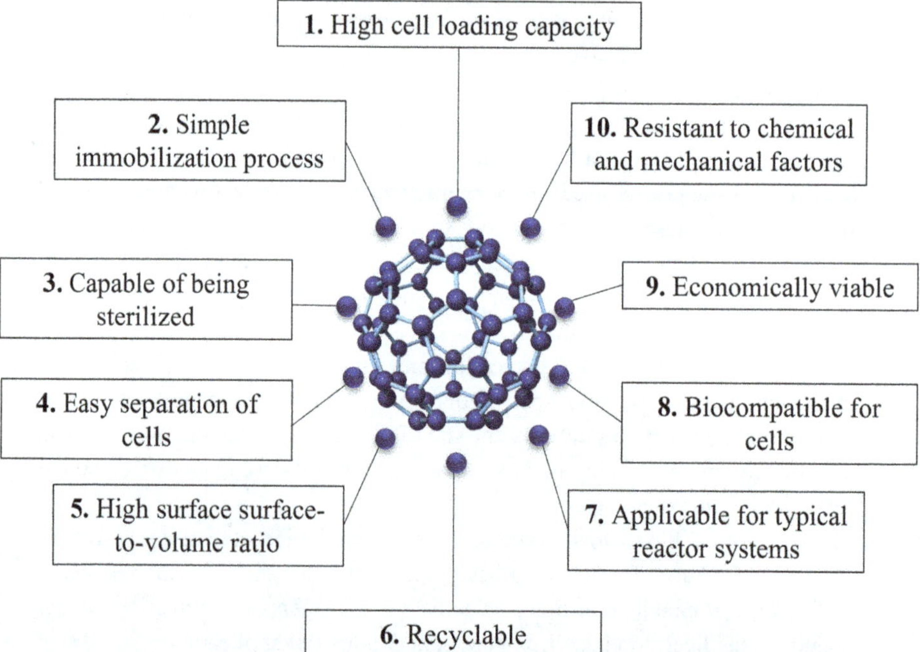

Fig. 7.19 The main requirements for the carriers used in cell immobilization

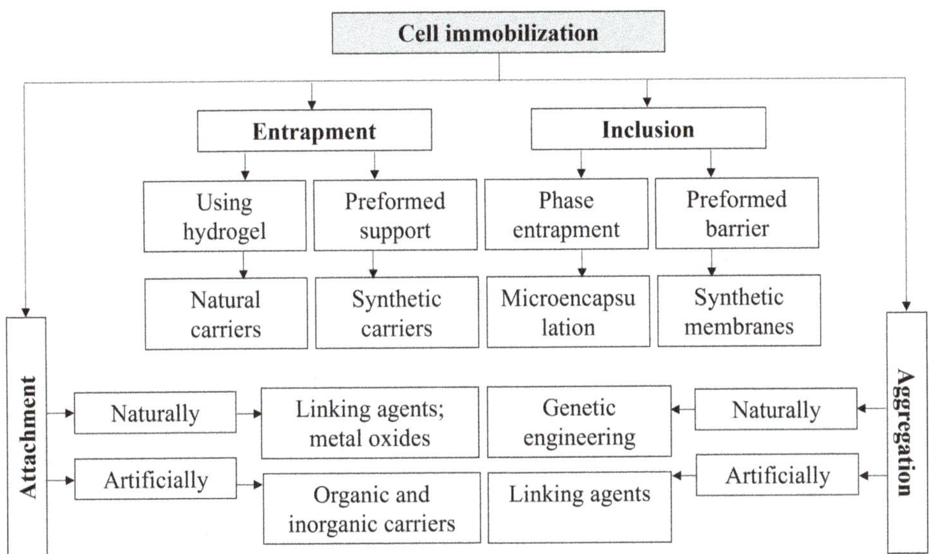

Fig. 7.20 Categorization of immobilized cell systems

7.7.1 Surface Attachment

Cell immobilization by attachment to a carrier can be accomplished naturally or induced artificially with the help of linking agents on natural, organic, and inorganic matrices (Fig. 7.20). A suitable carrier for attachment or adsorption must possess a high affinity toward the immobilizing cell. The attachment of cells to the surface of a material is gained by the ionic interactions and Van der Waals forces (Fig. 7.21).

Entrapment Within Supports

Entrapment of cells can be conducted by the porous matrix or self-agitation (Fig. 7.22). In practice, gel entrapment is widely used because of its simplicity and high cell containment.

7.7.2 Containment or Inclusion of Cells

The process is used when specific target products or cells need to be separated. A particular barrier can be formed around the cells to be immobilized, e.g., synthetic membranes (micro-, ultrafiltration, and ion exchange membranes), and microcapsules are useful. Different microencapsulation techniques have been advanced to entrap whole cells; some methods are united with gel immobilization within the microcapsule.

Membranes available for cell immobilization can be in three forms: cell immobilization in/on the membrane, loaded biofilm on the membrane, and immobilization in a cell compartment, which is allocated by a membrane (Fig. 7.23).

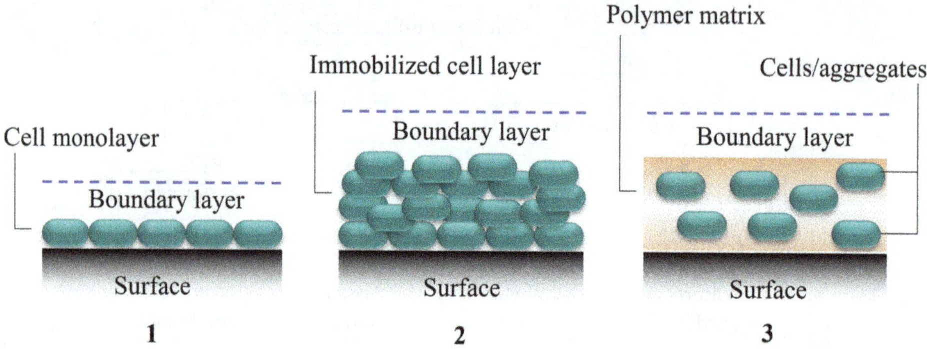

Fig. 7.21 Immobilization of whole cells to a surface by attachment. (1) monolayer adsorption; (2) biofilm (multilayer) adsorption; (3) artificial biofilm adsorption using a loaded matrix, gel, sponge, etc

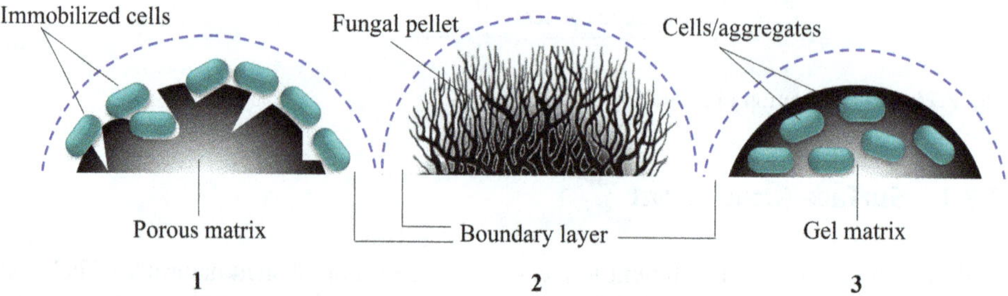

Fig. 7.22 Immobilization of whole cells within a spherical carrier: (1) entrapment within a preformed porous support; (2) self-aggregation immobilization, i.e., a fungal pellet; (3) immobilization in a gel matrix

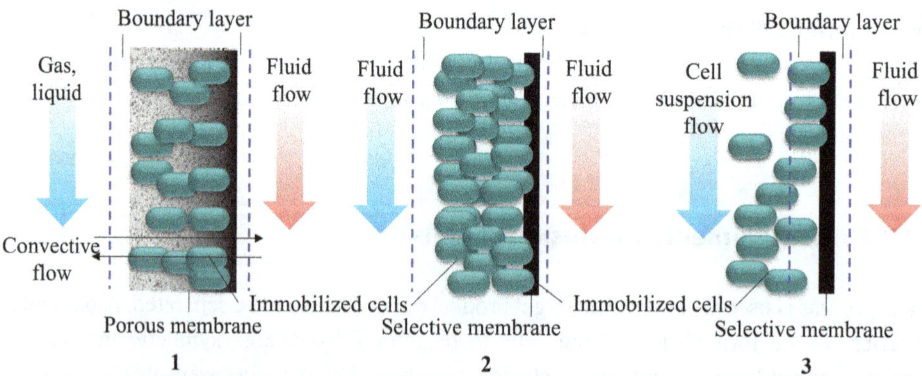

Fig. 7.23 Types of synthetic membranes for living cell immobilization. (1) immobilization in the porous structure of a membrane; (2) biofilm attachment on a permeable membrane; (3) a membrane to separate the cell layer from a cell-free liquid

Immobilized whole cells are used in such areas as the production of medicines, environmental protection, industry, and various analyses. For industrial purposes, immobilized cells and enzymes served as the sources of amino acids, antibiotics and the production of sugar syrups, the transformation of steroids, etc. Environmental applications include the treatment of sewage and the production of methane (Table 7.9).

To ensure the catalysis of multistage reactions, for example, the conversion of glucose to ethanol, immobilization is carried out under mild conditions, particularly providing good gas exchange conditions. One of the typical immobilization methods is the inclusion of biocatalysts in gels. Alginate is used as a gel-forming compound, as it forms stable gel beds by interacting with calcium ions. The substrate, like glucose, can pass through the pores of alginate-cell beds to reach the yeasts to produce metabolites, such as ethanol (Fig. 7.24).

Table 7.9 Examples of immobilized whole cells and their application

Carrier	Cell type	Application
Attachment		
Resin, coke, seashell	Bacteria (Bacillus sp., Aeromonas sp., Alcaligenes sp., Zymomonas mobilis)	Ethanol, amylase production, decolorization
Celite, fiber cloth, bagasse, straw, wood	Fungi (Penicillium sp., Saccharomyces cerevisiae, Candida sp., Aspergillus sp.	Penicillin, beer, xylitol, laccase and itaconic acid production
Entrapment		
Alginate, Cellulose acetate phthalate, NPAE alginate, cryogel	Bacteria (Bifidobacterium sp., Lactococcus sp., Lactobacillus sp., Bacillus agaradhaerens)	Lactic acid, nisin, probiotics and β-cyclodextrin production
Calcium alginate, carrageenan, chitosan	Microalgae (Chlorella vulgaris, Scenedesmus acutus, Scenedesmus bicellaris)	Removal of the pollutants, heavy metals, nitrogen, phosphorus and biocides
Calcium alginate, polyethylene oxide	Fungi (Saccharomyces cerevisiae, Candida versatilis, Zygosaccharomyces rouxii)	Ethanol and bioflavor production
Microcapsulation		
Alginate-PLL-alginate	Bacteria (Escherichia coli, Erwinia herbicola, Lactococcus lactis)	Tyrosine and bacteriocin production, urea removal, therapeutic delivery of whole cells
Membrane immobilization		
Polyvinyl chloride, polyacrylonitrile, ceramic	Bacteria (Escherichia coli, Bacillus stearothermophilus)	Production of β-lactamase and lactic acid
Polysulfone, silicon carbide, polypropylene	Funfi (Saccharomyces cerevisiae, Aspergillus niger)	Ethanol, beer and citric acid production
Self-aggregation of cells		
No carrier used	Bacteria (Zymomonas mobilis)	Ethanol production
No carrier used	Fungi (Aspergillus awamori sp., Penicillium chrysogenum, Saccharomyces cerevisiae, Trichoderma reesei)	Enzymes, penicillin and beer production

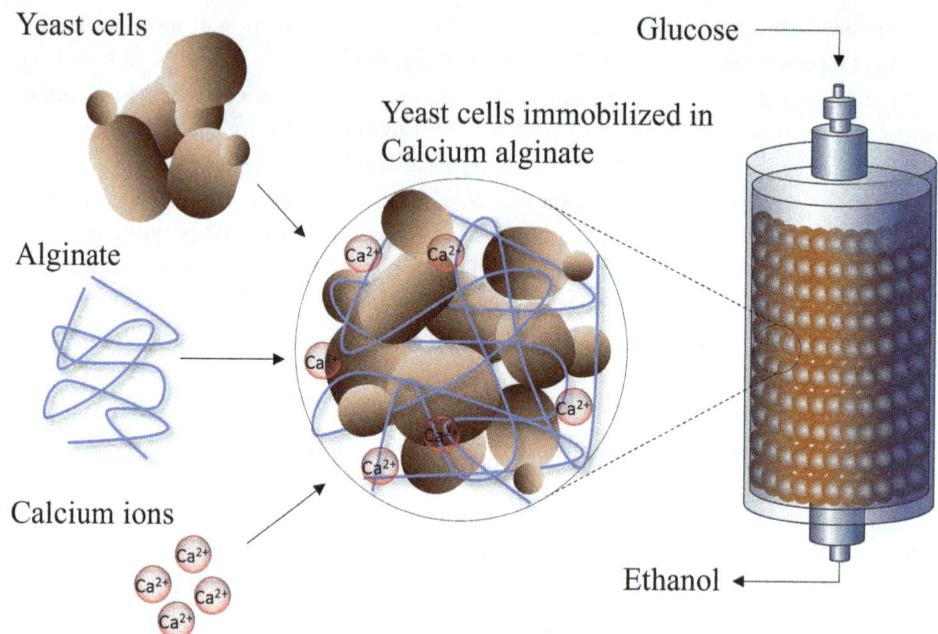

Fig. 7.24 Ethanol production using immobilized yeast cells

Immobilized Plant Cell Systems

At present, many essential compounds used in the pharmaceutical, food, and perfume industries are isolated from plant cells. However, these substances have a complex chemical structure and are difficult to synthesize.

The solution to this problem is characterized by the search for alternative techniques to produce such valuable compounds; immobilized plant cell cultures, therefore, have been gaining great attention.

As a rule, immobilization methods should be sufficiently accurate, as plant cells are susceptible to environmental changes. Therefore, various approaches are being adopted to conduct appropriate immobilization (Fig. 7.25).

The most suitable way for immobilization is the inclusion of plant cells in the gels under strict sterile conditions. Calcium alginate or other carriers are widely used for the support of immobilization.

Areas of application of immobilized plant cells to produce biologically active substances are quite diverse (Table 7.10).

Plant cells immobilized in the alginate gel are used to synthesize or transform various substances. The productivity of the cells included in the gel is ten times higher than that of the same cells in the suspension. With prolonged cultivation of cell cultures in suspension, they undergo significant functional changes, and the cells, in this case, lose their high initial productivity. This problem can be solved based on the immobilization of cells. The immobilized cells were found to prolong the stationary phase, and the synthesis of many compounds occurs when the cells do not divide.

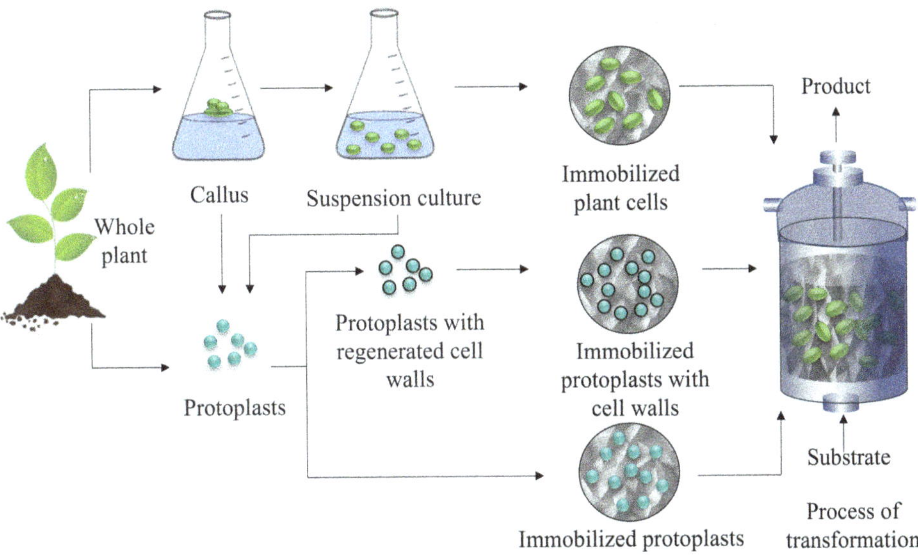

Fig. 7.25 The combination of the immobilization with the plant culture techniques

Table 7.10 Examples of immobilized plant cells

Plant species	Substrate	Product	Carrier
Bioconversion			
Catharanthus roseus	Cathenamine	Ajmalicine (antihypertensive drug)	Agarose
Digitalis lanata	Digitoxin	Digoxin (medication)	Alginate
Daucus carota		Periplogenin (clinical factors)	Alginate
Mentha spicata	−/− menthone	+/− neomenthol (perfumery)	Polyacrylamide
Papaver somniferum	Codeinone	Codeine (pain reliever)	Alginate
Synthesis from precursors			
Capsicum frutescens	Isocapric acid, valine	Capsaicin (topical use)	Polyurethane foam
Datura innoxia	Ornithine	Hyoscine (medication)	Alginate
Coffea arabica	Theobromine	Caffeine (stimulant)	Membranes
De novo synthesis			
Solanum aviculare		Steroid glycosides	Polyphenylene oxide
Glycine max		Phenolics	Hollow fibers
Catharanthus roseus		Enzymes	Agarose, polyurethane
Vicia faba		Ethane	Alginate
Lavandula vera		Pigments	Polyurethane gel
Ginkgo biloba		Ginkgolides (clinical applications)	Polyester fibre

7.7.3   Immobilized Animal Cell Systems

The vital activity of the cell cultures is mainly determined by their interaction with the substrate or matrix for growth. All cells derived from mammalian tissues have a negative charge on the surface, and they attach well to negatively charged surfaces, such as glass or polystyrene. The cells are treated with glycoproteins (obtained from blood serum or cells) for a solid attachment to the surfaces. Polylysine, collagen, and other substances are also quite common that promote cell adhesion.

The ideal carrier for immobilizing animal cell cultures should meet as many of the requirements as possible. Porous support spheres have the most desirable characteristics (nontoxic, high loading capacity, maximum surface area to valium ratio, various shapes and sizes, stable, reusable, suitable to use, etc.) and the potential to overcome many of the restrictions of existing immobilized cell systems (Fig. 7.26).

Mammalian cells, capable of immobilization on the surface, are widely used in various fields of biology, biotechnology, bioengineering, medicine, and clinical studies. For example, immobilized cell culture is essential in investigating growth processes and cell differentiation, mechanisms of cell aging, and the metabolism of various foreign compounds (Table 7.11).

7.7.4   Co-immobilized Biocatalysts

Co-immobilized biocatalysts are complex agents that compose various biological agents (enzymes-cells, coenzymes-enzymes, cells-cells, cells-compounds, etc.). The support materials-carriers and binding methods applied for co-immobilization differ significantly

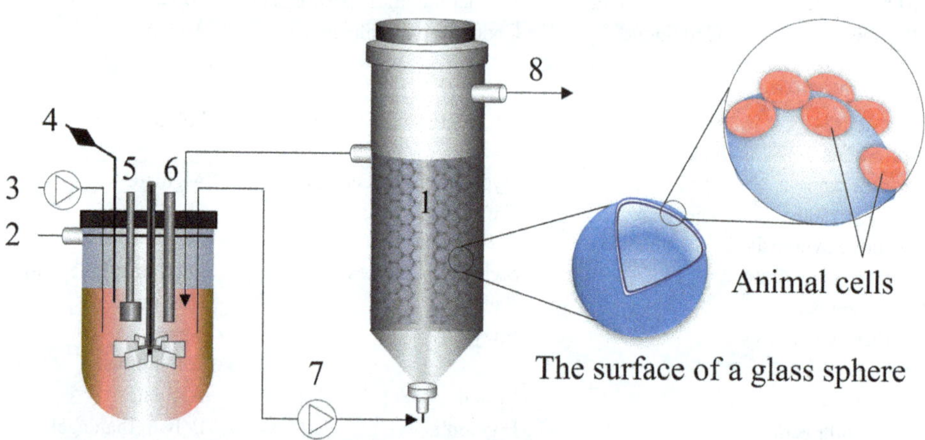

Fig. 7.26 Diagram of immobilized cell culture system: (1) packed bed with porous glass balls; (2) sampling port; (3) media feed; (4) gas exchange; (5) pH probe; (6) dissolved oxygen probe; (7) peristaltic pump; (8) used media out

Table 7.11 Examples of immobilized animal cell cultures and their application

Carrier	Cell type	Application
Attachment		
Polyethylene film	Midbrain cells	Neural differentiation
PLGA	Chondrocyte	Tissue engineering
Entrapment		
Barium alginate	Engineered NIH/3 T3	Cancer therapy
Calcium alginate	Engineered HEK 293	Cancer therapy
Chitosan	Rat osteosarcoma cells	Tissue engineering
Microcapsulation		
Alginate-PLL-alginate	BHK fibroblast	Viability assessment
Alginate-PLL-alginate	Engineered mouse myoblast	Tumor suppression
Alginate-PLL-alginate	Islets of Langerhans	Diabetes treatment
Collagen-based support	Rat hepatocytes	Tissue engineering
Membrane immobilization		
Polypropylene	BAEC	Artificial lung
Polysulfone	H1 fibroblast	Aerobic cell growth
Self-aggregation		
No carrier	BHK cells	Recombinant protein
No carrier	Vero cells	Recombinant protein
No carrier	293 cells	Recombinant protein

according to which biological agent is to be used (Fig. 7.27). The comprehensive set of reactions cannot be achieved due to the pH difference, temperature, and nature of individual biocatalysts. To meet this challenge, it is desirable to benefit from a mixture of individually immobilized bio-agents.

Typically, co-immobilized bio-agents are exploited with mixtures of substrates that a microorganism cannot fully utilize because of its lack of a proper enzyme spectrum. However, multistage enzyme synthesis of bulk compounds is relevant to the modern food and pharmaceutical industry (Table 7.12).

Overall, the possibilities and prospects for using enzymes and cells in the immobilized state in practice are much broader than those achieved to date; just on this path, industry, medicine, and research are expecting the creation of new, highly effective methods and technologies for implementation.

Brainstorming

1. What are immobilized cells?
2. How does the metabolism of immobilized cells differ from that of free cells?
3. Describe the main requirements for the carriers used in cell immobilization.
4. Name 4 ways of cell immobilization.
5. How are cells incorporated into the spherical shell?
6. Name the types of synthetic membranes for living cell immobilization.

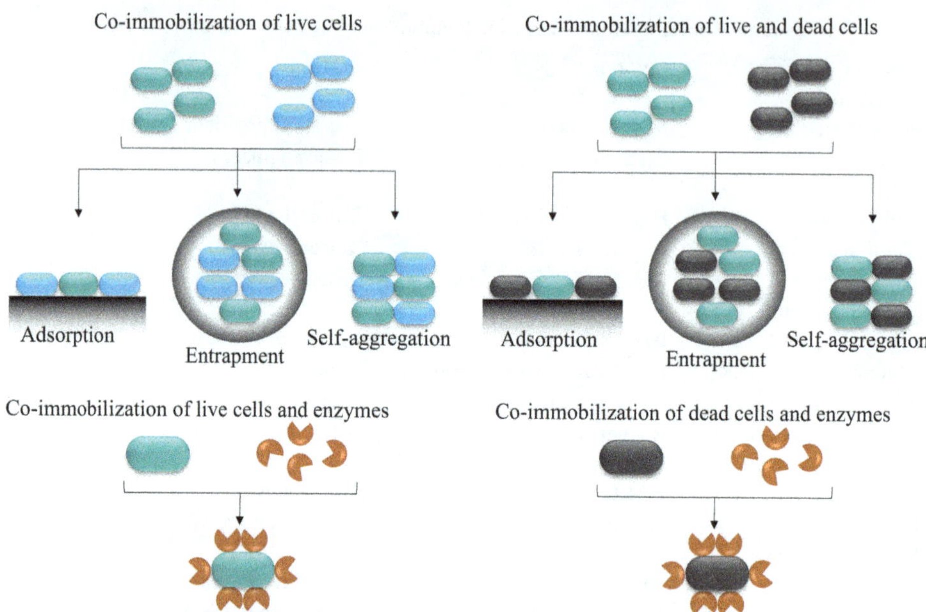

Fig. 7.27 Main types of co-immobilized biocatalysts

Table 7.12 Co-immobilized biocatalysts (cells-enzymes)

Biocatalyst 1	Biocatalyst 2	Application
Saccharomyces cerevisiae	β-glucosidase	Cellobiose fermentation
Saccharomyces cerevisiae	Glucoamylase	Low dextrin beer
Saccharomyces cerevisiae	β-galactosidase	Lactose fermentation
Saccharomyces cerevisiae	Pepsin	Low-protein wine
Saccharomyces cerevisiae	Glucose oxidase	Gluconic acid and fructose
Aspergillus niger	Catalase	Gluconic acid and fructose
Aspergillus niger	Glucoamylase	ORR_2RR removal from beverages
Zymomonas mobilis	β-glucosidase	Cellobiose fermentation

7. Where are immobilized bacterial cells used?
8. Where are immobilized yeast cells used?
9. How are immobilized plant cells obtained and used?
10. Describe the immobilized animal cell systems.

Take-Home Messages
- Whole-cell immobilization allows the use of complex enzyme systems and cofactor-dependent reactions that are difficult to replicate with isolated enzymes alone. Immobilized cells offer greater metabolic stability, can perform multistep conversions, and often show higher product yields than free cells or enzymes.

- Immobilized cells can be physically confined through entrapment, attachment, inclusion, or aggregation, while maintaining biological activity over time. Entrapment in gels (e.g., calcium alginate) is one of the most widely used and effective methods due to its simplicity, biocompatibility, and cell containment.
- Plant cell immobilization enables stable and high-yield production of pharmaceutically valuable secondary metabolites, overcoming the limitations of long-term suspension cultures. Animal cells can be immobilized using positively charged or protein-coated surfaces, which enhances adhesion and supports applications in medical and pharmaceutical research.
- Co-immobilized biocatalysts combine different enzymes and/or cells into a single system, enabling complex bioconversions not achievable by individual components alone.

Evaluating the Effectiveness of Biotechnological Production

8

One of the advantages of biotechnological processes over traditional technologies is that they can be classified as controlled processes. The ability to manage biological agents relies on a fundamental property of all living systems—their capacity to adapt to environmental conditions. Adaptation is essentially a dynamic reorganization of metabolism that ensures the preservation of the organism's structural and functional integrity under constantly changing environmental influences.

By regulating the conditions under which a biotechnological process is carried out—through adjustments of physical and chemical parameters—it is possible to obtain the desired product. In other words, the regulation of a biotechnological process is grounded in certain core principles. It is quite evident that the management of biotechnological processes is only possible when sufficient information is available about the biological agents in use. In addition, it is crucial to understand how these bio-agents behave under specific conditions, which are often unique and may lead to atypical reaction patterns.

Biotechnological processes can be implemented in either batch (periodic) or continuous cultivation modes. In batch fermentation, a single-cell culture goes through all phases of development: lag phase, exponential (log) phase, stationary phase, and death phase. The yield of the desired product depends on the timing of product collection. In contrast, continuous cultivation typically maintains the culture in the stationary phase. By keeping key parameters in the bioreactor at a constant, stationary level, it becomes possible to regulate the productivity of the system. Thus, production management is essentially a matter of controlling and maintaining the required steady-state cultivation conditions.

To achieve this, various analytical tools are employed—such as sensors for detecting medium parameters, electrodes, flow meters, and more. These devices are integrated into unified systems that include dispensers, microprocessors, and computers. In enzymatic processes in particular, microprocessors are widely used to monitor and control operations. A microprocessor is capable of executing several million operations per second.

Control processes in biotechnology can be divided into three levels of complexity: the first (relatively simple), the second, and the third (the highest level of complexity). Despite the rapid advancement of control systems and automation technologies, many biotechnological processes still cannot be fully automated. This is due to several characteristics inherent to biological systems.

Biological systems are self-regulating, which can result in uncontrolled changes during the cultivation process. On one hand, this property is the basis for process control; on the other hand, it means that biological systems may shift their behavior in response to environmental parameters that are not under control. For instance, seasonal variations can affect productivity, even when growth conditions are strictly regulated.

8.1   Biotechnological Process Assessment Criteria

When setting the task of obtaining the target product, it is vital to proceed not only from the capabilities of biological systems to produce useful compounds, but also from the potential costs of producing this product in various ways, i.e., from its profitability. The second condition for the success of setting a biotechnological task is the precision and correctness of its formulation.

The statement of the problem cannot be successful without a comprehensive and in-depth analysis of the state of the matter. So, when setting the task, it is necessary to have deep knowledge of the question, knowledge of potential producers and a preliminary assessment of the profitability of possible production, which depends on the productivity of the system used.

The productivity of a manufacturing system refers to the amount of product obtained per unit volume of the bioreactor per unit time, based on the substrate consumed (Fig. 8.1).

The productivity of a biotechnological system is influenced by several factors, the most important of which are:
- The activity of the producer organism.
- The yield coefficient of the product from the substrate.
- The quantity of active biomass in the bioreactor (i.e., the portion of biomass directly involved in product formation).

The generalized dependence of productivity P allows solving the question of increased efficiency in the given process. So, if we try to increase productivity by enhancing the amount of the producer in the reactor, then we will have to solve the problem of intensification of mass transfer. If we intend to increase productivity by enhancing the rate of substrate consumption, then we need to further investigate the mechanisms for regulating the consumption of the product by the producer's cells.

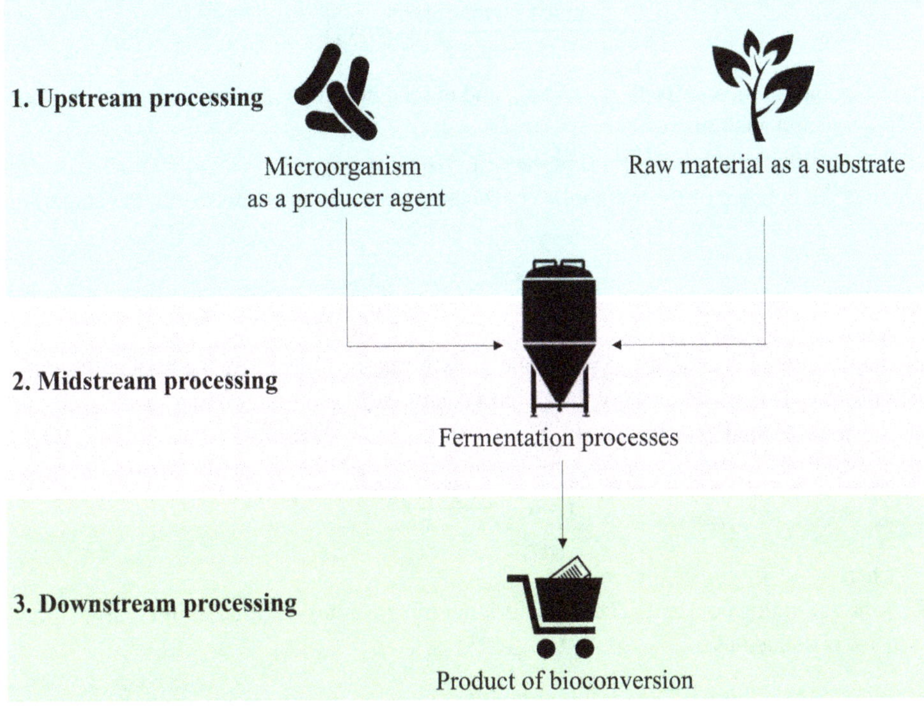

Fig. 8.1   Schematic representation of the target product bioconversion

The *economic coefficient* (*Y*):

$$Y = \frac{\Delta X}{\Delta S}$$

where

ΔX—increase in biomass (g/L), corresponding to substrate consumption in the amount of ΔS (mol/L). In this case, Y is the yield of biomass (V_S). Currently, the economic coefficient is used more widely. There are three classes of economic coefficients:

1. Providing information about the technical process, the cost and process efficiency, and combining indicators such as Y_S, Y_{O2} and Y_{kj}.

$$Y_S = \frac{\text{biomass formation}}{\text{substrate consumption}} \left[kg \, / \, kg \right]$$

$$Y_{O_2} = \frac{\text{biomass formation}}{\text{oxygen consumption}} \left[kg \, / \, kg \right]$$

$$Y_{kj} = \frac{\text{biomass formation}}{\text{heat formation}}\left[kg/kg\right]$$

2. Including indicators Y_C, Y_P and Y_N, and describing catabolism, the amphibolic pathway and anabolism.
3. Describing the energy ratio in the cell (Y_{ATP}).

There is the concept of the metabolic coefficient (q), which is expressed by the equation:

$$p = \frac{\mu}{Y}$$

where

Y—the economic coefficient. Always the growth rate is affected by the concentration of substrate. This is reflected by the Monod equation (Monod correlation) and is valid for substrate-limited growth:

$$\mu = \mu_{max} \cdot \frac{S}{K_S + S}$$

where

K_S—the saturation constant. The concentration of substrate—S (mol/L) in the flow bioreactor is defined as:

$$S = \frac{DK_S}{\mu_{max} - D}$$

where

D—the bulk velocity of the flow through the entire diameter of bioreactor.

The productivity (P) is found as follows:

$$P = \frac{\text{product concentration}}{L \cdot \text{fermentation time}}\left[\frac{\text{units}}{L \cdot h}\right]$$

The output of the product is influenced by organizational factors (bioreactor type, qualification of specialists) and the genetic and biochemical characteristics of the producer. At the upstream processing stage, the product yield and productivity of the selected process are analyzed.

Such indicators as specific energy inputs and unproductive expenditures of substrate and energy in bioconversion processes are important indicators in assessing the efficiency of production.

8.2   Design and Optimization of Biotechnological Processes

The broad application of biotechnological processes has sparked considerable interest in modeling for the purposes of design, management, and optimization. The increasing availability of powerful software and utility tools has enabled a more detailed and accurate approach to designing and applying models for biotechnology-based industrial processes.

The effective implementation of biotechnological processes is closely linked to proper design and management (Fig. 8.2). In the early stages of biotechnology, individual efforts were made to regulate producer development by adjusting environmental parameters. At that time, regulation was largely empirical—without understanding the underlying mechanisms, effective monitoring and control were impossible. Typically, the subject of control was an extensive, periodic microbial culture, which had significant limitations: a lack of control tools and dynamic variability of the producer.

In recent years, with the introduction of controlled cultures, biotechnologists have shifted from simply maintaining environmental parameters to managing the process as a whole. Implementing controlled cultivation requires the construction of control algorithms based on models of the biotechnological process. Modeling—both experimental and mathematical—has become one of the most important directions in biotechnology, as it enables the investigation and development of new processes, the improvement of equipment, and the refinement of production schemes.

Experimental modeling makes it possible to investigate and optimize processes whose nature is not yet fully understood. Often, it is the only viable method for studying a biotechnological process. The first stage typically takes place at the laboratory level, where new producers are examined and initial process designs are developed. Results are then transferred to experimental, pilot-scale, and industrial-scale operations. In experimental setups, all the technological details of the future process are worked out: personnel are

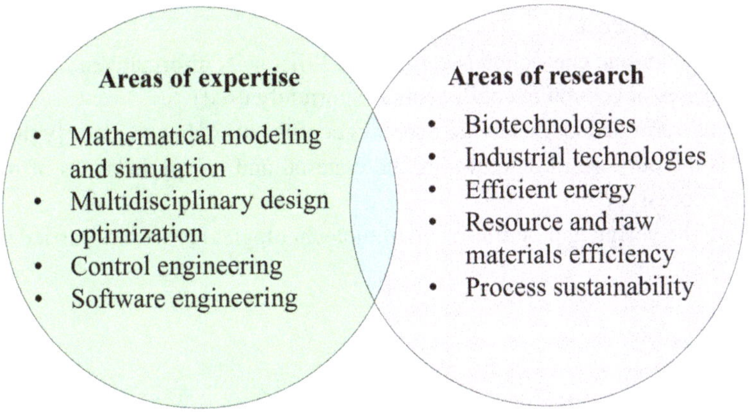

Fig. 8.2 Design areas of the biotechnological process

trained, equipment is developed, and technical and economic indicators are refined. Only after this phase are large-scale, costly industrial experiments and testing undertaken.

However, experimental modeling comes with challenges. It is labor-intensive and complex to implement for new processes. Scaling technology and equipment is particularly difficult. Unlike chemical processes, the development of biological agents depends not only on fluid dynamics and reagent behavior but also on the agents' own metabolism. As a result, biological scaling requires unique solutions, and no universal approach exists to date.

The optimization of biotechnological processes involves a combination of experimental and mathematical modeling, supported by modern optimization techniques. Both approaches are essential for efficient process development. Experimental modeling often precedes mathematical modeling, providing a foundation of information. Mathematical models serve to generalize experimental results and enhance their interpretation.

There are various types of models. Unstructured models use simplified representations of cell growth, considering only the amount of the biological agent. In contrast, structured metabolic models account for the composition of the biological phase and offer greater detail. Compartmental models are simpler, involving fewer variables, while metabolic models are more complex and describe key aspects of cellular metabolism. The availability of mathematical models enables more rational experimental planning, improves data analysis, and significantly reduces the volume of experimental work required.

8.2.1 Brainstorming

1. What criteria are used to evaluate the effectiveness of biotechnological processes?
2. What factors determine the productivity of biotechnological systems?
3. What is the rate of biomass growth and generation time?
4. How do you determine the overall productivity of a biotechnological process?
5. Why and how to determine the rate of substrate consumption and the rate of biosynthesis of the product?
6. How are economic coefficients determined? What is its informativeness?
7. What classes of economic coefficients are currently used?
8. What is determined by metabolic (trophic) coefficients? How to identify them?
9. What is the purpose of determining the material and energy balances of biotechnological processes?
10. How are the design and optimization of biotechnological processes carried out?

Take-Home Messages

- Biotechnological processes are controllable due to the adaptive nature of biological systems, allowing for dynamic adjustment of environmental and operational conditions to optimize product yields. Process evaluation in biotechnology must consider both biological productivity and economic feasibility. A thorough analysis of potential producers and profitability is crucial when formulating a biotechnological objective.
- Continuous cultivation offers more stable control than batch processes, as cultures can be maintained in a stationary phase with constant parameters, enabling more predictable production outcomes. Automation and control systems play a central role in modern biotechnology, using integrated sensors, dispensers, and microprocessors to regulate and monitor production in real time.
- Modeling is fundamental for the design and optimization of biotechnological processes, helping to predict system behavior, guide experimental planning, and scale up from lab to industrial levels. Experimental modeling is indispensable for exploring poorly understood processes and developing new technologies, though it is often labor-intensive and faces scaling challenges.
- Mathematical models—both unstructured and structured—complement experimental work by enabling efficient data processing, optimizing process variables, and reducing experimental workload. A combined approach of experimental and mathematical modeling enhances process efficiency, precision, and scalability, which are essential for industrial implementation of biotechnological innovations.

Biological Safety in Microbiological Production

9

9.1 Sanitary, Hygienic, Microbiological, and Production Control

The specific nature of biotechnology—namely, its reliance on processes involving biological agents—determines the unique technogenic factors present in biotechnological production. The concept of biological safety in biotechnology encompasses two key aspects: biosafety and biosecurity.

The first aspect, biosafety, refers to ensuring the safety of biotechnological processes for human health, including the protection of personnel from pathogens and other harmful effects of biological agents. The second aspect, biosecurity, has gained growing political and ethical significance in recent years. It involves protecting society from the malicious use of biological agents. This issue is addressed at the international level: several documents are being developed under the auspices of the United Nations to ensure that the benefits of biotechnology are used for the common good—not for terrorism or criminal activities. Accordingly, global control and regulation of biotechnology are emphasized within the UN counter-terrorism strategy.

The potential impact of biological factors on humans and the environment represents a specific hygienic, epidemiological, toxicological, and ecological challenge in biotechnology. A biological factor is defined as a group of biological agents whose effects on humans or the environment are related to their ability to:

- Reproduce under natural or artificial conditions
- Produce biologically active substances
- Exert harmful effects through their presence, activity, or by-products in production or the environment
- The biological factors associated with biotechnological production include:
- Viable cells of biological agents

- Inactivated cells
- Metabolic products released by biological agents into the environment
- Products of biological synthesis extracted from biomass
- At key technological stages in biotechnological production, biological factors may be released into the environment. Primary sources include:
- Aerosols from gas-air emissions during fermentation or biomass separation and concentration stages that contain viable cells
- Aerosols from drying systems that may include inactivated cells
- Aerosols generated during the extraction of extracellular metabolites from the culture medium
- Biosynthesis products during their separation from biomass
- Wastewater containing viable or inactivated cells, as well as metabolic by-products

Each component of the biological factor is characterized by its specific potential impact on human health. The use of biological agents in biotechnological processes not only shapes the production technology itself but also dictates the framework for hygienic and sanitary assessment.

Sanitary and hygienic assessment of a biological object and finished products, including living cells of the producer.

All commercial microbial strains permitted for use in biotechnology belong to non-pathogenic or opportunistic groups (excluding strains used in vaccine production). According to regulatory and technical documentation, these strains are classified as hazard class 3, which indicates moderate individual risk and limited risk to the general population. In case of process disruptions or inadequate sanitary and hygienic working conditions, such microorganisms can adversely affect human health—potentially causing microbial carriage and allergic reactions (Fig. 9.1).

A particular feature of the hygienic assessment in biotechnological production is that not only pathogenic microorganisms are evaluated but also non-pathogenic saprophytic strains. Hygienic standards apply to all microbial strains used or intended for use in biotechnological processes. This also includes final products whose active ingredients are living microorganisms or their spores.

A comprehensive evaluation of industrial strains requires a step-by-step assessment (Fig. 9.2).

In the toxicological and hygienic evaluation of the biological activity of strains and their metabolic products, the following factors are taken into account:

- Toxicity (acute, subacute, and chronic): the ability of the strain to cause harmful effects, including intoxication or death, in living organisms
- Carcinogenic properties: the ability to induce tumors
- Mutagenic properties: the capacity to cause mutations in somatic and germ cells
- Teratogenic properties: the ability to cause malformations or developmental defects in embryos
- Gonadotoxic properties: the impact on the development and function of sex cells

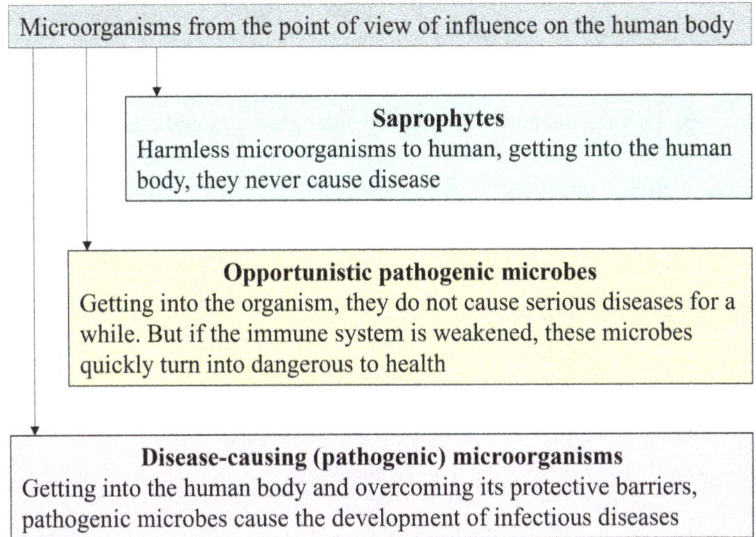

Fig. 9.1 Classification of microorganisms according to the pathogenicity categories

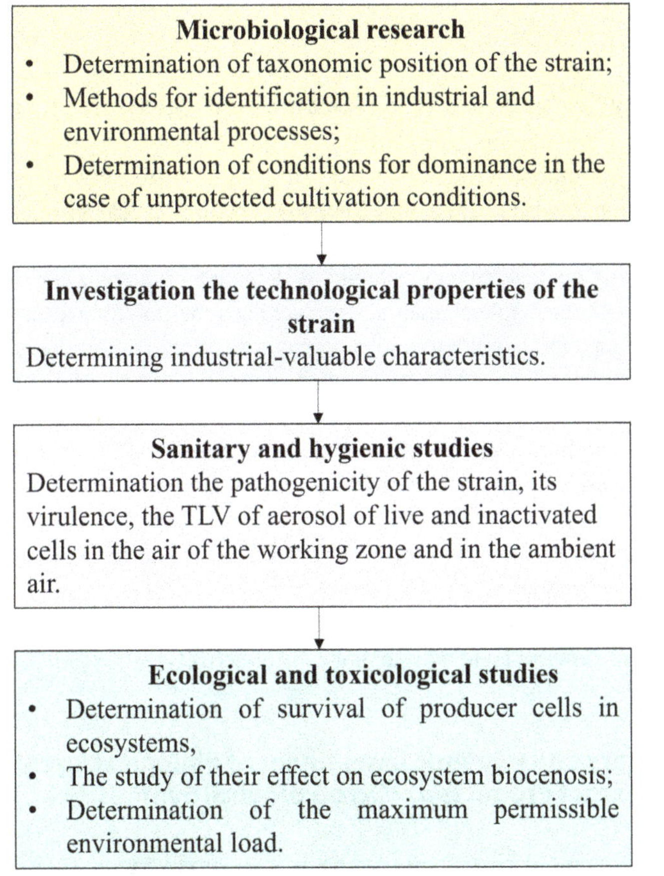

Fig. 9.2 Evaluation stages of microbial strains used in industry

- Embryotoxic properties: any toxic effect on the embryo or fetus
- Allergenic properties: the ability to alter immune function and cause sensitization

Additional factors considered include the strain's persistence in the external environment, its accumulation in macroorganisms, and overall environmental stability. Even non-pathogenic microorganisms, which do not cause infectious diseases, are evaluated for the following potential hazards:

- Toxicity and toxigenicity
- Transient colonization or dissemination within host organs
- Immunogenicity (specific effects on the immune system)
- Dysbiotic effects (disruption of normal microflora)

When determining the threshold limit value (TLV) for final products containing live cells or spores, further assessments are conducted. These include evaluating skin irritation potential and determining the thresholds for acute and chronic toxicity. The most limiting and harmful effect observed is used as the basis for establishing hygienic standards.

9.1.1 Safety Considerations for Recombinant Microorganisms

In recent years, recombinant microorganisms have been increasingly used in biotechnological production, requiring strict adherence to specialized biosafety rules:

- The genetically engineered strain should be defective or *auxotrophic*, meaning it cannot survive without supplementation (e.g., lacking genes to produce an essential amino acid or vitamin). In natural conditions, without the required nutrient, the microorganism cannot replicate or synthesize the target product. Such strains are thus fragile and non-viable in changing environments.
- During fermentation, negative pressure is maintained inside the bioreactor. If any microbial release occurs, only small amounts of non-viable, auxotrophic microorganisms escape.
- After production, equipment is sterilized without causing corrosion of materials used in the production apparatus.
- Exhaust and ejection systems are equipped with special filters that prevent microbial release into the environment by trapping cells effectively.

9.1.2 Sanitary and Hygienic Assessment of Biological Agents and Target Products of Microbiological Synthesis

Products of microbiological synthesis can be divided into two groups:

The *first group* is the products of synthesis, exometabolites released into the environment (soluble or gaseous compounds). Exometabolites include compounds that are both target products of the biotechnological process (such as antibiotics, enzymes, organic acids, certain vitamins, etc.) and compounds that are not targeted but are intermediate metabolic products (for example, acidic and neutral compounds of hydrocarbon oxidation) or intermediate synthesis products (for example, amid-phenylacetic acid, a precursor of penicillin synthesis). They can usually be found in industrial wastewater and in the form of aerosol emitted into the air with gas emissions.

The *second group* is the target microbiological synthesis products accumulated in the cell. Such products include hormones extracted from mammalian tissues, recombinant proteins formed by cultured animal cells, nucleic components, certain enzymes and coenzymes of the microbial cell, toxins, and other biologically active substances. These products are extracted from the cell by chemical technology methods (aqueous extraction, extraction with organic solvents, ion exchange, chromatographic and other methods). In small concentrations, they can be contained in liquid production streams and in the form of an aerosol in gas-air emissions.

The influence of products from microbiological synthesis and their impact on working personnel and the environment depends on the production volume, the perfection of the technological process, and the nature of metabolites. The largest production volumes and biological activity are characteristic of antibiotics and enzymes. Enzyme products, like other protein substances can have allergic effects and may cause skin and mucous membrane irritations, including contact allergic dermatitis.

In manufacturing, products of microbiological synthesis as a «biological factor» can be present in liquid process streams, in the form of an aerosol in gas-air emissions, as well as in solid industrial wastes (especially with the surface method of their production). With this method of cultivating producers, the air of factory premises is contaminated with organic dust, represented by the components of nutrient medium (particles of bran, sawdust, etc.), the intermediate products of the production (the producer's culture), the smallest particles of powdered products, etc., which poses a particular danger to the personnel.

Regular microbiological control of the technological process is of great importance for ensuring safe working conditions for personnel and ensuring the dominance of production strains-producers in these processes.

The following main factors provide microbiological safety of biotechnological productions:

- Application of a permitted strain or associative culture in the technological process.
- The presence of concomitant microflora in the production process and the finished product in an amount authorized by regulations.
- The absence of pathogenic microorganisms among the concomitant microflora that infects the production process and the finished product.

- Compliance with the developed hygienic standards for the content of living cells of microorganisms in the production and residential areas.

An essential task of microbiological control in biotechnological productions based on processes of microbial synthesis is to ensure a strict orientation and activity of biosynthetic processes that guarantee the production of the desired product in the required quality. This task is solved based on the control in the preparation stages of the inoculum and the physiological state of the culture in the cultivation process.

Most biotechnological processes are conducted under aseptic conditions during the periodic cultivation of microorganisms with protection against the penetration of foreign microorganisms. Microbiological control of such productions includes:

- Control of inoculum purity and activity.
- Control of prepared culture medium sterility for cultivation.
- Control of air sterility arriving at all stages of the technological process.
- Control of microbiological purity of the finished product.

In the technical regulations of any biotechnological production, there is a section including the main methods of control, terms, and controlled stages of production, considering the technology features.

Assessment of the sanitary and microbiological state of the environment in biotechnological productions.

Biological aerosols, which are released during biotechnological processes, represent the primary source of potential biological hazards. A biological aerosol is an aerosol in which the dispersed phase includes viable or inactivated microbial cells, microbial metabolic products, spores, or plant dust.

In sanitary-microbiological studies of air environments, the number of culture-producer cells, the total contamination of microorganisms and their group composition, the amount of protein, and the content of other chemicals are determined. The criterion for assessing the sanitary and microbiological state of the air in the working area is TLV or tentative safe exposure levels (TSELs) of the strain used in production.

Based on air quality monitoring results—both inside production facilities and from controlled emissions—compliance with hygienic norms is determined. This includes the TLV for microbial cells in workplace air and the efficiency of equipment in capturing and inactivating airborne microflora.

Thus, the main sanitary-microbiological and hygienic characteristics of air in biotechnological production are the concentrations of production strain, extraneous microflora, and the species composition of the latter. Critical is the control of extraneous microflora for technologies based on unprotected fermentation processes, using raw materials well assimilated by various microorganisms.

The source of air pollution in the working area by microorganisms is technological equipment. Therefore, in addition to the air environment, sanitary and microbiological investigations of technological equipment surfaces, premises walls, and working clothes are conducted in each control point. At the same time, dust samples are taken from the air to determine the protein content.

A properly conducted sanitary and hygienic assessment of the enterprise, in particular, studies of the air environment at the industrial site, in the working and sanitary protection zones, is of great importance for ensuring the microbiological safety of biotechnological productions.

Hygienic provision of biological safety in biotechnological productions:

- The elimination of possible adverse effects of «biological factors» on human health and the quality of the environment is decided both at the stages of technology development, the enterprise's design, and the stages of their operation.
- Safety measures when working with biological agents should ensure the prevention of the occurrence of harm to workers and the public:
- Diseases, carrier conditions, intoxication caused by microorganisms and products of their vital activity, as well as cell and tissue cultures.
- Sensitization of the organism is caused by microorganisms and the products of their vital activity.

The production process technology, equipment, and special preventive measures ensure safety when working with dangerous biological agents. *Production processes must*:

- Allow the possibility of disinfection or neutralization of the territory, premises, equipment, vehicles, clothing, and protective equipment, with the specifics of working with a biological agent
- Allow control over working conditions and hygiene requirements
- Exclude the adverse impact of working methods with biological agents on personnel
- Exclude the occurrence of fires and explosive situations in the allocation of products of vital activity and the decay of biological agents
- Exclude the possibility of environmental pollution

The production equipment must correspond to the sanitary-hygienic and ergonomic requirements, namely:

- To provide the possibility of monitoring the measurement with specific biological hazard parameters to compare them with the corresponding maximum permissible values
- To allow the possibility of monitoring the physiological state and behavior of biological agents.
- To allow the possibility of disinfection and neutralization.

The system of preventive measures should:

- Provide the possibility of creating specific active or passive immunity when working with pathogenic microorganisms
- Ensure the normalization of the duration of work in harmful conditions
- Provide an opportunity to increase the body's resistance (preventive nutrition)

To prevent the harmful effects of microorganisms and products of their vital activity on personnel, safety requirements are imposed on the following types of activity:

- Production and control of biological products, the basis or producers of which are microorganisms, biological fluids, tissues, and organs, as well as cell and tissue cultures
- Study of drugs for prevention, treatment, diagnosis, and other purposes in medicine, veterinary, agriculture, etc.

In occupational safety standards, for each type of activity with biological agents, the biological hazard parameters and their permissible values are established, as well as methods for their measurement and control; that is, hygienic standards must be developed. The hygienic standard is a quantitative indicator of TLV of harmful production and environmental factors, identified by the intensity or duration of exposure, which does not adversely affect a person.

Toxicological and hygienic evaluations of microbial strains and their products include assessments of acute, subacute, and chronic toxicity, focusing on their ability to cause intoxication or mortality in test organisms. The principles of toxicological evaluation of products of microbiological synthesis based on inactivated microbial cells or their metabolic products (fodder protein products, amino acids, antibiotics, enzymes, etc.) are based on methods similar to those used to justify the hygienic regulations of chemicals.

The study surveyed the toxicological, embryotropic, carcinogenic, teratogenic, sensitizing, and other product properties. In addition, in the toxicological assessment of the finished dry product to substantiate its TLV in the air of the working area, experimental studies, hygienic assessment of working conditions, a survey of the health of workers is conducted, and a method for determining the content of the product in the air is being developed.

In the hygienic study of working conditions and the health status of workers in the production of microbial preparations, particular attention is paid to the following:

- The duration of exposure to the dust of the product (short-term, intermittent, continuous) for workers
- The aggregate state of the dust of the product
- The content of dust and microbial cells in the workplace, considering the technological process and time

The safety of biotechnological processes and productions for personnel and the environment is ensured by the observance of design documentation for the facility's construction and the technical regulations for production. Furthermore, product safety is ensured by the product quality, technical conditions, and compliance with the instructions or technical regulations for their application.

Biotechnological plants use various raw materials and chemicals, producing a wide range of products in various forms. Several components of raw materials and some types of finished products, in the case of airborne industrial premises, can adversely affect workers' health, create fire and explosive aerosols, and, if there is insufficient purification of technological emissions into the environment, pollute the landscape at considerable distances from the enterprise. The harmfulness of biotechnological production can be determined by both chemical and biological factors. Therefore, when designing biotechnological enterprises, special attention is paid to the issues of safety engineering, industrial sanitation, and environmental protection.

Concerning safety engineering, industry standards have been developed. In addition, all enterprises must have working safety documentation inside the plant. In general, factory and general production instructions for workplaces for all stages and types of work are reflected in the production regulations.

The regulations must contain complete data on waste, technological emissions, and effluents, as well as specific recommendations for their purification, disposal, or destruction. Of course, to the same extent, it applies to all types of primary and secondary commercial products produced and marketed by a biotechnological enterprise. The most reliable way to ensure the biosafety of biotechnological productions is for the production organization to comply with the asepsis rules.

To ensure healthy and safe working conditions, it is essential to observe the system of labor safety standards, strict management of technological production regimes, the implementation of recommendations developed based on studying and operating existing biotechnological productions, the safe organization of workplaces and production in general, the correct behavior of personnel, and personal hygiene. In accordance with sanitary requirements, the personnel of biotechnological production should be provided with overalls, special footwear, and individual means of protection.

Thus, biotechnological production can be a sources of «biological factors», bioaerosols, and sewage containing living and inactivated cells, organisms, as well as products of their synthesis and metabolism.

9.1.3   Disinfection of Microbiological Waste Products

Biotechnology production, like other industrial activities, can negatively affect human health and the environment. The nature and extent of this impact depend on the composition and volume of gas-air emissions, wastewater, and solid waste generated during production.

Methods for Removing Suspended Impurities from the Air

- "Dry" cleaning is based on the subsidence of particles under the action of gravity or on the action of inertial forces when the direction of air movement changes.
- "Wet" cleaning is based on the irrigation of air with water or passing it through a layer of water.
- Air passage through filters in which dust, moisture, and microorganisms are trapped (including sterilization).
- Electrical gas cleaning precipitation of particles suspended in gas in an electric field.

To neutralize exhaust gases from gaseous or vaporous toxic substances, the following methods are used:

- Absorption (water, organic solvents, not interacting with extracted gas).
- Adsorption (porous materials).
- Catalytic methods (based on the chemical transformation of toxic components into non-toxic ones on the surface of solid catalysts).
- Thermal methods for neutralizing gases from easily oxidized, toxic, and foul-smelling impurities (based on direct combustion).

Methods of Sewage Treatment

1. Mechanical: separating insoluble coarse-dispersed impurities by straining, sedimentation, or filtration. Large, easily precipitated impurities are filtered through mechanized gratings. Granular mineral contamination is deposited in sedimentation tanks. More fine particles and oil films are separated on filters with granular material—sand filters or ultrafiltration (reverse osmosis).
2. Physical-chemical: based on changes in the physical state of contaminants, which facilitates their removal from wastewater (coagulation, flotation, ion exchange method, etc.).
3. Chemical: based on the chemical interaction of reagents with substances dissolved in wastewater (condensation, oxidation, neutralization reactions), resulting in the formation of non-toxic substances, soluble compounds becoming insoluble, etc.
4. Thermal: This wastewater is destroyed at high temperatures to produce non-toxic combustion products and solid residue. They are used to neutralize industrial wastewater containing toxic organic and mineral substances.
5. Biological: Based on the ability of microorganisms to use as a nutrient substrate, many organic and some inorganic compounds are contained in wastewater. For microbiological industry enterprises, this method is the main one.

Biological purification can be carried out under aerobic and anaerobic conditions with continuously cultivated microorganisms in submerged or superficial ways. For this, biological filters and aero tanks are used.

Methods of Processing Solid Industrial Waste

- Mechanical, mechanical-thermal, and thermal processing. During such processing, crushing is carried out—reduction of the sizes of processed materials (jaw, cone, roll, and other crusher structures). The mechanical pretreatment of solid waste includes: Grinding into particles no larger than 5 mm using mills or disintegrators; Classification and sorting by separating the waste into fractions using gratings, sieves, or wire mesh; Granulation, i.e., the formation of spherical aggregates from pastes or powders, using drum, plate, or centrifugal granulators or tablet machines.
- Pyrolysis, re-melting, or burning occurs during the heat treatment of the waste. In this case, the organic part of the waste passes into CO_2, N_2, NO_3, H_2O, HCl, SO_2, etc., and the inorganic remains in the ash.
- Enrichment, used for solid industrial waste containing ferrous and non-ferrous metals (gravity, electrical, flotation, special methods).
- Physical-chemical isolation: leaching (extraction), dissolution, crystallization.
- Burial place. Particularly toxic, including radiation, non-recyclable solid industrial waste is to be buried in metal capsules, then into cubes of cured liquid glass, designed for an indefinitely long shelf life. They are placed underground in geological workings or deep depressions of the seabed. This is one of the most severe and intractable problems of environmental protection. There are safe places for the isolation of solid industrial waste.

As a result of special processing methods, including biotechnological, solid industrial waste can be used:

- For the amelioration of acidic (calcium-containing industrial waste) and solonetzic soils (iron and calcium sulfates, sulfuric acid production wastes).
- As organic fertilizers (sewage sludge, not less than 3 years lying on silt areas).
- As additives in cattle feed.

Brainstorming

1. What is meant by the term "biological factor" in biotechnological production?
2. What concerns and risks are associated with biological factors?
3. What are the main sources of emissions in biotechnological production?
4. How is the phased, integrated assessment of industrial microbial strains conducted?
5. What types of studies are involved in the comprehensive toxicological and hygienic assessment of industrial strains and their metabolic products?
6. How are microbiological synthesis products classified according to sanitary regulations?
7. What is the purpose of regular microbiological control of the technological process, and how is it implemented?
8. How is the sanitary and microbiological state of the environment in biotechnological facilities assessed?

9. What measures ensure the hygienic provision of biological safety in biotechnological production?
10. What practical methods are used to neutralize microbiological production waste?

Take-Home Messages

- Biological factors in biotechnology pose sanitary, toxicological, or ecological risks. The biosafety of biotechnological production involves protecting human health and the environment, while biosecurity ensures that biological agents are not misused or weaponized. A comprehensive assessment of industrial strains includes toxicological testing for acute, chronic, mutagenic, allergenic, and carcinogenic properties, ensuring that strains used are safe for humans and the environment.
- Emissions from fermentation equipment, drying units, extraction processes, and wastewater systems are major sources of environmental and occupational exposure to biological agents.
- Microbiological synthesis products are classified into extracellular (exometabolites) and intracellular products, each requiring different safety and purification protocols due to their potential biological activity.
- Regular microbiological monitoring ensures that the desired production strains dominate the process and that contamination with harmful or uncontrolled microflora is avoided. Assessing the sanitary-microbiological state of air, equipment, and surfaces is essential in preventing the spread of microbial contaminants and maintaining safe working environments.
- Hygienic safety measures include engineering controls, sterilization systems, strict process design, and personal protective equipment to minimize exposure to biohazards.
- Waste neutralization methods—including mechanical, chemical, biological, and thermal treatments—are critical for preventing environmental contamination from biotechnological operations.

9.2   Quality Assurance of Biotechnology Product

9.2.1   GMP System in Production and Quality Control

In the current global landscape, the pharmaceutical market operates in a truly international context. A medicine may be produced on one continent and consumed on another. The diversity of available medicines is steadily growing. These products are manufactured in countries with different social systems, varying national traditions and customs, and often in sharply contrasting climatic zones.

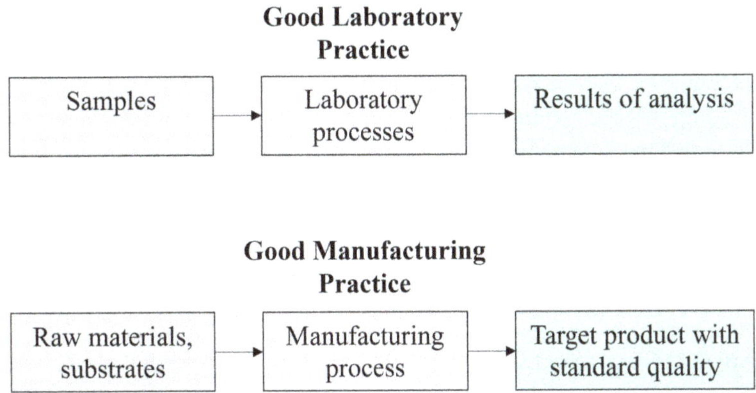

Fig. 9.3 GMP vs. GLP

Medicines are a type of product whose quality, efficacy, and safety cannot always be evaluated by the consumer—hence the necessity for specialized control laboratories. It is well known that the quality of a medicine is ensured through pharmacopoeias, which serve a legislative function. However, since the mid-twentieth century, an additional framework became necessary. This framework, or more precisely a set of documents, is known as Good Manufacturing Practice (GMP). GMP does not replace the pharmacopoeia but rather complements it, serving the ultimate goal of guaranteeing high-quality medicines for the consumer.

While the pharmacopoeia pertains directly to the medicinal product, GMP—officially titled "Rules for the Organization of Production and Quality Control of Medicinal Products"—applies to the manufacturing enterprise itself. It is important to distinguish GMP from GLP (Good Laboratory Practice) (see Fig. 9.3).

The GMP rules are the requirements of the regulations to produce medicines, which ensure a high quality of work in the enterprise concerning all manufactured medicinal products that have an official nature and constitute a list of regulatory documents. They are mandatory for all enterprises that produce ready-made medicines and biologically active substances intended to manufacture finished medicines. Failure to comply with these rules leads to sanctions imposed on the enterprise to rally until its closure.

GMP consists of a set of regulatory requirements for medicine production that ensure high quality throughout all stages of work at the enterprise. These rules have an official status and form a comprehensive list of normative documents. They are mandatory for all enterprises that produce finished medicines or active pharmaceutical ingredients (APIs) used in such medicines. Non-compliance with GMP can result in sanctions, including the possible shutdown of the enterprise.

The first GMP regulations were adopted in the United States in 1963. They gained international relevance in 1969, when, under the auspices of the World Health Organization (WHO), around 100 countries signed a multilateral agreement. This agreement introduced the Certification Scheme on the Quality of Pharmaceutical Products Moving in International

Commerce, aimed at assisting the health authorities of importing countries in assessing the production standards and quality of pharmaceutical imports.

Manufacturers of pharmaceuticals and medical devices must adhere to GMP principles. Achieving compliance is significantly facilitated when facilities and equipment are appropriately designed and constructed from the outset. The following section outlines the key considerations necessary for GMP compliance.

9.2.2   Rule #1—Get the Facility Design Right from the Start

Facility Layout The production area should be organized to align with the sequence of operations, minimizing the risk of cross-contamination, mix-ups, and errors.   Final products should not pass through or near areas containing intermediate products or raw materials.

A well-planned layout enhances productivity by eliminating unnecessary traffic, segregating materials to prevent confusion, mix-ups, and errors. For example: A company's poor planning resulted in a non-linear manufacturing process, causing the near-final product to be exposed to an early-stage product, increasing contamination risks. A comprehensive review of the layout should precede any corrective measures.

Environment:
  Control air quality, water, lighting, ventilation, temperature, and humidity to avoid
    impacting product quality.
  Design facilities to minimize contamination risks from the environment.
  Ensure appropriate lighting, temperature, humidity, and ventilation.
  Interior surfaces (walls, floors, and ceilings) should be smooth, crack-free, and non-
    particulate shedding.
  Easy-to-clean surfaces for walls, floors, ceilings, pipework, light fittings, and ventila-
    tion points.
  Adequately sized drains with trapped gullies.
Equipment:
  Design, locate, and maintain equipment according to its intended use.
  Ensure equipment is easy to repair, maintain, and clean.
  Select equipment suitable for its intended use.
  Verify that equipment is not reactive, additive, or absorptive.
  Calibrate equipment at defined intervals, as necessary.
  Clearly label equipment.

Adhering to these principles in facility design and equipment selection contributes to a GMP-compliant operation in the pharmaceutical and medical device manufacturing industry.

9.2.3   Rule #2—Validate Processes

Ensuring the controlled and consistent operation of a state-of-the-art facility involves rigorous testing and documentation to demonstrate that equipment and processes consistently meet their intended functions. This commitment to consistent performance is crucial for product safety, efficacy, and building a reputation for quality and reliability.

Validation, as defined, is a documented program ensuring a high degree of assurance that a specific process, method, or system consistently produces results meeting predetermined acceptance criteria. Key aspects of validation include.

Well-planned and clearly defined validation activities, considering the life cycle of facilities, equipment, utilities, processes, and products.

Preparation of a Validation Master Plan (VMP) containing the company's validation policy, roles and responsibilities, facility and equipment summaries, guidance on developing acceptance criteria, and the validation strategy.

Three main phases:

1. Installation Qualification (IQ) to verify correct installation
2. Operational Qualification (OQ) to verify correct operation
3. Performance Qualification (PQ) to verify consistent production to specification

Protocols detailing each test and acceptance criteria, followed by comprehensive reports upon completion.

After successful testing and confirmation of control, maintaining the validated state involves:

- Adherence to written procedures and proper equipment maintenance and calibration
- Implementation of change control procedures for any required changes to the validated state
- Regular evaluation of equipment, facilities, utilities, and systems to confirm their continued state of control

For instance, in a scenario where a computer system was validated upon installation, an auditor later discovered undocumented changes, indicating the system was no longer in its validated state. The oversight was attributed to a weak change control system that allowed changes without formal evaluation and re-validation

9.2.4   Rule #3—Write Good Procedures and Follow Them

In the pharmaceutical and medical device industries, the establishment of robust procedures is crucial to ensure processes are executed in a controlled and consistent manner.

Effective documentation serves as an integral component of the quality assurance system and is fundamental for adhering to Good Manufacturing Practice (GMP) requirements.

- In these industries, the following documents are commonly required.
- Detailed requirements that products or materials must meet, serving as the basis for quality evaluation.
- Providing details on material and equipment requirements, outlining steps to complete a task.
- Offering guidance for performing specific tasks and providing higher-level instructions compared to operating or work instructions.

Establishing effective procedures is vital in pharmaceutical and medical device industries to ensure consistent and controlled processes. Robust documentation is integral for compliance with Good Manufacturing Practice (GMP) requirements.

Ensuring the usability of procedures can be validated through a "usability test," involving individuals unfamiliar with the task following the procedure. Observing and noting areas where the document proves challenging provides valuable feedback. Regular review of documentation, a GMP requirement, is necessary to maintain currency. Companies often implement a three-year review cycle, but this duration can be adjusted based on the potential for change or the criticality of the associated process.

While having well-crafted procedures is essential, adherence is equally crucial, mandated by GMP requirements. Although written procedures might not always seem like the most efficient approach, deviation without supervisor or Quality Department approval is not permitted. This is driven by two key considerations.

Risk of Pitfalls: Shortcuts may introduce pitfalls that could incur significant costs in the long run.

Purposeful Steps: Each step in a procedure serves a specific purpose, even if its relevance isn't immediately apparent. Steps may act as checks for subsequent stages of the process.

While fostering ideas for improvement is encouraged, altering the entire process requires a thorough impact assessment before implementation.

9.2.5　Rule #4—Identify Who Does What

The accurate manufacturing of medicinal products hinges on the competence of individuals. It is imperative that all staff members possess a clear understanding of their daily responsibilities to mitigate misunderstandings and minimize the risk to product quality.

To establish clarity, a comprehensive job description should be developed for each role, outlining.

Furthermore, specific task responsibilities should be explicitly outlined in procedures to avoid any ambiguity. Certain areas, such as cleaning, validation, and calibration, are prone to potential overlaps in responsibilities.

An organizational chart can be made visible, either on an intranet or a local notice board, ensuring transparency about individual roles and responsibilities throughout the organization.

9.2.6   Rule #5—Keep Good Records

Maintaining accurate records is crucial in tracking every activity conducted during batch manufacturing, spanning from raw material receipt to the final product release. Good Manufacturing Practice (GMP) emphasizes the significance of meticulous record-keeping, serving as evidence during audits that procedures are being followed and processes are well-understood and controlled.

To instill a culture of good record-keeping in daily practices, consider the following guidelines:

- Record all essential information immediately after completing a task.
- Avoid relying on memory or jotting down results on loose sheets; use a systematic approach.
- Clearly write your name in ink when signing records.
- Understand that your signature certifies the accuracy of the record and adherence to defined procedures.
- If a mistake occurs, draw a single line through it, initial and date the correction, and provide a reason for the correction at the page's bottom.
- Document any deviations from procedures, seeking guidance from supervisors or the Quality Department.
- Refrain from documenting someone else's work unless designated and trained to do.
- Don't assume that undocumented work has been completed properly; if it's not written down, it didn't happen.

Be cautious of signature fatigue, and review procedures to limit the number of signatures to critical steps. Include a "checked by" signature only when required by a predicate rule.

Retention requirements mandate keeping records for each stage of the manufacturing process, encompassing product master records, batch or manufacturing records, material/component control records, personnel records, training records, equipment logs, and cleaning logs. Clearly define the relationship between records and manufacturing activities, and implement secure controls to maintain record integrity throughout their retention period.

For batch documentation, specific retention requirements dictate preservation for one year after the batch's expiry or at least five years after certification by the Authorized Person, depending on the longer duration. Other documentation's retention periods vary based on the supported business activity.

9.2.7   Rule #6—Train and Develop Staff

Ensuring compliance with Good Manufacturing Practice (GMP) requirements hinges on having the right individuals in the right roles. Are your employees adequately skilled and knowledgeable for their tasks? Have you provided them with the necessary tools? If so, you can take pride in the fact that your workforce is actively contributing to fostering a GMP culture.

Training plays a pivotal role in achieving this goal. It is imperative to offer training to all employees whose responsibilities involve working in production and storage areas or laboratories, encompassing maintenance and cleaning staff. This training is especially critical when their activities could impact the product's quality.

In addition to foundational training covering the theory and practice of GMP, employees should receive role-specific training tailored to their responsibilities. While it might be unavoidable to have untrained visitors in production areas at times, proper measures should be taken. In such instances, providing visitors with relevant information in advance, especially regarding personal hygiene, and closely supervising them throughout their time in these areas becomes essential.

9.2.8   Rule #7—Practice Good Hygiene

Establishing an effective sanitation program is crucial to minimizing the risk of product contamination. The program should align with the required cleanliness standards for the specific product, considering variations such as sterile products for operating theatres versus those injected into the bloodstream.

The ongoing fight against contamination demands the constant attention of every employee, every day. Always bear in mind these fundamental practices:

- Maintain good personal hygiene by adhering to handwashing protocols and wearing necessary protective garments.
- Report any illness to your supervisor promptly, as entering the manufacturing area may be restricted until recovery.
- Minimize contact with product, product contact surfaces, and equipment.
- Strictly avoid eating, drinking, smoking, or chewing in manufacturing areas.
- Adhere to prescribed cleaning and sanitation procedures consistently.

- Report any conditions that may pose a risk of product contamination.
- Dispose of trash and waste materials appropriately and store them according to guidelines.

9.2.9 Rule #8—Maintain Facilities and Equipment

Having a well-defined maintenance schedule for facilities, utilities, and equipment is paramount. Regular maintenance serves as a proactive measure to prevent breakdowns, mitigate the risk of product contamination, and ensure the equipment remains in its "validated state." In case of unexpected events, immediate repairs should be conducted.

Written procedures for all scheduled and emergency maintenance activities are essential. These procedures should outline the responsible personnel and tasks involved and specify any required lubricants, coolants, cleaning agents, etc.

When crafting maintenance procedures, it's beneficial to assess whether the work can be conducted outside the manufacturing area to minimize impact. If this is not feasible, ensure that the cleaning requirements to restore the plant to a GMP standard are clearly detailed.

Accurate record-keeping for maintenance activities is a GMP requirement. Utilizing equipment logs can facilitate this process, recording information such as the last usage of the equipment, its purpose, cleaning history, inspection and repair dates, and calibration history.

9.2.10 Rule #9—Design Quality into the Whole Product Lifecycle

Ensuring the health and safety of customers in the pharmaceutical and medical device industries hinges on product quality. While the Quality Control (QC) department can inspect for quality, it is crucial to integrate quality into the entire product lifecycle.

Control measures should be implemented at various stages:

Controlling Components:
- Rigorous checks on materials and components upon entry to the plant are essential to verify compliance with defined specifications.
- Approval of materials and components is a prerequisite before utilization in manufacturing.
- Rejected items must be clearly identified and securely stored to prevent inadvertent use.

Controlling the manufacturing process:
- Implementation of records and procedures ensures consistent job performance by employees.
- Each product requires:

- A master record delineating specifications and manufacturing procedures
- Individual batch or history records documenting conformity to the master record
- Written schedules and procedures for the cleaning and maintenance of equipment and areas
- Packaging and labeling controls:
  - Packaging and labeling are prone to mix-ups and errors; thus, meticulous controls are essential.
  - Assigning a unique batch or lot number to each product facilitates traceability.
  - Inspection of packaging and labeling areas before processing a new batch ensures the absence of material from previous batches.
- Holding and distribution controls:
  - Controls against contamination, mix-ups, and errors should be established.
  - Provision of separate areas for quarantine and finished product testing.
  - Implementation of procedures for handling, storage, and distribution records aids in tracing shipments.

9.2.11 Rule #10—Perform Regular Audits

Regulatory authorities, such as the Food and Drug Administration (FDA) or the Therapeutic Goods Association (TGA), will perform audits to evaluate your adherence to GMP regulations. Apart from these external audits, it is advisable to conduct internal audits, also known as self-inspections, to confirm GMP compliance. Performing self-inspections several times a year is a best practice, with a focus on different manufacturing areas and departments during each inspection.

Adopting such a document meant that the pharmacopeia, as a single barrier guaranteeing the efficacy and safety of drugs, needed to be revised. Therefore, it was required to strengthen the control of manufactured pharmaceutical products. In addition, the agreement gave certain rights to exporters of medicines, which are trying to expand the sales market. However, in order to take advantage of these rights, exporters had to fulfill three conditions:

- Registration of medicinal products in the country where the products are manufactured.
- State inspection of pharmaceutical enterprises.
- GMP rules should be formally adopted.

The GMP rules in their current form have a similar rubrication, regardless of national, regional, or international, and consist of eight sections.

GMP rules, in their modern form, follow a similar structure across national, regional, and international frameworks and are typically organized into eight main sections:

Terminology

This section defines key terms used throughout the document. Clarifying terminology is essential to avoid misinterpretation or dispute. It begins with a definition of pharmaceutical production and continues with explanations of concepts such as production processes, operations, materials, and premises.

Quality Assurance

This section outlines mandatory measures to ensure product quality. These include staffing requirements, the presence of a designated person responsible for quality, proper documentation of all production stages, protocols for handling product recalls in case of quality issues, and procedures for investigating the causes of such deviations.

Personnel

GMP rules specify requirements related to company staff. The head of a pharmaceutical enterprise (not necessarily the owner) must have a relevant educational background. A clear distinction of responsibilities among senior management is required. Provisions for staff training, personal hygiene, and conduct—especially in cleanrooms—are detailed. Guidelines for the use and maintenance of functional clothing are also included, along with the obligation to inform employees of relevant instructions.

Buildings and Facilities

This section, comprising over 40 clauses, sets requirements for the layout and construction of pharmaceutical facilities. While some criteria—such as locating enterprises away from residential areas or polluting industries—are difficult to fulfill, they are important for quality assurance. For biotechnological facilities, additional safety instructions apply, particularly regarding recombinant producers and ventilation systems. In antibiotic production, stricter rules apply. Due to the high allergenic potential of β-lactam antibiotics, processes involving penicillin and its derivatives must be carried out in isolated rooms to avoid cross-contamination—even trace amounts may cause allergic reactions.

Equipment

GMP regulations cover all equipment used—from fermenters to purification systems. Equipment must be suitable for the intended processes, regularly calibrated, and easy to clean. Surfaces that come into contact with raw materials, intermediates, or finished products must be non-reactive, corrosion-resistant, and compatible with disinfectants.

Production Process

This section emphasizes the importance of raw materials, including their documentation, storage, and testing. In biotechnology, raw materials such as corn extract, cottonseed, or pea flour must be checked for microbial contamination and stored in designated, dry, non-production areas. Sampling and control of raw materials are especially important due to the complexity of nutrient media. In-process control must be carried out both by staff and a technical control department, and all results must be documented and preserved for at least one year after the product's expiration date.

Technical Control Department

A dedicated quality control unit is mandatory. This department evaluates raw materials, intermediates, and finished products. It monitors product stability and retains reference samples from each batch for further testing or verification.

Validation

Validation involves a thorough and documented confirmation that a process consistently produces products meeting predefined quality standards. This includes evaluation of the process, the equipment used, and the final product. The validation report determines whether a process is accepted or rejected. It is mandatory for all new technological processes, especially those involving sterile products. Currently, GMP rules are adopted in 140 countries around the world. "Good Laboratory Practice (GLP)" rules—properly or correctly organized laboratory tests—require not only strict adherence to a set of tests but also the maximum possible standardization of conditions for drug testing. The reliability of the results when working with laboratory animals should be conditioned by the observance of many requirements, for example, the selection of linear animals and their maintenance on a standard diet. In addition, animals in the vivarium should be placed so that various kinds of stress effects are minimized to avoid distortion of the test results.

Today, GMP regulations have been adopted in over 140 countries. Complementary to GMP are the GLP (Good Laboratory Practice) rules, which ensure the proper design and conduct of laboratory tests. These include strict protocols for working with laboratory animals, including standardized diets, genetically similar animals, and controlled housing to minimize stress and ensure consistent results.

New substances—whether discovered intentionally or accidentally—must undergo rigorous preclinical and clinical testing before being introduced into medical practice. These procedures ensure the safety and efficacy of pharmaceuticals for human use. Preclinical testing is governed by GLP, while clinical trials follow the GCP (Good Clinical Practice) framework. All stages of drug development are tightly regulated and integrated under GLP, GMP, and GCP standards..

GLP stands for Good Laboratory Practice , and it refers to a set of principles and standards that ensure the quality and integrity of non-clinical laboratory studies. GLP is particularly important in industries such as pharmaceuticals, chemicals, agrochemicals, and biotechnology where laboratory studies are conducted to support regulatory submissions and ensure the safety and efficacy of products.   Key principles and rules associated with GLP include the following:

- GLP emphasizes the establishment of a Quality Assurance (QA) program to oversee all aspects of the study, ensuring that procedures are followed and data are generated, recorded, and reported accurately.
- Laboratories must have suitable facilities and equipment to conduct the planned studies. These facilities should be properly maintained and calibrated to ensure the reliability of results.
- Qualified and trained personnel should conduct studies. Roles, responsibilities, and qualifications of all individuals involved in the study must be documented.
- GLP requires the development and adherence to SOPs for all aspects of the laboratory work. SOPs outline step-by-step procedures for conducting studies and ensure consistency.
- Thorough documentation is a key aspect of GLP. All phases of a study, including study design, execution, and results, must be documented in a clear, complete, and traceable manner.
- Raw data generated during a study must be preserved and made available for inspection. Data integrity, accuracy, and reliability are paramount.
- Records and raw data should be retained for a specified period after the completion of a study. This archival period is typically defined by regulatory authorities.
- Test and control articles used in studies must be characterized, stored, and handled appropriately. The identity, purity, strength, and composition of these articles must be known.
- Quality control samples are included in studies to monitor the performance of the analytical methods and the overall conduct of the study.
- GLP requires the preparation of a final report that accurately reflects the conduct and results of the study. The report should be written in a clear and concise manner.

Compliance with GLP is crucial for organizations involved in research and development to ensure the acceptance of their studies by regulatory authorities and to demonstrate the reliability of their data. Non-compliance with GLP can result in the rejection of studies and may impact the approval of products.

GCP stands for Good Clinical Practice, and it is a set of international ethical and scientific quality standards for designing, conducting, recording, and reporting clinical trials that involve human participants. GCP ensures the protection of participants' rights, well-being, and the credibility of clinical trial data. These rules are essential for maintaining the integrity and reliability of clinical research. Here are the key principles and rules associated with GCP. GCP emphasizes that clinical trials should be conducted in accordance with ethical principles, ensuring the rights, safety, and well-being of trial participants. Informed consent, confidentiality, and respect for individuals are fundamental.

The design, conduct, and documentation of clinical trials should adhere to scientific and ethical principles. Protocols must be written clearly, detailing the study objectives, design, methodology, statistical considerations, and the rights and safety of participants.

Participants must provide voluntary and informed consent before participating in a clinical trial. Investigators must provide participants with detailed information about the trial, including potential risks and benefits, in a language they can understand.

Investigators (doctors, scientists, and other healthcare professionals) are responsible for the conduct of the trial at their respective sites. They must be qualified by education, training, and experience.

Ethical review by an Independent Review Board (IRB) or Independent Ethics Committee (IEC) is mandatory before initiating a clinical trial. The IRB/IEC ensures that the rights and safety of participants are protected.

Adverse events, including serious adverse events, must be promptly reported to regulatory authorities, the sponsor, and the ethics committee. Safety monitoring is an integral part of GCP.

Accurate, complete, and verifiable records must be maintained for the duration of the trial. Source documents and case report forms (CRFs) should reflect the participants' clinical status and provide evidence of trial conduct.

Quality control and assurance measures should be implemented to ensure that the data generated and reported are reliable and accurate. Regular monitoring visits and audits are conducted to assess compliance.

Investigational products (study drugs, devices, or treatments) must be manufactured, handled, and administered in accordance with GCP and regulatory requirements.

Monitoring of clinical trials is essential to verify the protection of participants' rights and the reliability of trial results. Monitoring activities include site visits, source data verification, and compliance assessments.

Compliance with GCP is a regulatory requirement for conducting clinical trials, and adherence to these rules is critical for obtaining regulatory approval and ensuring the acceptance of clinical trial data by health authorities worldwide.

In conclusion, Good Manufacturing Practice (GMP), Good Clinical Practice (GCP), and Good Laboratory Practice (GLP) are integral frameworks that establish and uphold quality standards in different domains of the life sciences industry. Each practice serves a distinct purpose, contributing to the overall integrity, safety, and reliability of products and research outcomes.

International Organization for Standardization in Industrial Biotechnology

ISO, or the International Organization for Standardization, plays a crucial role in industrial biotechnology by establishing and promoting international standards. These standards provide a framework for consistency, interoperability, and quality in various aspects of industrial biotechnology processes, products, and services. The application of ISO standards in industrial biotechnology helps enhance efficiency, ensure safety, and facilitate international trade. Specific areas within industrial biotechnology where ISO standards may be applicable include bioprocessing, bio-based products, and quality management systems tailored to the biotechnology sector. These standards contribute to the overall advancement and harmonization of practices in industrial biotechnology on a global scale.

While international standardization is no stranger to the field of biotechnology (ISO/TC 276, Biotechnology, was created back in 2013), the pace of development in this field and the breadth of its applications means that this is an area to watch for emerging-market needs.

ISO, or the International Organization for Standardization, is an independent, non-governmental international organization that develops and publishes standards to ensure the quality, safety, and efficiency of products, services, and systems. ISO standards cover a wide range of industries and sectors globally. Here is an overview of ISO rules and regulations.

ISO, established on February 23, 1947, has an extensive track record, having issued over 24,500 international standards covering nearly every facet of technology and manufacturing as of November 2022. With 811 technical committees (TCs) and subcommittees (SCs) overseeing standards development, the organization addresses standardization across both technical and non-technical domains, excluding electrical and electronic engineering, a responsibility handled by the IEC. Headquartered in Geneva, Switzerland, ISO operates in 167 countries as of 2023. The official languages of ISO are English, French, and Russian.

The International Organization for Standardization is an independent, non-governmental organization whose membership consists of different national standards bodies. As of 2022, there are 167 members representing ISO in their country, with each country having only one member. ISO is responsible for creating and issuing international standards across a broad spectrum of technical and non-technical domains, excluding electrical and electronic engineering, a domain overseen by the International Electrotechnical Commission.

As of February 2023, ISO has crafted more than 24,676 standards, encompassing areas ranging from manufactured products and technology to food safety, agriculture, and healthcare.

Genome editing technology, a rapidly advancing bioscience field with diverse applications in biotechnology, involves modifying nucleic acids in a site-specific manner, including DNA or RNA alterations such as insertion, deletion, or modification. This technology, applicable to various cell types, finds global significance in areas such as human cell-based therapeutics, agriculture, microbial-based therapeutics, synthetic biology, and biomanufacturing.

Given the active utilization of genome editing technology, the lack of standardized terms and definitions poses a challenge to the interpretation and communication of concepts, data, and results. This document aims to address this need by providing a unified set of standards for terms and definitions, catering to the requirements of biotechnology stakeholders and serving as a reference for genome editing technology.

The standards outlined in this document are designed to harmonize and expedite effective communication, technology development, qualification, and evaluation of genome editing products. Its objective is to enhance confidence and clarity in scientific communication, data reporting, and data interpretation within the genome editing field. Notably, specific requirements for the application of genome editing technologies in agriculture and food are not covered, and users seeking such specifications can refer to standards developed by relevant ISO Technical Committees, like ISO/TC 34/SC 16 for horizontal methods in molecular biomarker analysis or ISO/TC 215 for health informatics.

Further sub-categories within this document include terms specific to different types of genome editing technology, such as "CRISPR specific," "Meganuclease specific," "mega-TAL specific," "TALEN specific," and "ZFN specific." The index provides an alphabetical list of all these terms. Definitions are presented in English word order whenever possible.

Acknowledging the rapid development and evolution of genome editing as a biotechnology field, it is anticipated that additional terms and definitions will be required to accommodate the ongoing maturation of genome editing technologies. As the field progresses, the continuous expansion of this standardized vocabulary will be essential to accurately represent and communicate the nuanced terminology associated with emerging advancements in genome editing.

9.2.12 The Codex Alimentarius

The Codex Alimentarius, established by the Food and Agriculture Organization (FAO) and the World Health Organization (WHO), is an international food standards body. It develops harmonized international food standards, guidelines, and codes of practice to protect the health of consumers and ensure fair practices in the food trade. The term "Codex Alimentarius" is Latin for "Food Code."

The Codex Alimentarius Commission was established under the influence of the pharmaceutical industry in 1963, in compliance with resolutions adopted in 1961 at the eleventh session of the FAO Conference at the UN and the sixteenth assembly of the WHO. However, the historical roots of the name "Codex Alimentarius" can be traced back to the Codex Alimentarius Austriacus (the collection of standards and descriptions of various food products developed in the Austro-Hungarian Empire from 1897 to 1911). Later, from 1954 to 1958, Austria actively contributed to the creation of a regional European code known as Codex Alimentarius Europaeus. It was the Council of Codex Alimentarius Europaeus in 1961 that adopted a resolution proposing to arm the FAO and WHO with the standards they had developed.

9.2.13 Key Points About Codex Alimentarius

Creation and Purpose: The Codex Alimentarius Commission was created in 1963 by FAO and WHO to develop food standards and related texts, such as codes of practice and guidelines.

Scope: The Codex covers a wide range of issues related to food safety, quality, and fairness in international trade. It addresses areas such as food additives, contaminants, labeling, hygiene, methods of analysis, and more.

International Collaboration: The standards developed by Codex Alimentarius are the result of international collaboration. Member countries contribute to the development and review of standards to ensure broad consensus.

Protecting Consumer Health: One of the primary objectives is to protect the health of consumers and ensure fair practices in the food trade. Codex standards aim to provide a basis for national legislation and regulations.

Trade Facilitation: Codex standards facilitate international trade by providing a common ground for the assessment of food products. When countries adopt Codex standards, it helps prevent technical barriers to trade.

Consumer Protection: The establishment of maximum residue limits for pesticides and veterinary drugs in food is an example of how Codex contributes to consumer protection.

Committees and Task Forces: The work of Codex is carried out through various committees and task forces, each dealing with specific aspects of food standards. These include committees on food hygiene, food additives, contaminants, nutrition, labeling, and more.

Adoption of Standards: While Codex standards are not legally binding, many countries adopt them as a basis for their national food regulations. The World Trade Organization (WTO) recognizes Codex standards as a reference point for international trade disputes related to food safety and quality.

Codex Alimentarius Meetings: The Commission holds regular meetings where member countries discuss and adopt standards. These meetings provide a forum for negotiations and consensus-building.

Overall, Codex Alimentarius plays a crucial role in promoting international cooperation, protecting consumer health, and facilitating fair practices in the global food trade.

The Codex regulates all food products, both processed and raw. In addition to standards for individual types of products, the Codex includes general standards that regulate issues such as product labeling, food hygiene, food additives, pesticide content, and procedures for the safety assessment of food products and biotechnologies.

The Codex is published in several languages, including English, French, and Spanish. Some standards within the Codex are also available in Russian and Arabic, and certain standards have been translated into Chinese.

The Codex Regulates
- Food labeling (general rules, guidelines for indicating nutritional value, guidelines for various statements)
- Food additives (basic standards, rules for use, requirements for food chemicals)
- Food contamination (general requirements, maximum contamination levels of food by radionuclides, aflatoxins, and others)
- Maximum allowable levels of pesticides and veterinary drugs in food
- Methods for assessing biotechnological risks
- Hygiene in food production
- Methods of analysis and selection of samples
- Specific standards exist for various types of food, including:
- Meat and meat products
- Fish and fish products
- Milk and dairy products
- Dietetic and infant food
- Fresh and processed vegetables, fruits, and juices
- Cereals and legumes and products
- Fats, oils, and products
- Various food products (chocolate, sugar, honey, mineral water, and others)

9.2.14 Hazard Analysis and Critical Control Points (HACCP)

The HACCP system is an organizational structure of production that consists of documents, production processes, and resources necessary for implementing HACCP.

The HACCP system is an effective management tool used to protect a company's brand when promoting food products in the market and safeguarding production processes from biological (microbiological), chemical, physical, and other contamination risks.

International organizations, such as the Codex Alimentarius Commission, have endorsed the use of HACCP as the most effective way to prevent diseases caused by poor-quality food products. The application of HACCP can be beneficial for confirming compliance with legislative and regulatory requirements.

Hazard Analysis and Critical Control Points (HACCP) is a systematic preventive approach to food safety that addresses biological, chemical, and physical hazards in the production processes that can cause the finished product to be unsafe. Here are the key principles and rules of HACCP: Hazard analysis:

- Identify potential hazards associated with the food production process.
- Determine the severity and likelihood of these hazards.
- Identify critical control points (CCPs):

- Identify the points in the production process where control can be applied to prevent, eliminate, or reduce a hazard to an acceptable level.
- These are the critical stages where control is crucial for ensuring food safety.
- Establish critical limits:
  - Set measurable criteria for each CCP to ensure that hazards are controlled within acceptable levels.
  - Critical limits define the maximum or minimum values to which biological, chemical, or physical parameters must be controlled.
- Establish monitoring procedures:
  - Develop procedures to monitor and record the parameters at each CCP.
  - Monitoring ensures that the process is under control and that critical limits are being met.
- Establish corrective actions:
  - Define actions to be taken when monitoring indicates a deviation from the critical limit.
  - Corrective actions are necessary to bring the process back under control and prevent unsafe products from reaching consumers.
- Establish verification procedures:
  - Establish activities, other than monitoring, to verify that the HACCP system is working effectively.
  - Verification involves validating, auditing, and reviewing the HACCP system.
- Establish record-keeping and documentation:
  - Maintain documentation of the HACCP plan and all related records.
  - Proper documentation provides evidence that the HACCP system is implemented and effective.
- Establish procedures for verification:
  - Implement regular reviews and audits to verify that the HACCP system is functioning as intended.
  - This involves confirming that the plan is being followed and is effective in ensuring food safety.
- Implement a HACCP training program:
  - Ensure that personnel involved in the HACCP plan are trained in the principles of HACCP and their role in implementing the plan.
  - Training is crucial for the successful implementation of the system.
- Review and update the HACCP plan:
  - Periodically review and reassess the HACCP plan, especially when there are changes in the process, product, or regulations.
  - Update the plan to reflect any modifications necessary for maintaining food safety.
  - By following these HACCP principles, food producers can systematically control and reduce the risks associated with the production process, ensuring the safety of the final food products.

Brainstorming

1. What are the reasons for introducing international rules into biotechnology practice?
2. What is the content and purpose of GMP rules?
3. What are the key principles of GCP and GLP, and how do they differ from GMP?
4. Why can testing the quality of a drug according to GMP rules provide more reliable results?
5. What are the reasons for conducting validation when replacing producer strains in production?
6. What are the specific GMP requirements for biotechnology-based production?
7. What is the Codex Alimentarius, and what is its main purpose within the framework of international food standards?
8. When and why was the Codex Alimentarius Commission established, and what historical events influenced its formation?
9. How does the historical background of the Codex Alimentarius—such as the *Codex Austriacus* and *Codex Europaeus*—contribute to its development?
10. What are the key points regarding the creation, scope, and objectives of the Codex Alimentarius Commission?
11. How does Codex Alimentarius contribute to consumer health protection and ensuring fair practices in the global food trade?
12. In what ways does Codex Alimentarius facilitate international trade, and how are its standards adopted or used by member countries?
13. Can you provide examples of specific areas covered by Codex standards, such as food additives, contaminants, labeling, and hygiene?
14. While Codex standards are not legally binding, how do they influence national food regulations and international law?
15. How does the Codex Alimentarius Commission carry out its work through committees and task forces, and what specific issues do these groups address?
16. What role does Codex play in consumer protection, especially in establishing maximum residue limits for pesticides and veterinary drugs?
17. What is the significance of Codex meetings, and how do member countries use them for negotiations and consensus-building?
18. What is the primary purpose of Hazard Analysis and Critical Control Points (HACCP) in the food industry?
19. How does HACCP address different types of biological, chemical, and physical hazards in food production?
20. What are the key steps involved in conducting a hazard analysis within the HACCP system?
21. Why is it important to identify Critical Control Points (CCPs) in the production process?
22. How do Critical Limits help ensure food safety within the HACCP framework?
23. What role do Monitoring Procedures play in maintaining control at CCPs?

24. Why are Corrective Actions crucial in the HACCP system, and how are they determined?
25. What is the significance of Verification Procedures in HACCP, and what activities do they involve?
26. Why is record-keeping and documentation essential for the effective implementation of HACCP?
27. How does a HACCP training program contribute to food safety, and who should participate in it?
28. In what ways should a company regularly review and update its HACCP plan to maintain food safety?
29. How does HACCP help prevent unsafe food products from reaching consumers?

Take-Home Messages
- GMP (Good Manufacturing Practice) ensures that medicines and biotechnological products are consistently produced and controlled according to quality standards, covering all aspects from raw materials to final packaging and staff training.
- GLP (Good Laboratory Practice) and GCP (Good Clinical Practice) complement GMP by establishing rigorous standards for laboratory research and clinical testing, ensuring the reliability, traceability, and ethical integrity of pharmaceutical development.
- Validation is a critical GMP component that confirms a production process consistently yields products meeting predefined quality standards, especially when introducing new technologies or changing producer strains.
- The Codex Alimentarius provides international food safety and quality standards to protect consumer health and facilitate fair trade. Although not legally binding, its influence shapes national regulations and global food policy.
- HACCP (Hazard Analysis and Critical Control Points) is a systematic, science-based approach for identifying and controlling food safety hazards, ensuring preventive action rather than reactive response in food production.
- The effective implementation of GMP, Codex, and HACCP principles requires ongoing documentation, staff training, and regular audits, which together uphold public trust, product safety, and international trade compatibility.

Appendix 1: Exponential Notation for Base 10

Number	Quantity	Exponential notation	One followed by
1	One	10^0	No zeros
10	Ten	10^1	One zero
100	Hundred	10^2	Two zeros
1,000	Thousand thousand	10^3	Three zeros
10,000	Ten thousand	10^4	Four zeros
100,000	Hundred	10^5	Five zeros
1,000,000	Million	10^6	Six zeros
1,000,000,000	Billion	10^9	Nine zeros
1,000,000,000,000	Trillion	10^{12}	Twelve zeros
1,000,000,000,000,000	Quadrillion	10^{15}	Fifteen zeros
1,000,000,000,000,000,000	Quintillion	10^{18}	Eighteen zeros

Length

    1 millimeter = 1 mm = 0.001 m = 10^{-3} m

    1 micrometer = 1 μm = 0.000001 m = 10^{-6} m

    1 nanometer = 1 nm = 0.000000001 m = 10^{-9} m

Weight

    1 milligram = 1 mg = 0.001 g = 10^{-3} g

    1 microgram = 1 μg = 0.000001 g = 10^{-6} g

    1 nanogram = 1 ng = 0.000000001 g = 10^{-9} g

    1 picogram = 1 pg = 0.000000000001 g = 10^{-12} g

Appendix 2: Microbial Terminology: Singulars and Plurals

Singular	Plural	Singular	Plural	Singular	Plural
alga	algae	diagnosis	diagnoses	mucosa	mucosae
ameba	amebae	fimbria	fimbriae	mycelium	mycelia
bacillus	bacilli	flagellum	flagella	mycosis	mycoses
bacterium	bacteria	focus	foci	phylum	phyla
cilium	cilia	fungus	fungi	pilus	pili
clostridium	clostridia	hydrolysis	hydrolyses	nucleus	nuclei
coccus	cocci	hypha	hyphae	septum	septa
conidium	conidia	inoculum	inocula	synthesis	syntheses
datum	data	medium	media		

© The Editor(s) (if applicable) and The Author(s), under exclusive license to
Springer Nature Switzerland AG 2025

I. Digel et al., *Introduction to Industrial Biotechnology*, Learning Materials in
Biosciences, https://doi.org/10.1007/978-3-032-07918-3

Appendix 3: Prefixes and Suffixes in Microbiology

Prefix or suffix	Meaning	Example
a-, an-	not, without	avirulent (lacking virulence), anaerobic (without air)
aer-	air	aerobic
anti-	against	antiseptic
-ase	enzyme	penicillinase
chlor-	green	chlorophyll
-chrom-	color	metachromatic (staining differently with the same dye)
-cide	causing death	germicide
co-, com-, con-	together	coenzyme
-cyan-	blue	pyocyanin (a blue bacterial pigment)
de-	down, from	dehydrate (remove water)
-dem-	people, district	epidemic
endo-	within	endospore (spore within a cell)
-enter-	intestine	enteritis (inflammation of the intestine)
epi-	upon	epidermis
erythro-	red	erythocyte (red blood cell)
eu-	well, normal	eukaryotic (true nucleus)
exo-	outside	exoenzyme (enzyme that acts outside the cell that produced it)
extra-	outside of	extracellular
flav-	yellow	flavoprotein (a protein containing a yellow enzyme)
-gen	produce, originate	antigen (a substance that induces the production of antibodies)
glyc-	sweet	glycemia (the presence of sugar in the blood)
hetero-	other	heterotroph (organism that obtains carbon from organic compounds)
homo-	common, same	homologous (similar in structure or origin)
hydr-	water	dehydrate
hyper-	excessive, above	hypersensitive
hypo-	under	hypotonic (having low osmotic pressure)

(conitnued)

Prefix or suffix	Meaning	Example
iso-	same, equal	isotonic (having the same osmotic pressure)
-itis	inflammation	appendicitis, meningitis
leuko-	white	leukocyte (white blood cell)
ly-, -lys, -lyt-	loosen; dissolve	bacteriolysis (dissolution of bacteria)
meso-	middle	mesophilic (preferring moderate temperatures)
meta-	changed	metachromatic (staining differently with the same dye)
micro-	small; one-millionth part	microscopic
milli-	one-thousandth part	millimeter (10^{-3} meter)
mito-	thread	mitochondrion (small, rod-shaped or granular organelle)
mono-	single	monotrichous (having a single flagellum)
multi-	many	multinuclear (having many nuclei)
myc-	fungus	mycotic (caused by fungus)
myx-	mucus	myxomycete (slime mold)
-oid	resembling	lymphoid (resembling lymphocytes)
-ose	a sugar	lactose (milk sugar)
-osis	disease of	coccidioidomycosis (disease caused by *Coccidioides*)
pan-	all	pandemic (widespread epidemic)
para-	beside	parasite (an organism that feeds in and at expense of the host)
patho-	disease	pathogenic (producing disease)
peri-	around	peritrichous (having flagella on all sides)
-phag-	eat	phagocyte (a cell that ingests other cell substances)
-phil	like, having affinity for	eosinophilic (staining with the dye eosin)
-phot-	light	photosynthesis
-phyll	leaf	chlorophyll (green leaf pigment)
pleo-	more	pleomorphic (occurring in more than one form)
poly-	many	polymorphonuclear (having a many-shaped nucleus)
post	after	postnatal (after birth)
pyo-	pus	pyogenic (producing pus)
-sta-	stop	bacteriostatic (inhibiting bacterial multiplication)
sym-, syn-	together	symbiosis (life together)
thermo-	heat	thermophilic (liking heat)
trans-	through, across	transfusion
-trich-	hair	monotrichous (having a single flagellum)
-troph	nourishment	autotroph (organism that obtains carbon from CO_2)
tox-	poison	toxin
zym-	ferment	enzyme

Appendix 4: Pronunciation Key for Bacterial, Fungal, Protozoan, and Viral Names

A

Acetobacter (a-see-toe-back-ter)

Achromobacter (a-krome-oh-back-ter)

Acinetobacter calcoaceticus (a-sin-et-oh-back-ter kal-koh-ahsee-ti-kus)

Actinomyces israelii (ak-tin-oh-my-seez iz-ray-lee-ee)

Actinomycetes (ak-tin-oh-my-seats)

Adenovirus (ad-eh-no-vi-rus)

Agrobacterium tumefaciens (ag-rho-bak-teer-ee-um too-mehfaysh-ee-enz)

Alcaligenes (al-ka-li-jen-ease)

Amoeba (ah-mee-bah)

Arbovirus (are-bow-vi-rus)

Aspergillus niger (ass-per-jill-us nye-jer)

Aspergillus oryzae (ass-per-jill-us or-eye-zee)

Azolla (aye-zol-lah)

Azotobacter (ay-zoh-toe-back-ter)

B

Bacillus anthracis (bah-sill-us an-thra-siss)

Bacillus cereus (bah-sill-us seer-ee-us)

Bacillus coagulans (bah-sill-us coh-ag-you-lans)

Bacillus fastidiosus (bah-sill-us fas-tid-ee-oh-sus)

Bacillus stearothermophilus (bah-sill-us steer-oh-ther-maw-fill-us)

Bacillus subtilis (bah-sill-us sut-ill-us)

Bacillus thuringiensis (bah-sill-us thur-in-jee-en-sis)

Bacteroides (back-ter-oid-eez)

Baculovirus (back-you-low-vi-rus)

Beggiatoa (beg-gee-ah-toe-ah)

Beijerinckia (by-yer-ink-ee-ah)

Bordetella pertussis (bor-deh-tell-ah per-tuss-iss)

Borrelia burgdorferi (bor-real-ee-ah berg-dor-fir-ee)

I. Digel et al., *Introduction to Industrial Biotechnology*, Learning Materials in Biosciences, https://doi.org/10.1007/978-3-032-07918-3

Bradyrhizobium (bray-dee-rye-zoe-bee-um)

Branhamella (bran-ham-el-lah)

Brucella abortus (bru-sell-ah ah-bore-tus)

C

Campylobacter jejuni (kam-peh-low-back-ter je-june-ee)

Candida albicans (kan-did-ah al-bi-kanz)

Caulobacter (caw-loh-back-ter)

Ceratocystis ulmi (see-rah-toe-sis-tis ul-mee)

Chlamydia trachomatis (klah-mid-ee-ah trah-ko-ma-tiss)

Claviceps purpurea (kla-vi-seps purr-purr-ee-ah)

Clostridium acetobutylicum (kloss-trid-ee-um a-seat-tow-bu-till-i-kum)

Clostridium botulinum (kloss-trid-ee-um bot-you-line-um)

Clostridium difficile (kloss-trid-ee-um dif-fi-seal)

Clostridium perfringens (kloss-trid-ee-um per-frin-gens)

Clostridium tetani (kloss-trid-ee-um tet-an-ee)

Coccidioides immitis (cock-sid-ee-oid-eez im-mi-tiss)

Coronavirus (kor-oh-nah-vi-rus)

Corynebacterium diphtheriae (koh-ryne-nee-bak-teer-ee-um diftheer -ee-ee)

Coxsackievirus (cock-sack-ee-vi-rus)

Cryptococcus neoformans (krip-toe-cock-us knee-oh-for-manz)

Cytophaga (sigh-taw-fa-ga)

D

Desulfovibrio (dee-sul-foh-vib-ree-oh)

E

Eikenella corrodens (eye-keh-nell-ah kor-roh-denz)

Entamoeba histolytica (en-ta-mee-bah his-toh-lit-ik-ah)

Enterobacter (en-ter-oh-back-ter)

Enterococcus faecalis (en-ter-oh-kock-us fee-ka-liss)

Enterovirus (en-ter-oh-vi-rus)

Epidermophyton (eh-pee-der-moh-fy-ton)

Escherichia coli (esh-er-ee-she-ah koh-lee)

F

Filobasidiella neoformans (fee-loh-bah-si-dee-ell-ah knee-ohfor-manz)

Flavivirus (flay-vih-vi-rus)

Flavobacterium (flay-vo-back-teer-ee-um)

Francisella tularensis (fran-siss-sell-ah tu-lah-ren-siss)

Frankia (frank-ee-ah)

Fusobacterium (fu-zoh-back-teer-ee-um)

G

Gallionella (gal-ee-oh-nell-ah)

Gardnerella vaginalis (gard-nee-rel-lah va-jin-al-is)

Giardia intestinalis (jee-are-dee-ah in-test-tin-al-is)

Giardia lamblia (jee-are-dee-ah lamb-lee-ah)
Gluconobacter (glue-kon-oh-back-ter)
Gonyaulax (gon-ee-ow-lax)
Gymnodinium breve (jim-no-din-i-um brev-eh)
H
Haemophilus influenzae (hee-moff-ill-us in-flew-en-zee)
Helicobacter pylori (he-lih-koh-back-ter pie-lore-ee)
Hepadnavirus (hep-ad-nah-vi-rus)
Hepatitis virus (hep-ah-ti-tis vi-rus)
Herpes simplex (her-peas sim-plex)
Herpes zoster (her-peas zoh-ster)
Histoplasma capsulatum (his-toh-plaz-mah cap-su-lah-tum)
Hyphomicrobium (high-foh-my-krow-bee-um)
I
Influenza virus (in-flew-en-za vi-rus)
K
Klebsiella pneumoniae (kleb-see-ell-ah new-moan-ee-ee)
L
Lactobacillus brevis (lack-toe-ba-sil-lus bre-vis)
Lactobacillus bulgaricus (lack-toe-ba-sil-lus bull-gair-i-kus)
Lactobacillus casei (lack-toe-ba-sil-us kay-see-ee)
Lactobacillus plantarum (lack-toe-ba-sil-us plan-tar-um)
Lactobacillus thermophilus (lack-toe-ba-sil-us ther-mo-fil-us)
Lactococcus lactis (lack-toe-kock-us lak-tiss)
Legionella pneumophila (lee-jon-ell-ah new-moh-fill-ah)
Leptospira interrogans (lep-toe-spire-ah in-ter-roh-ganz)
Leuconostoc citrovorum (lew-kow-nos-tok sit-ro-vor-um)
Listeria monocytogenes (lis-tear-ee-ah mon-oh-sigh-tojen-eze)
M
Malassezia (mal-as-seez-e-ah)
Methanobacterium (me-than-oh-bak-teer-ee-um)
Methanococcus (me-than-oh-ko-kus)
Microsporum (my-kroh-spore-um)
Mobiluncus (moh-bi-lun-kus)
Moraxella catarrhalis (more-ax-ell-ah kah-tah-rah-liss)
Moraxella lacunata (more-ax-ell-ah lak-u-nah-tah)
Mucor (mu-kor)
Mycobacterium leprae (my-koh-bak-teer-ee-um lep-ree)
Mycobacterium tuberculosis (my-koh-bak-teer-ee-um too-berkew-loh-siss)
Mycoplasma pneumoniae (my-koh-plaz-mah newmoan-ee-ee)

N

Neisseria gonorrhoeae (nye-seer-ee-ah gahn-oh-ree-ee)

Neisseria meningitidis (nye-seer-ee-ah men-in-jit-id-iss)

Neurospora sitophila (new-rah-spor-ah sit-oh-phil-ah)

O

Orthomyxovirus (or-thoe-mix-oh-vi-rus)

Oscillatoria (os-sil-la-tor-ee-ah)

P

Papillomavirus (pap-il-oh-ma-vi-rus)

Parainfluenza virus (par-ah-in-flew-en-zah vi-rus)

Paramecium (pair-ah-mee-see-um)

Paramyxovirus (par-ah-mix-oh-vi-rus)

Parvovirus (par-vo-vi-rus)

Pasteurella multocida (pass-ture-ell-ah mul-toe-sid-ah)

Pediococcus soyae (ped-ih-oh-ko-kus soy-ee)

Penicillium camemberti (pen-eh-sill-ee-um cam-em-bare-tee)

Penicillium roqueforti (pen-eh-sill-ee-um rok-e-for-tee)

Peptostreptococcus (pep-to-strep-to-ko-kus)

Phytophythora infestans (fy-toe-fy-thor-ah in-fes-tanz)

Picornavirus (pi-kor-na-vi-rus)

Plasmodium falciparum (plaz-moh-dee-um fall-sip-air-um)

Plasmodium malariae (plaz-moh-dee-um ma-lair-ee-ee)

Plasmodium ovale (plaz-moh-dee-um oh-vah-lee)

Plasmodium vivax (plaz-moh-dee-um vye-vax)

Pneumocystis carinii (new-mo-sis-tis car-i-nee-ee)

Poliovirus (poe-lee-oh-vi-rus)

Polyoma virus (po-lee-oh-mah vi-rus)

Propionibacterium acnes (proh-pee-ah-nee-bak-teer-ee-umak_-neez)

Propionibacterium shermanii (proh-pee-ah-nee-bak-teer-ee-umsher-man-ee-ee)

Proteus mirabilis (proh-tee-us mee-rab-il-us)

Pseudomonas aeruginosa (sue-dough-moan-ass aye-rue-gin-o-sa)

R

Rabies virus (ray-bees vi-rus)

Retrovirus (re-trow-vi-rus)

Rhabdovirus (rab-doh-vi-rus)

Rhinovirus (rye-no-vi-rus)

Rhizobium (rye-zoh-bee-um)

Rhizopus nigricans (rise-oh-pus nye-gri-kanz)

Rhizopus stolon (rise-oh-pus stoh-lon)

Rhodococcus (roh-doh-koh-kus)

Rickettsia rickettsii (rik-kett-see-ah rik-kett-see-ee)

Rotavirus (row-tah-vi-rus)

Rubella virus (rue-bell-ah vi-rus)
Rubeola virus (rue-bee-oh-la vi-rus)
S
Saccharomyces carlsbergensis (sack-ah-row-my-sees karls-bergen-siss)
Saccharomyces cerevisiae (sack-ah-row-my-sees sara-vis-ee-ee)
Saccharomyces rouxii (sack-ah-row-my-sees roos-ee-ee)
Salmonella enteritidis (sall-moh-nell-ah en-ter-it-id-iss)
Salmonella typhi (sall-moh-nell-ah tye-fee)
Salmonella typhimurium (sall-moh-nell-ah tye-fe-mur-ee-um)
Serratia marcescens (ser-ray-sha mar-sess-sens)
Shigella dysenteriae (shig-ell-ah diss-en-tair-ee-ee)
Spirillum minus (spy-rill-um my-nus)
Sporothrix schenckii (spore-oh-thrix shenk-ee-ee)
Staphylococcus aureus (staff-ill-oh-kok-us aw-ree-us)
Staphylococcus epidermidis (staff-ill-oh-kok-us epi-der-mid-iss)
Streptobacillus moniliformis (strep-tow-bah-sill-us mon-ill-i-form-is)
Streptococcus agalactiae (strep-toe-kock-us a-ga-lac-tee-ee)
Streptococcus cremoris (strep-toe-kock-us kre-more-iss)
Streptococcus mutans (strep-toe-kock-us mew-tanz)
Streptococcus pneumoniae (strep-toe-kock-us new-moan-ee-ee)
Streptococcus pyogenes (strep-toe-kock-us pie-ah-gen-ease)
Streptococcus salivarius (strep-toe-kock-us sal-ih-vair-ee-us)
Streptococcus sanguis (strep-toe-kock-us san-gwis)
Streptococcus thermophilis (strep-toe-kock-us ther-moh-fill-us)
Streptomyces griseus (strep-toe-my-seez gree-see-us)
T
Thiobacillus (thigh-oh-bah-sill-us)
Torulopsis (tore-you-lop-siss)
Treponema pallidum (tre-poh-nee-mah pal-ih-dum)
Trichomonas vaginalis (trick-oh-moan-as vag-in-al-iss)
Trichophyton (trick-oh-phye-ton)
Trypanosoma brucei (tri-pan-oh-soh-mah bru-see-ee)
V
Varicella-zoster virus (var-ih-sell-ah zoh-ster vi-rus)
Veillonella (veye-yon-ell-ah)
Vibrio anguillarum (vib-ree-oh an-gwil-air-um)
Vibrio cholerae (vib-ree-oh kahl-er-ee)
Y
Yersinia enterocolitica (yer-sin-ee-ah en-ter-oh-koh-lih-tih-kah)
Yersinia pestis (yer-sin-ee-ah pess-tiss)

Appendix 5: Overview of Bergey's Manual of Systematic Bacteriology

Taxonomic rank	Representative genera
Volume 1. The Archaea and the Deeply Branching and Phototrophic Bacteria	
Domain Archaea	Thermoproteus, Pyrodictium, Sulfolobus
Phylum Crenarchaeota	Methanobacterium
Phylum Euryarchaeota	Methanococcus
Class I. Methanobacteria	Halobacterium, Halococcus
Class II. Methanococci	Thermoplasma, Picrophilus, Ferroplasma
Class III. Halobacteria	Thermococcus, Pyrococcus
Class IV. Thermoplasmata	Archaeoglobus
Class V. Thermococci	Methanopyrus
Class VI. Archaeoglobi	
Class VII. Methanopyri	
Domain Bacteria	Aquifex, Hydrogenobacter
Phylum Aquifi cae	Thermotoga, Geotoga
Phylum Thermotogae	Thermodesulfobacterium
Phylum Thermodesulfobacteria	Deinococcus, Thermus
Phylum "Deinococcus-Thermus"	Chrysogenes
Phylum Chrysiogenetes	Chlorofl exus, Herpetosiphon
Phylum Chlorofl exi	Thermomicrobium
Phylum Thermomicrobia	Nitrospira
Phylum Nitrospira	Geovibrio
Phylum Deferribacteres	Prochloron, Synechococcus, Pleurocapsa,
Phylum Cyanobacteria	Oscillatoria, Anabaena, Nostoc,Stigonema
Phylum Chlorobi	Chlorobium, Pelodictyon
Volume 2. Domain Bacteria: The Proteobacteria	

© The Editor(s) (if applicable) and The Author(s), under exclusive license to
Springer Nature Switzerland AG 2025
I. Digel et al., *Introduction to Industrial Biotechnology*, Learning Materials in
Biosciences, https://doi.org/10.1007/978-3-032-07918-3

Taxonomic rank	Representative genera
Phylum Proteobacteria	Rhodospirillum, Rickettsia, Caulobacter,
Class I. Alphaproteobacteria	Rhizobium, Brucella, Nitrobacter,
Class II. Betaproteobacteria	Methylobacterium, Beijerinckia,
Class III. Gammaproteobacteria	Hyphomicrobium
Class IV. Deltaproteobacteria	Neisseria, Burkholderia, Alcaligenes,
Class V. Epsilonproteobacteria	Comamonas, Nitrosomonas,Methylophilus,
Volume 3. Domain Bacteria: The Low G 1 C	Thiobacillus
Gram-Positive Bacteria	Chromatium, Leucothrix, Legionella,
Phylum Firmicutes	Pseudomonas, Moraxella, Acinetobacter,
Class I. Clostridia	Azotobacter, Vibrio, Escherichia, Klebsiella,
Class II. Mollicutes	Proteus, Salmonella, Shigella, Yersinia,
Class III. Bacilli	Haemophilus
Volume 4. Domain Bacteria: The High G 1 C	Desulfovibrio, Bdellovibrio, Myxococcus,
Gram-Positive Bacteria	Polyangium
Phylum Actinobacteria	Campylobacter, Helicobacter
Class Actinobacteria	Clostridium, Peptostreptococcus, Eubacterium,
Volume 5. Domain Bacteria: The	Desulfotomaculum, Heliobacterium,
Planctomycetes, Spirochaetes, Fibrobacteres,	Veillonella
Bacteriodetes, and Fusobacteria	Mycoplasma, Ureaplasma, Spiroplasma,
Phylum Planctomycetes	Acholeplasma
Phylum Chlamydiae	Bacillus, Caryophanon, Paenibacillus,
Phylum Spirochaetes	Thermoactinomyces, Lactobacillus,
Phylum Fibrobacteres	Streptococcus, Enterococcus, Listeria,
Phylum Acidobacteria	Leuconostoc, Staphylococcus
Phylum Bacteriodetes	Actinomyces, Micrococcus, Arthrobacter,
Phylum Fusobacteria	Corynebacterium, Mycobacterium, Nocardia,
Phylum Verrucomicrobia	Actinoplanes, Propionibacterium,
Phylum Dictyoglomus	Streptomyces, Thermomonospora, Frankia,
	Actinomadura, Bifi dobacterium
	Planctomyces, Gemmata
	Chlamydia, Chlamydiophila
	Spirochaeta, Borrelia, Treponema, Leptospira
	Fibrobacter
	Acidobacterium
	Bacteroides, Porphyromonas, Prevotella,
	Flavobacterium, Sphingobacterium,
	Flexibacter, Cytophaga
	Fusobacterium, Streptobacillus
	Verrucomicrobium
	Dictyoglomus

Recommended Literature

Ahmet J, Tiwari BK, Imam SH, Rao MA. Starch-based polymeric materials and nanocomposites. Chemistry, processing, and applications; 2012. ISBN: 978-1-4398-5117-3.

Alexandre G, et al. Advances in applied microbiology; 2009. ISBN: 978-0-12-374788-4.

Antranikian G, Egorova K. Extremophiles, a unique resource of biocatalysts for industrial biotechnology. In: Physiology and biochemistry of extremophiles. Gerday, C; n.d.

Baker-Austin C, Dopson M. Life in acid: PH homeostasis in acidophiles. Trends Microbiol. 2007;15:165–71.

Barbosa-Cánovas GV, Gutiérrez-López GF. Food science and food biotechnology. 2003. ISBN 1-56676-892-6

Bhandari B, Bansal N, Zhang M, Schuck P. Handbook of food powders. Processes and properties; 2013. ISBN: 978-0-85709-513-8

Brahmachari G. Biotechnology of microbial enzymes: production, biocatalysis, and industrial applications. 2nd ed; 2023. ISBN: 0443190593

Brakstad OG. Natural and stimulated biodegradation of petroleum in cold marine environments. In: Margesin R, Schinner F, Marx J-C, Gerday C, editors. Psychrophiles: from biodiversity to biotechnology. Heidelberg: Springer; 2008. p. 389–407.

Brock TD. Life at high temperatures. Science. 1985;230:132–8.

Burns RG, Dick RP. Enzymes in the environment. Activity, ecology, and applications; 2002. ISBN: 0-8247-0614-5

Carter J, Saunders V. Virology: principles and applications. 2007. ISBN: 978-0-470-02387-7.

Cheung PCK, Mehta BM. Handbook of food chemistry. 2015. ISBN: 978-3-642-36604-8.

Collins T, D'Amico S, Marx J-C, et al. Cold-adapted enzymes. In: Gerday C, Glansdorff N, editors. Physiology and biochemistry of extremophiles. Washington, DC: ASM Press; 2007. p. 165–79.

Collins T, Roulling F, Piette F, et al. Fundamentals of cold-adapted enzymes. In: Margesin R, Schinner F, Marx J-C, Gerday C, editors. Psychrophiles: from biodiversity to biotechnology. Heidelberg: Springer; 2008. p. 211–28.

Dias D, Urban S. Phytochemical analysis of the southern Australian marine alga, Plocamium mertensii using HPLC-NMR. Phytochem Anal. 2008;19:453–70.

Dias DA, Urban S. Application of HPLC-NMR for the rapid chemical profiling of a Southern Australian sponge, Dactylospongia sp. J Sep Sci. 2009;32:542–8.

Dias DA, White JM, Urban S. Laurencia filiformis: phytochemical profiling by conventional and HPLC-NMR approaches. Nat Prod Commun. 2009;4:157–72.

Eriksson LA. Theoretical biochemistry: process and properties of biological systems. 2001. ISBN: 1380-7323 (Series).

I. Digel et al., *Introduction to Industrial Biotechnology*, Learning Materials in Biosciences, https://doi.org/10.1007/978-3-032-07918-3

Flickinger MC. Downstream industrial biotechnology. Recovery and purification; 2013a. ISBN: 978-1-118-13124-4

Flickinger MC. Upstream industrial biotechnology. Expression systems and process development; 2013b. ISBN: 978-1-118-13123-7

Glansdorff N., ASM Press: Washington, DC, 2008; pp 361–421.

Golyshina OV, Timmis KN. Ferroplasma and relatives, recently discovered cell wall-lacking archaea making a living in extremely acid, heavy metal-rich environments. Environ Microbiol. 2005;7:1277–88.

Grunwald P Industrial biocatalysis. 2015.] ISBN 978-981-4463-88-1 (Hardcover), 978-981-4463-89-8 (eBook).

Guerrero-Lemus R, Martinez-Duart JM. Renewable energies and CO_2. 2013. ISBN: 978-1-4471-4384-0.

Hagihara H, Igarashi K, Hayashi Y, et al. Novel alpha-amylase that is highly resistant to chelating reagents and chemical oxidants from the alkaliphilic bacillus isolate KSM-k38. Appl Environ Microbiol. 2001;67:1744–50.

Haider SI, Ashok A. Biotechnology: a comprehensive training guide for the biotechnology industry. 2009. ISBN: 978-1-4200-8448-1.

Himmel ME. Biomass conversion. Methods and Protocols. 2012. ISBN: 978-1-61779-955-6.

Huston AL. Biotechnological aspects of cold-adapted enzymes. In: Margesin R, Schinner F, Marx J-C, Gerday C, editors. Psychrophiles: from biodiversity to biotechnology. Heidelberg: Springer; 2008. p. 347–63.

Kayser O et al. Pharmaceutical biotechnology, drug discovery and clinical applications. 2004. ISBN: 3-527-30554-8.

Kent JA. Industrial chemistry and biotechnology. 2012. ISBN: 978-1-4614-4258-5.

Krulwich TA, Hicks DB, Swartz T, Ito M. Bioenergetic adaptations that support alkaliphily. In: Gerday C, Glansdorff N, editors. Physiology and biochemistry of extremophiles. Washington, DC: ASM Press; 2008. p. 311–29.

Laskin AI, Sariaslani S, Gadd GM. Advances in applied microbiology. ISBN: 978-0-12-002663-0; 2007.

Liese A, Seelbach K, Wandrey C. Industrial biotransformations. 2006. ISBN: 978-3-527-31001-2.

Madsen EL. Environmental microbiology. 2008a; ISBN-13: 978-1-4051-3647-1

Eugene L. Madsen. Environmental microbiology, from genomes to biogeochemistry. 2008b. ISBN-13: 978-1-4051-3647-1.

Margesin R, Schinner F, editors. Cold-adapted organisms. Heidelberg: Springer; 1999.

Margesin R, Schinner F, Marx J-C, Gerday C, editors. Psychrophiles: from biodiversity to biotechnology. Heidelberg: Springer; 2008.

Montville TJ, Matthews. Food microbiology: an introduction. 2nd ed; 2008. ISBN: 978-1-55581-396-3

Mosier NS, Ladisch MR. Modern biotechnology: connecting innovations in microbiology and biochemistry to engineering fundamentals 2009. ISBN 978-0-470-11485-8.

Nakano S, Fukaya M. Analysis of proteins responsive to acetic acid in acetobacter: molecular mechanisms conferring acetic acid resistance in acetic acid bacteria. Int J Food Microbiol. 2008;125:54–9.

Nester EW, et al. Microbiology: a human perspective. 6th ed; 2011. ISBN 978-0-07-299543-5

Padan E, Bibi E, Ito M, Krulwich TA. Alkaline PH homeostasis in bacteria: new insights. Biochim Biophys Acta. 2005;1717:67–88.

Prescott LM. Microbiology. 2002. ISBN: 0-07-282905-2.

Prescott, Harley, and Klein's microbiology, 7th ed. 2008. ISBN 978-0-07-299291-5.

Price G. Biology: an illustrated guide to science. 2006. ISBN-10: 0-8160-6162-9.

Ratledge C Kristiansen B. Basic biotechnology. 2nd. 2001. ISBN 978-0-521-779170.

Rawlings DE. Heavy metal mining using microbes. Annu Rev Microbiol. 2002;56:65–91.

Renneberg R. Biotechnology for beginners. 2007. ISBN: 9780123735812.

Richard F. Grossman and domasius nwabunma. Poly (Lactic Acid): Synthesis, Structures, Properties, Processing, and Application; 2010. ISBN: 978-0-470-29366-9

Saiki RK, Gelfand DH, Stoffel S, et al. Primer-directed enzymatic amplification of DNA with a thermostable DNA polymerase. Science. 1988;239:487–91.

Sambamurthy K, Kar A. Pharmaceutical biotechnology, introduction, theory, classification. 2006. ISBN: 978-81-224-2424-9

Sangeetha J, Thangadurai D, Islam S, Hospet R. Innovations in biotechnology: algal metabolites: biotechnological, commercial, and industrial applications. 2023. ISBN: 1774912732.

Sayyed RZ, El-Enshasy HA. Industrial biotechnology: microbial surfactants. 2022. ISBN: 9781032162478.

Schaechter M. Encyclopedia of microbiology. 3rd ed; 2009. ISBN: 9780123749802

Rolf D. Schmid. Pocked guide to biotechnology and genetic engineering. 2003. ISBN: 3-527-30895-4.

Shimasaki CD. The business of bioscience. What Goes into Making a Biotechnology Product; 2009. ISBN: 978-1-4419-0063-0

Simonato F, Campanaro S, Lauro FM, et al. Piezophilic adaptation: a genomic point of view. J. Biotechnol. 2006;126:11–25.

Singleton P, Sainsbury D. Dictionary of microbiology and molecular biology. 3rd ed; 2006. ISBN: 978-0-470-03545-0

Sinha RK Sinha R. Environmental biotechnology. 2008. ISBN 978-81-7910-229-9.

Soetaert W, Vandamme EJ. Industrial biotechnology. Sustainable Growth and Economic Success; 2010. ISBN: 978-3-527-31442-3

Stevens CV. Renewables-based technology. 2006. ISBN: 978-0-470-02241-2.

Strausberg SL, Ruan B, Fisher XE, et al. Directed evolution of stability and catalytic activity in calcium-free subtilisin. Biochemistry. 2005;44:3272–9.

Talaro KP. Foundations in microbiology. 8th ed; 2012. ISBN 978-0-07-337529-8

Talaro-Talaro. Foundations in microbiology. 4th ed; 2011. ISBN: 978-0072320428

Tortora GJ. Microbiology: an introduction. 2010; ISBN-13: 978-0-321-55007-1

Turner P, Mamo G, Karlsson EN. Potential and utilization of thermophiles and thermostable enzymes in biorefining. Microb Cell Fact. 2007;6:9–32.

Varfolomeev SD, Krylova LP. Biotechnology in agriculture, industry and medicine. 2012. ISBN: 978-1-62100-576-6.

Willey JM, Sherwood LM, Woolverton CJ. Prescott's principles of microbiology. 2009. ISBN: 978-0-07-337523-6.

Wittmann C, Liao JC. Industrial biotechnology. Microorganisms. 2017;1. ISBN: 978-3-527-34179-5

Xia L, Ruppert M, Wang M, Panjikar S, et al. Structures of alkaloid biosynthetic glucosidases decode substrate specificity. ACS Chem Biol. 2012;7:226–34.

Yamazaki M, Mochida K, Asano T, Nakabayashi R, et al. Coupling deep transcriptome analysis with untargeted metabolic profiling in Ophiorrhiza pumila to further the understanding of the biosynthesis of the anti-cancer alkaloid camptothecin and anthraquinones. Plant Cell Physiol. 2013:686–96.

Index

© The Editor(s) (if applicable) and The Author(s), under exclusive license to
Springer Nature Switzerland AG 2025
I. Digel et al., *Introduction to Industrial Biotechnology*, Learning Materials in
Biosciences, https://doi.org/10.1007/978-3-032-07918-3

GPSR Compliance
The European Union's (EU) General Product Safety Regulation (GPSR) is a set
of rules that requires consumer products to be safe and our obligations to
ensure this.

If you have any concerns about our products, you can contact us on

ProductSafety@springernature.com

In case Publisher is established outside the EU, the EU authorized
representative is:

Springer Nature Customer Service Center GmbH
Europaplatz 3
69115 Heidelberg, Germany